Sea-Level Rise for the Coasts of California, Oregon, and Washington:

Past, Present, and Future

Committee on Sea Level Rise in California, Oregon, and Washington

Board on Earth Sciences and Resources and Ocean Studies Board

Division on Earth and Life Studies

NATIONAL RESEARCH COUNCIL
OF THE NATIONAL ACADEMIES

THE NATIONAL ACADEMIES PRESS
Washington, D.C.
www.nap.edu

THE NATIONAL ACADEMIES PRESS • 500 Fifth Street, NW • Washington, DC 20001

NOTICE: The project that is the subject of this report was approved by the Governing Board of the National Research Council, whose members are drawn from the councils of the National Academy of Sciences, the National Academy of Engineering, and the Institute of Medicine. The members of the committee responsible for the report were chosen for their special competences and with regard for appropriate balance.

This study was supported by the California Department of Water Resources, Contract No. 4600008602; the National Oceanic and Atmospheric Administration, Contract No. DG133R08CQ0062; the U.S. Army Corps of Engineers, Contract No. W912HQ-09-P-0155; and the United States Geological Survey, Grant/Cooperative Agreement No. G09AP00152. Any opinions, findings, or conclusions expressed in this publication are those of the author(s) and do not necessarily reflect the views of the organizations or agencies that provided support for the project.

International Standard Book Number-13: 978-0-309-25594-3
International Standard Book Number-10: 0-309-25594-5

Additional copies of this report are available from the National Academies Press, 500 Fifth Street, NW, Keck 360, Washington, DC 20001; (800) 624-6242 or (202) 334-3313 (in the Washington metropolitan area); http://www.nap.edu.

Cover: Lighthouse Point, Santa Cruz, California. SOURCE: Courtesy of Shmuel Thaler, Santa Cruz Sentinel.

Printed in the United States of America.

THE NATIONAL ACADEMIES

Advisers to the Nation on Science, Engineering, and Medicine

The **National Academy of Sciences** is a private, nonprofit, self-perpetuating society of distinguished scholars engaged in scientific and engineering research, dedicated to the furtherance of science and technology and to their use for the general welfare. Upon the authority of the charter granted to it by the Congress in 1863, the Academy has a mandate that requires it to advise the federal government on scientific and technical matters. Dr. Ralph J. Cicerone is president of the National Academy of Sciences.

The **National Academy of Engineering** was established in 1964, under the charter of the National Academy of Sciences, as a parallel organization of outstanding engineers. It is autonomous in its administration and in the selection of its members, sharing with the National Academy of Sciences the responsibility for advising the federal government. The National Academy of Engineering also sponsors engineering programs aimed at meeting national needs, encourages education and research, and recognizes the superior achievements of engineers. Dr. Charles M. Vest is president of the National Academy of Engineering.

The **Institute of Medicine** was established in 1970 by the National Academy of Sciences to secure the services of eminent members of appropriate professions in the examination of policy matters pertaining to the health of the public. The Institute acts under the responsibility given to the National Academy of Sciences by its congressional charter to be an adviser to the federal government and, upon its own initiative, to identify issues of medical care, research, and education. Dr. Harvey V. Fineberg is president of the Institute of Medicine.

The **National Research Council** was organized by the National Academy of Sciences in 1916 to associate the broad community of science and technology with the Academy's purposes of furthering knowledge and advising the federal government. Functioning in accordance with general policies determined by the Academy, the Council has become the principal operating agency of both the National Academy of Sciences and the National Academy of Engineering in providing services to the government, the public, and the scientific and engineering communities. The Council is administered jointly by both Academies and the Institute of Medicine. Dr. Ralph J. Cicerone and Dr. Charles M. Vest are chair and vice chair, respectively, of the National Research Council.

www.national-academies.org

COMMITTEE ON SEA LEVEL RISE IN CALIFORNIA, OREGON, AND WASHINGTON

ROBERT A. DALRYMPLE, *Chair*, Johns Hopkins University, Baltimore, Maryland
LAURENCE C. BREAKER, Moss Landing Marine Laboratories, California
BENJAMIN A. BROOKS, University of Hawaii, Manoa
DANIEL R. CAYAN, Scripps Institution of Oceanography, San Diego, California
GARY B. GRIGGS, University of California, Santa Cruz
WEIQING HAN, University of Colorado, Boulder
BENJAMIN P. HORTON, University of Pennsylvania, Philadelphia
CHRISTINA L. HULBE, Portland State University, Oregon
JAMES C. MCWILLIAMS, University of California, Los Angeles
PHILIP W. MOTE, Oregon State University, Corvallis
WILLIAM TAD PFEFFER, University of Colorado, Boulder
DENISE J. REED, University of New Orleans, Louisiana
C.K. SHUM, Ohio State University, Columbus

Ocean Studies Board Liaison

ROBERT A. HOLMAN, Oregon State University, Corvallis

National Research Council Staff

ANNE M. LINN, Study Director, Board on Earth Sciences and Resources
MARTHA MCCONNELL, Program Officer, Ocean Studies Board (through September 2011)
COURTNEY R. GIBBS, Program Associate, Board on Earth Sciences and Resources
JASON R. ORTEGO, Research Associate, Board on Earth Sciences and Resources

BOARD ON EARTH SCIENCES AND RESOURCES

OCEAN STUDIES BOARD

National Research Council Staff

Preface

Projections of sea-level rise are increasingly being incorporated into coastal planning at national, state, and local levels. This assessment of sea-level rise for the California, Oregon, and Washington coasts was requested by 10 state and federal agencies:

- California Department of Water Resources
- California Energy Commission
- California Department of Transportation
- California State Water Resources Control Board
- California Ocean Protection Council
- Oregon Watershed Enhancement Board
- Washington Department of Ecology
- National Oceanic and Atmospheric Administration (NOAA)
- U.S. Army Corps of Engineers (USACE)
- U.S. Geological Survey (USGS)

At the committee's first meeting, each agency described its needs for sea-level information.[1] The state agencies need estimates and projections of sea-level rise in their areas to assess coastal risk; to plan investments in water, transportation, energy, and pollution-control infrastructure; to modify design and construction standards; to develop adaptation strategies that will protect the environment and infrastructure against increased salt-water intrusion, coastal erosion, and inundation; and to identify necessary changes in state law or policy. NOAA and the USGS need sea-level information at state, national, and global scales to assess coastal vulnerability and response to sea-level rise; to improve models and forecasts; to develop research priorities; and to develop decision support tools for a variety of users, including the public. Finally, the USACE needs sea-level information to guide water resource investment decisions.

Assessments of sea-level rise at state and regional levels are challenging because data on the geophysical processes involved are relatively sparse and there are no agreed-upon models or approaches for projecting future sea-level rise. Consequently, in addition to searching the scientific literature, it was necessary to consult widely with colleagues and to carry out original data analyses. The results were discussed during four committee meetings in 2011 and countless teleconference and email discussions.

The committee used standard statistical techniques to calculate means, trends, and uncertainties associated with sea-level rise, and to extrapolate recent data into the future. To ensure that the calculated results were sound, the committee verified its results in several ways. Calculations performed using standard statistical packages or the equations and data presented in the report were cross-checked by one or two committee members. This process was used to check the means and uncertainties of the various components of sea-level rise, the tide gage and satellite altimetry measurement errors and corrections, vertical land motion observations and models, and estimates of the effect of gravitational attraction. Calculations that required specialized software, including extracting the steric contribution from model results, calculating trends from satellite measure-

[1] Presentations to the committee by the 10 sponsor agencies on January 12, 2011.

ments and glacial isostatic adjustment (GIA) models, and projecting future sea-level rise, were carried out or checked by a colleague or student of the lead committee member. The method for extrapolating the cryospheric contribution to sea-level rise was developed in collaboration with a statistician, who also verified the results. Where possible, the data and equations for these calculations are provided in the report or the public-access file, enabling an independent check from reviewers.

The committee would like to thank the individuals who briefed the committee; supplied data, figures, or model results; or provided other input or feedback: Jonathan Allan, Brian Atwater, Patrick Barnard, Laura Brophy, John Church, Abe Doherty, Catia Domingues, Peter Gleckler, Chris Goldfinger, Dominic Gregorio, Jonathan Gregory, Eric Grossman, Junyi Guo, Erica Harris, Greg Hood, Masayoshi Ishii, Ian Joughin, Jeanine Jones, Tom Kendall, Paul Komar, Eli Levitt, Sydney Levitus, Becky Lunde, Anne Pardaens, Archie Paulson, Stephan Rahmstorf, Eric Rignot, Peter Ruggiero, Carl Safina, Ingo Sasgen, Armand Thibault, Wouter van der Wal, Hansheng Wang, Kelin Wang, Jeff Weber, Josh Willis, Frank Wu, Patrick Wu, Jianjun Yin, and Phoebe Zarnetske. Special thanks go to Balaji Rajagopalan, who developed the statistical approach for the ice extrapolations; James Foster, who compiled and analyzed leveling data in California; Richard Peltier, who provided details of his GIA models and computed past and future predictions of relative sea-level changes in Washington, Oregon, and California; and Jerry Mitrovica, who provided gravity fingerprints along the U.S. west coast for Alaska, Greenland, and Antarctica. The committee also thanks the students, postdocs, and colleagues who crunched numbers, validated results, and created (and recreated) figures, including Jianbin Duan, Zhenwei Huang, Chungyen Kuo, Darrin Sharp, Scott Waibel, and Yuchan Yi. Without the hard work and contributions of all these individuals, it would have been difficult to complete this report.

Finally, I thank all the members of the committee for their service, some of whom had to go way beyond that usually required for an NRC committee because of the short study period and the complexity of the task. Finally, I thank Anne Linn for her tireless efforts as Study Director and for bringing the report to fruition.

Robert A. Dalrymple, *Chair*
Committee on Sea Level Rise in
California, Oregon, and Washington

Acknowledgments

This report has been reviewed in draft form by individuals chosen for their diverse perspectives and technical expertise, in accordance with procedures approved by the NRC's Report Review Committee. The purpose of this independent review is to provide candid and critical comments that will assist the institution in making its published report as sound as possible and to ensure that the report meets institutional standards for objectivity, evidence, and responsiveness to the study charge. The review comments and draft manuscript remain confidential to protect the integrity of the deliberative process. We wish to thank the following individuals for their participation in the review of this report:

Linda K. Blum, University of Virginia, Charlottesville
Roland Bürgmann, University of California, Berkeley
John A. Church, Centre for Australian Weather and Climate Research, Hobart, Tasmania
Peter J. Gleckler, Lawrence Livermore National Laboratory, California
Peter H. Gleick, Pacific Institute for Studies in Development, Environment, and Security, Oakland, California
Mark F. Meier, emeritus, University of Colorado, Boulder
Jerry X. Mitrovica, Harvard University, Cambridge, Massachusetts
Antony R. Orme, University of California, Los Angeles
W. Richard Peltier, University of Toronto, Canada
Stephen Price, Los Alamos National Laboratory, New Mexico
Claudia Tebaldi, Climate Central, Princeton, New Jersey, and Palo Alto, California
John M. Wallace, University of Washington, Seattle
Joshua K. Willis, Jet Propulsion Laboratory, Pasadena, California

Although the reviewers listed above have provided many constructive comments and suggestions, they were not asked to endorse the conclusions or recommendations nor did they see the final draft of the report before its release. The review of this report was overseen by Ken Brink, Woods Hole Oceanographic Institution, and Warren Washington, National Center for Atmospheric Research. Appointed by the National Research Council, they were responsible for making certain that an independent examination of this report was carried out in accordance with institutional procedures and that all review comments were carefully considered. Responsibility for the final content of this report rests entirely with the authoring committee and the institution.

Contents

Summary

Sea level rose during the 20th century, and observations and projections suggest that it will rise at a higher rate during the 21st century. Rising seas increase the risk of coastal flooding, storm surge inundation, coastal erosion and shoreline retreat, and wetland loss. The cities and infrastructure that line many coasts are already vulnerable to damage from storms, which is likely to increase as sea level continues to rise and inundate areas further inland.

Global mean sea level is rising primarily because global temperatures are rising, causing ocean water to expand and land ice to melt. However, sea-level rise is not uniform; it varies from place to place. Sea-level rise along the coasts of California, Oregon, and Washington (referred to hereafter as the U.S. west coast) depends on the global mean sea-level rise and also on regional factors, such as ocean and atmospheric circulation patterns in the northern Pacific Ocean, the gravitational and deformational effects of land ice mass changes, and tectonics along the coast. The comparative importance of these factors determines whether local sea level is higher or lower than the global mean, and how fast it is changing. Such information has enormous implications for coastal planning.

California Executive Order S-13-08 directed state agencies to plan for sea-level rise and coastal impacts, and it also requested the National Research Council (NRC) to establish a committee to assess sea-level rise to inform these state efforts. The states of Washington and Oregon, the U.S. Army Corps of Engineers, the National Oceanic and Atmospheric Administration, and the U.S. Geological Survey subsequently joined California in sponsoring this study to evaluate sea-level rise in the global oceans and along the coasts of California, Oregon, and Washington for 2030, 2050, and 2100. The charge to the committee is given in Box S.1.

The most comprehensive estimates of global sea-level rise are made by the Intergovernmental Panel on Climate Change (IPCC), which assesses the state of knowledge on climate change every 5 to 6 years. The last IPCC assessment, published in 2007, evaluated research results published until mid-2006. This report summarizes the IPCC (2007) findings on global sea-level change and updates them with more recent results. In contrast, no comprehensive assessments of the rate of sea-level rise off the coasts of California, Oregon, and Washington have been carried out. Consequently, this report summarizes published research results on the processes that contribute to sea-level change in the region and also presents the committee's analysis of relevant data and model results. Projections of global and local sea-level rise for 2030, 2050, and 2100 are based on model results and data extrapolations, as described below.

GLOBAL SEA-LEVEL RISE

Following a few thousand years of relative stability, global sea level has been rising since the late 19th or early 20th century, when global temperatures began to increase. The IPCC (2007) estimated that global sea level rose an average of 1.7 ± 0.5 mm per year over the 20th century, based on tide gage measurements from around the world. Rates for 1993–2003 were 3.1 ± 0.7 mm per year, based on precise satellite altimetry measurements and confirmed by tide gage

BOX S.1
Committee Charge

The committee will provide an evaluation of sea-level rise for California, Oregon, and Washington for the years 2030, 2050, and 2100. The evaluation will cover both global and local sea-level rise. In particular, the committee will

1. Evaluate each of the major contributors to global sea-level rise (e.g., ocean thermal expansion, melting of glaciers and ice sheets); combine the contributions to provide values or a range of values of global sea-level rise for the years 2030, 2050, and 2100; and evaluate the uncertainties associated with these values for each timeframe.
2. Characterize and, where possible, provide specific values for the regional and local contributions to sea-level rise (e.g., atmospheric changes influencing ocean winds, ENSO [El Niño-Southern Oscillation] effects on ocean surface height, coastal upwelling and currents, storminess, coastal land motion caused by tectonics, sediment loading, or aquifer withdrawal) for the years 2030, 2050 and 2100. Different types of coastal settings will be examined, taking into account factors such as landform (e.g., estuaries, wetlands, beaches, lagoons, cliffs), geologic substrate (e.g., unconsolidated sediments, bedrock), and rates of geologic deformation. For inputs that can be quantified, the study will also provide related uncertainties. The study will also summarize what is known about
   a. climate-induced increases in storm frequency and magnitude and related changes to regional and local sea-level rise estimations (e.g., more frequent and severe storm surges);
   b. the response of coastal habitats and geomorphic environments (including restored environments) to future sea-level rise and storminess along the west coast;
   c. the role of coastal habitats, natural environments, and restored tidal wetlands and beaches in providing protection from future inundation and waves.

records. More recent tide gage and altimetry data confirm that the higher rate of sea-level rise is continuing. However, because of natural climate variability, which affects sea level on decadal and longer timescales, more data are needed to determine whether the higher rates since the 1990s mark an acceleration in the long-term sea-level trend.

Components of Global Sea-Level Rise

A warming climate causes global sea level to rise by (1) warming the oceans, which causes sea water to expand, increasing ocean volume, and (2) melting land ice, which transfers water to the ocean. Human activities that transfer water between the land and ocean also affect global sea-level change. In particular, water withdrawn from aquifers eventually reaches the ocean, raising global sea level, whereas water stored behind dams effectively lowers global sea level.

The IPCC (2007) estimated that ice melt from glaciers, ice caps, and ice sheets contributed about 40 percent of the observed sea-level rise for 1961–2003 and that thermal expansion of ocean water contributed one-quarter of the observed rate for 1961–2003 and one-half for 1993–2003. Contributions from groundwater extraction and reservoir storage were poorly quantified but were thought to account for less than 10 percent of the observed rise. More recent data have changed these estimates. After the IPCC (2007) report was published, a bias was discovered in some ocean temperature measurements, which gave systematically warmer temperatures than the true values. Data sets corrected for this bias yield significantly lower rates of thermal expansion for the 1993–2003 period than were found by the IPCC (2007).

New research results also indicate that the relative contribution of land ice to global sea-level rise is increasing. Since 2006, the ice loss rate from the Greenland Ice Sheet has increased, and, according to most analyses, the contribution of Antarctic ice to sea-level change has shifted from negative (lowering sea level by accumulating ice) to positive (raising sea level). Ice loss rates from glaciers and ice caps have declined over the same period, but not enough to offset the increases in ice sheet melt. As a result of higher observed ice loss rates and a lower (corrected) contribution from thermal expansion, land ice is currently the largest contributor to global sea-level rise. In the most recent published estimate, land ice accounted for about 65 percent of the total sea-level rise from 1993 to 2008.

The contributions of groundwater withdrawal and reservoir storage to sea-level change remain poorly constrained, largely due to sparse data and inadequate models. Each process likely has a significant but opposite effect on sea-level change, on the order of 0.5 mm per year.

SEA-LEVEL RISE OFF CALIFORNIA, OREGON, AND WASHINGTON

The sea level at any particular place along the coast is commonly measured using tide gages, which record the height of the sea surface with respect to the land surface, both of which may change over time. Relative sea level will rise if ocean levels rise and/or land levels fall. Records from 12 west coast tide gages indicate local variability in sea-level change along the coast, although most of the gages north of Cape Mendocino, California, show that relative sea level has been falling over the past 6–10 decades, and most of the gages south of Cape Mendocino show that relative sea level has been rising.

Factors That Affect Northeast Pacific Ocean Levels

Along the west coast of the United States, climate patterns such as the El Niño-Southern Oscillation and, to a lesser extent, the Pacific Decadal Oscillation, affect winds and ocean circulation, raising local sea level during warm phases (e.g., El Niño) and lowering sea level during cool phases (e.g., La Niña). Large El Niño events can raise coastal sea levels by 10 to 30 cm for several winter months.

The large mass of glaciers and ice sheets exerts a gravitational pull that draws ocean water closer. As the ice melts, the gravitational pull decreases, ice melt enters the ocean, and the land and ocean basins both deform as a result of this loss of land ice mass. These gravitational and deformational effects produce a spatial pattern of regional sea-level change called a sea-level fingerprint. Melting from Alaska and, to a lesser extent, Greenland, causes relative sea level to fall at decreasing rates from northern Washington to southern California, whereas melting from Antarctica causes relative sea level to rise along all three states. The net effect is a reduction in the contribution of the three ice sources to relative sea-level rise by 42 percent along the north coast (Neah Bay), 24 percent along the central coast (Eureka), and 14 percent along the south coast (Santa Barbara) for 1992–2008.

Factors That Affect Land Elevation in California, Oregon, and Washington

Although modern melting of land ice has a significant effect on sea-surface heights in the northeast Pacific Ocean, the melting and eventual disappearance of North American ice sheets that began more than 20,000 years ago has a significant effect on land levels in California, Oregon, and Washington. The massive loss of ice from the ancient ice sheets continues to cause uplift of about 1 mm per year in northernmost Washington, which had been covered by an ice sheet, and subsidence of about 1–2 mm per year in areas at the ice margin and beyond, which includes the rest of Washington, Oregon, and California.

Tectonics causes substantial regional uplift along much of the Washington, Oregon, and northernmost California coast, where ocean plates are descending below North America at the Cascadia Subduction Zone. South of Cape Mendocino, California, the Pacific and North American plates are sliding past one another along the San Andreas Fault Zone, creating relatively little vertical land motion along the coast. Local tectonics, as well as compaction of sediments, pumping of water or hydrocarbons from subsurface reservoirs, and fluid recharge can produce locally high rates of land subsidence or uplift. Water or hydrocarbon extraction, which can lower surface elevations up to tens of centimeters per year if fluids are not returned to the subsurface, is most important in California.

The total vertical land motion from all of these geological processes and human activities can be estimated from Global Positioning System (GPS) measurements, which show that much of the coast is rising about 1.5–3.0 mm per year north of Cape Mendocino. The coast south of Cape Mendocino is sinking at an average rate of about 1 mm per year, although GPS-measured rates vary widely (-3.7–0.6 mm per year).

PROJECTIONS OF SEA-LEVEL RISE FOR 2030, 2050, AND 2100

Global Projections

Projections of global sea-level rise are generally made using models of the ocean-atmosphere-climate system, extrapolations, or semi-empirical methods.

Ocean-atmosphere models are based on knowledge of the physical processes that contribute to sea-level rise, and they predict the response of those processes to different scenarios of future greenhouse gas emissions. These models provide a reasonable estimate of the water density (steric) component of sea-level rise (primarily thermal expansion), but they underestimate the land ice contribution because they do not fully account for rapid changes in the behavior of ice sheets and glaciers as melting occurs (ice dynamics). The IPCC (2007) projections were made using this method, and they are likely too low, even with an added ice dynamics component. Estimates of the total land ice contribution can be made by extrapolating observations of recent ice loss rates from glaciers, ice caps, and ice sheets into the future. Extrapolations of future ice melt are most reliable for time frames in which the dynamics controlling behavior are stable, in this case, up to several decades. Semi-empirical methods, exemplified by Vermeer and Rahmstorf (2009), avoid the difficulty of estimating the individual contributions to sea-level rise by simply postulating that sea level rises faster as the Earth gets warmer. This approach reproduces the sea-level rise observed in the past, but reaching the highest projections would require acceleration of glaciological processes to levels not previously observed or understood as realistic.

Given the strengths and weaknesses of the different projection methods, as well as the resource constraints of an NRC study, the committee chose a combination of approaches for its projections. The committee projected the steric component of sea-level rise using output from global ocean models under an IPCC (2007) mid-range greenhouse gas emission scenario. The land ice component was extrapolated using the best available compilations of ice mass accumulation and loss (mass balance), which extend from 1960 to 2005 for glaciers and ice caps, and from 1992 to 2010 for the Greenland and Antarctic ice sheets. The contributions were then summed. The committee did not project the land hydrology contribution because available estimates suggested that the sum of groundwater extraction and reservoir storage is near zero, within large uncertainties.

Based on these calculations, the committee estimates that global sea level will rise 8–23 cm by 2030 relative to 2000, 18–48 cm by 2050, and 50–140 cm by 2100. The ranges reflect uncertainties related to the fit of the data; the level of future greenhouse emissions, which affects the steric component; and any future changes in the rate of ice flow, which affects the total ice contribution. These uncertainties, and hence the ranges, grow with the length of the projection period.

The committee's global projections for 2030 and 2050 are similar to the Vermeer and Rahmstorf (2009) projections for the same periods, but they have a wider range. For 2100, when IPCC (2007) projections are also available, the committee's projection is substantially higher than IPCC's projection (18–59 cm with an additional 17 cm if rapid dynamical changes in ice flow are included), mainly because of a faster growing cryosphere component, and lower than Vermeer and Rahmstorf's projection (78–175 cm).

Projections for California, Oregon, and Washington

Sea-level rise off the west coast of the United States is influenced by a variety of local factors; therefore, sea-level projections for California, Oregon, and Washington differ from global projections. The factors that affect local sea-level projections include steric variations; wind-driven differences in ocean heights; gravitational and deformational effects (sea-level fingerprints) of melting of ice from Alaska, Greenland, and Antarctica; and vertical land motions along the coast. The local steric and wind-driven components were estimated by extracting northeast Pacific data from the same ocean models used for the global projections. The cryosphere component was adjusted for gravitational and deformational effects and then extrapolated forward. Finally, vertical land motion was projected using continuous GPS measurements for two tectonically distinct areas: Cascadia, where the coastline is generally rising, and the San Andreas region, where the coastline is generally subsiding.

The projections for California, Oregon, and Washington are illustrated in Figure S.1. The steep change in projected sea-level rise at Cape Mendocino reflects the transition from land subsidence in California, which effectively increases sea-level rise, to land uplift in Oregon and Washington, which effectively decreases sea-level rise. The slight slope in the projection curves from north to south reflects the sea-level fingerprints, which lower relative sea level, especially along the Washington coast. For the California coast south of Cape

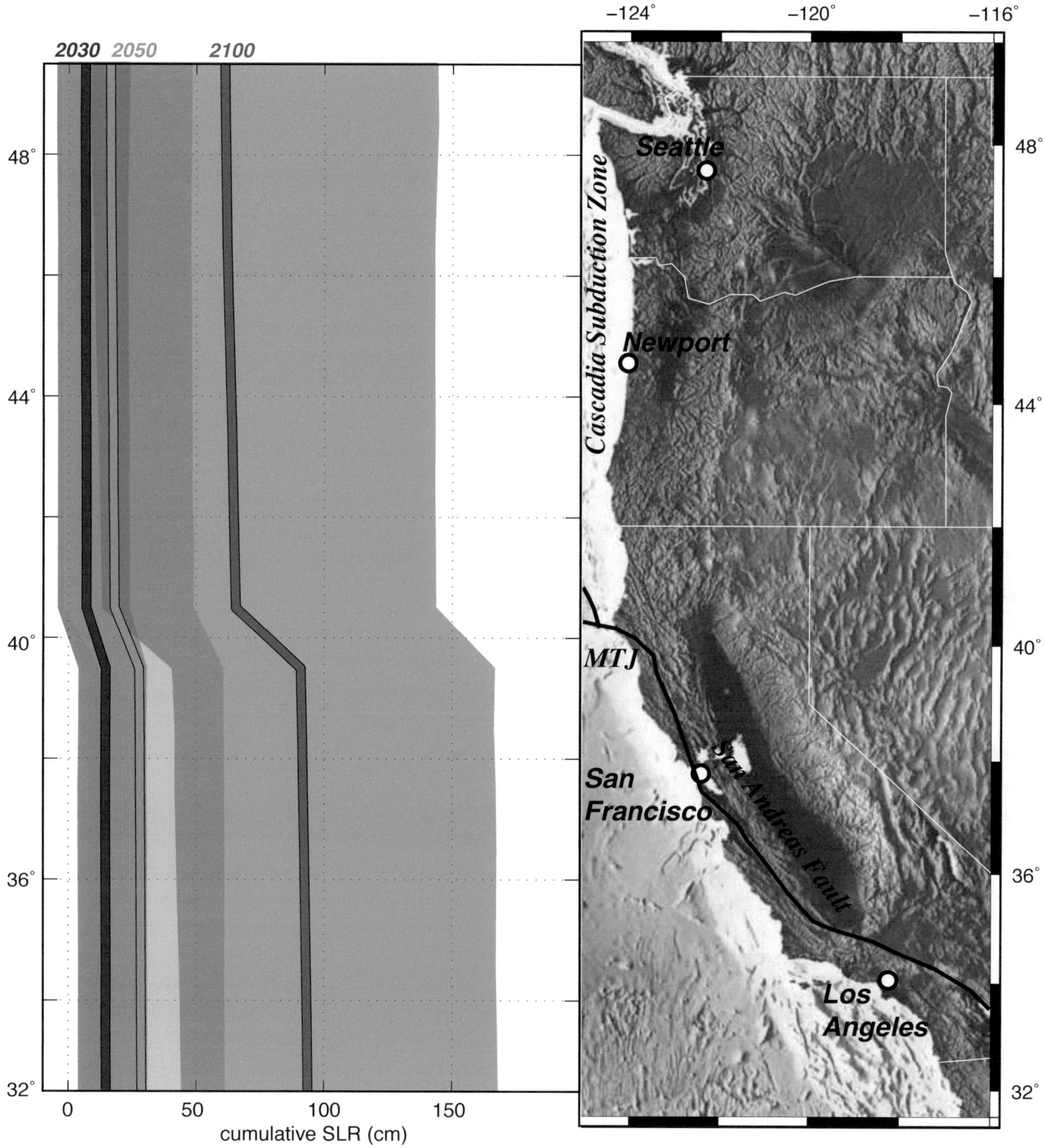

FIGURE S.1 Projected sea-level rise off California, Oregon, and Washington for 2030 (blue), 2050 (green), and 2100 (pink), relative to 2000, as a function of latitude. Solid lines are the projections, and shaded areas are the ranges. Ranges overlap, as indicated by the brown shading (low end of 2100 range and high end of 2050 range) and blue-green shading (low end of 2050 range and high end of 2030 range). MTJ = Mendocino Triple Junction, where the San Andreas Fault meets the Cascadia Subduction Zone.

Mendocino, the committee projects that sea level will rise 4–30 cm by 2030 relative to 2000, 12–61 cm by 2050, and 42–167 cm by 2100. For the Washington, Oregon, and California coasts north of Cape Mendocino, sea level is projected to change between -4 cm (sea-level fall) and +23 cm by 2030, -3 cm and +48 cm by 2050, and 10–143 cm by 2100. Major sources of uncertainty in the regional projections are related to assumptions about future ice losses and a constant rate of vertical land motion over the projection period. Uncertainties are larger for the regional projections than for the global projections because more components are considered and because uncertainties in the steric and ocean dynamic components are larger at a regional scale than at a global scale.

The combination of land uplift and gravitational and deformational effects reduces the threat of future sea-level rise for Washington and Oregon. However, the land is rising along the Washington and Oregon coasts likely because interseismic strain is building in the Cascadia Subduction Zone. A great earthquake (magnitude larger than 8), which has occurred in the area every few hundred to 1,000 years, would cause some coastal areas to immediately subside and relative sea level to suddenly rise. If this occurs, relative sea level could rise an additional meter or more over projected levels.

The committee's projections for the California coast are slightly higher than its global projections, primarily because much of the coastline is subsiding. The California projections are somewhat lower but have wider ranges than the Vermeer and Rahmstorf (2009) global projections, which are being used by California on an interim basis for coastal planning.

The projections of future sea-level rise have large uncertainties resulting from an incomplete understanding of the global climate system, the inability of global climate models to accurately represent all important components of the climate system at global or regional scales, a shortage of data at the temporal and spatial scales necessary to constrain the models, and the need to make assumptions about future conditions (e.g., greenhouse gas emissions, large volcanic eruptions) that drive the climate system. As the projection period lengthens, uncertainty in the projections grows. At short timescales (2030 and perhaps 2050), when the models more closely represent the future climate system, confidence in the global and regional projections is relatively high. By 2100, however, projections made using process-based numerical models, extrapolations, and semi-empirical methods all have large uncertainties. The actual sea-level rise will almost surely fall somewhere within the wide uncertainty bounds, although the exact value cannot be specified with high confidence.

SEA-LEVEL RISE AND STORMINESS

Most of the damage along the California, Oregon, and Washington coasts is caused by storms—particularly the confluence of large waves, storm surges, and high astronomical tides during a strong El Niño. The water levels reached during these large, short-term events have exceeded mean sea levels projected for 2100, so understanding their additive effects is crucial for coastal planning.

Changes in Storm Frequency and Magnitude

Climate change has been postulated to induce changes in storm frequency, magnitude, and direction. To date, there is no consensus among climate model simulations about whether the number and severity of storms will change in the northeast Pacific. A number of climate models predict a northward shift in the North Pacific storm track over the course of the 21st century, which could lessen the impact of winter storms in southern California and possibly increase their impact in Oregon and Washington. However, these changes may not emerge for a few decades, and most observational records are not yet long enough to determine conclusively whether storm tracks are moving north.

Several observational studies have reported that the largest waves have been getting higher and that winds have been getting stronger in the northeastern Pacific over the past few decades. Interpretation of these trends is controversial because wave and wind records are short, extending back only about 35 years. At least part of the observed increase likely reflects natural climate variability of the Pacific atmosphere-ocean system, particularly the occurrence of large El Niños and interdecadal fluctuations. If some or all of the increase represents a long-term trend, the frequency and magnitude of extremely high coastal wave events will likely increase.

Even if storminess does not increase in the future, sea-level rise will magnify the adverse impact of storm surges and high waves on the coast. For example, a model using the committee's sea-level projections predicts that the incidence of extreme high water events (1.4 m above historical mean sea level) in the San Francisco Bay area will increase substantially with sea-level rise, from less than 10 hours per decade today to a few hundred hours per decade by 2050 and to several thousand hours per decade by 2100.

Coastal Responses to Sea-Level Rise and Storminess

The natural shoreline can provide partial protection for coastal development against sea-level rise and storms. Coastal cliffs, beaches, and dunes take the brunt of storm waves and are therefore eroding over the long term. The net result of storms and sea-level rise is coastline retreat, with rates ranging from a few centimeters per year for cliffs made of resistant bedrock to several meters per year for beaches and dunes, which consist primarily of unconsolidated sand. These rates will increase with rising sea level and are likely to further increase if waves become higher. Although seawalls and revetments can make the shoreline more resistant to wave attack, they prevent beaches from migrating landward and will eventually be overwhelmed by sea-level rise.

Marshes and mudflats protect inland areas by storing flood waters and damping wave height and energy. To continue providing these services as sea level rises, marshes must be able to maintain their elevation relative to sea level and to move inland in places where they are subject to erosion at the seaward edge. Building elevation requires a sufficient supply of sediment and accumulation of organic material. Most studies of west coast marshes have focused on the supply of sediment. The frequent storms and associated floods in central and southern California potentially provide enough sediment for marshes to keep pace with the sea-level rise projected for 2030 and 2050 by the committee. In Oregon and Washington, rivers also potentially carry enough sediment for marshes to maintain elevation, despite upstream dams, especially because the projections of sea-level rise are lower. For 2100, marshes will need room to migrate, a high sediment supply, and uplift or low subsidence to survive the projected sea-level rise.

1

Introduction

Sea-level change is one of the most visible consequences of changes in the Earth's climate. A warming climate causes global sea level to rise principally by (1) warming the oceans, which causes sea water to expand, increasing ocean volume, and (2) melting land ice, which transfers water to the ocean. Tide gage and satellite observations show that global sea level has risen an average of about 1.7 mm yr^{-1} over the 20th century (Bindoff et al., 2007), which is a significant increase over rates of sea-level rise during the past few millennia (Shennan and Horton, 2002; Gehrels et al., 2004). Projections suggest that sea level will continue to rise in the future (Figure 1.1). However, the rate at which sea level is changing varies from place to place and with time. Along the west coast of the United States, sea level is influenced by changes in global mean sea level as well as by regional changes in ocean circulation and climate patterns such as El Niño; gravitational and deformational effects of ice age and modern ice mass changes; and uplift or subsidence along the coast. The relative importance of these factors in any given area determines whether the local sea level will rise or fall and how fast it will change.

Sea-level change has enormous implications for coastal planning, land use, and development along the 2,600 km shoreline of California, Oregon, and Washington (referred to hereafter as the U.S. west coast). Rising sea level increases the risk of flooding, inundation, coastal erosion, wetland loss, and saltwater intrusion into freshwater aquifers in many coastal communities (e.g., Heberger et al., 2009, 2011). Valuable infrastructure, development, and wetlands line much of the coast. For example, significant development along the edge of central and southern San Francisco Bay—including two international airports, the ports of San Francisco and Oakland, a naval air station, freeways, housing developments, and sports stadiums—has been built on fill that raised the land level only a few feet above the highest tides. The San Francisco International Airport will begin to flood with as little as 40 cm of sea-level rise (Figure 1.2), a value that could be reached in several decades (Figure 1.1).

Coastal infrastructure and ecosystems are already vulnerable to high waves during ocean storms (e.g., Figure 1.3), especially when storms coincide with high tides and/or El Niño events. For example, a strong El Niño, combined with a series of large storms at times of high astronomical tides, caused more than $200 million dollars in damage (in 2010 dollars) to the California coast during the winter of 1982–1983 (Griggs et al., 2005). Higher sea levels and heavy rainfall caused flooding in low-lying areas and increased the level of wave action on beaches and bluffs (Storlazzi and Griggs, 2000). More than 3,000 homes and businesses were damaged, 33 oceanfront homes were completely destroyed, and roads, parks, and other infrastructure was heavily damaged. The damage will likely increase as sea level continues to rise and more of the shoreline is inundated.

In November 2008, Governor Arnold Schwarzenegger issued Executive Order S-13-08 directing California state agencies to plan for sea-level rise and coastal impacts.[1] Included in the executive

[1] See <http://gov.ca.gov/executive-order/11036/>.

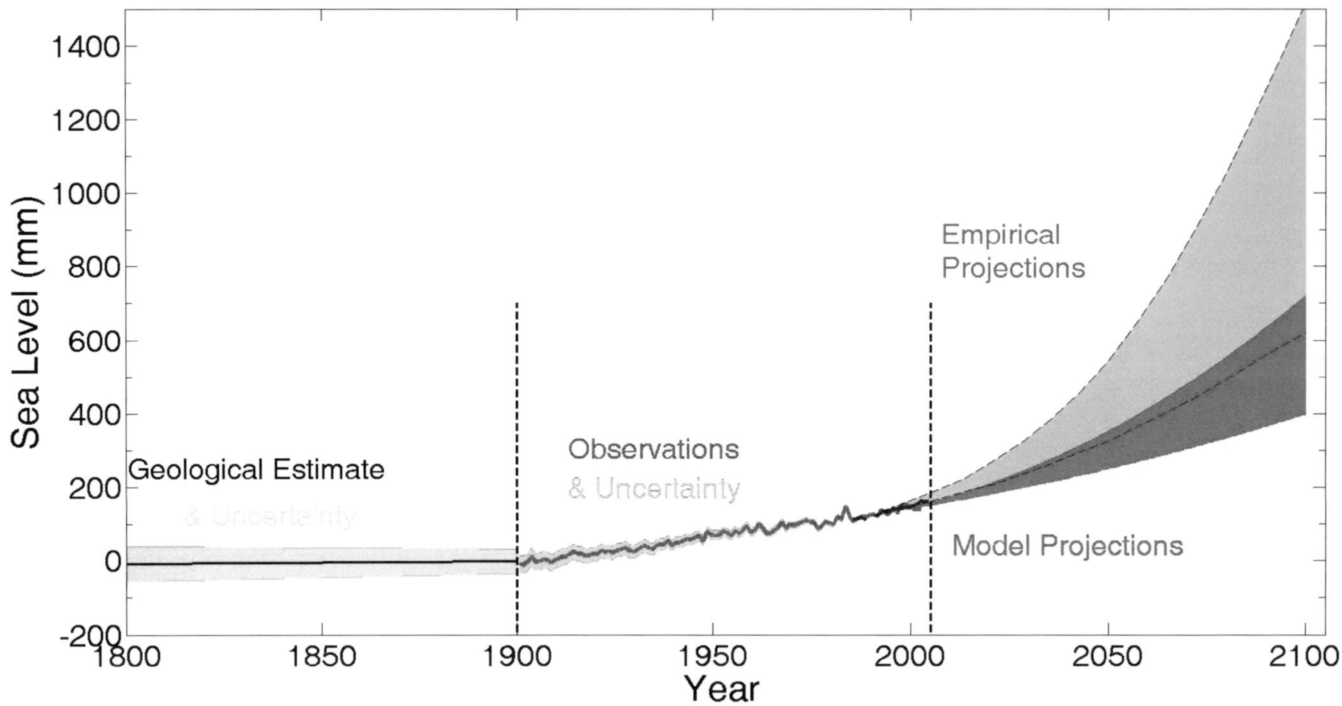

FIGURE 1.1 Estimated, observed, and projected global sea-level rise from 1800 to 2100. The pre-1900 record is based on geological evidence, and the observed record is from tide gages (red line) and satellite altimetry (blue line). Example projections of sea-level rise to 2100 are from IPCC (2007) global climate models (pink shaded area) and semi-empirical methods (gray shaded area; Rahmstorf, 2007). SOURCES: Adapted from Shum et al. (2008), Willis et al. (2010), and Shum and Kuo (2011).

order was a request that the National Research Council (NRC) establish a committee to assess sea-level rise in California to inform state planning and development efforts. Prior to release of the NRC report, the state agencies were instructed to incorporate sea-level-rise projections into their planning process. The range of projections adopted by California as interim values are 13–21 cm for 2030, 26–43 cm for 2050, and 78–176 cm for 2100 (CO-CAT, 2010).

Following the California executive order, the states of Oregon and Washington, the U.S. Army Corps of Engineers, the National Oceanic and Atmospheric Administration, and the U.S. Geological Survey joined California in sponsoring this NRC study. These agencies need sea-level information for a variety of purposes, including assessing coastal hazard vulnerability, risks, and impacts; informing adaptation strategies; and improving coastal hazard forecasts and decision support tools.

This report provides an assessment of current knowledge about changes in sea level expected in California, Oregon, and Washington for 2030, 2050, and 2100 (see Box 1.1 for the committee charge). The years for the assessment represent planning horizons: 2030 is a typical planning horizon for many local managers; 2050 is the latest date for which conventional population projections are available; and 2100 is the limit beyond which uncertainties become too high for planning.[2] The report primarily focuses on how much sea level is likely to rise globally (Task 1) and along the west coast of the United States (Task 2). Processes that have only transient effects on sea level (e.g., tides, tsunamis) were considered only if the nature of the process affects trends in sea level (e.g., changes in frequency of intensity of storms [Task 2a]). Coastal impacts or measures to lessen them were considered only in the context of summarizing what is known about how coastal habitats and natural and restored environments respond to and protect against future sea-level rise and storms (Tasks 2b and 2c).

[2] Jeanine Jones, California Department of Water Resources, personal communication, December 3, 2008.

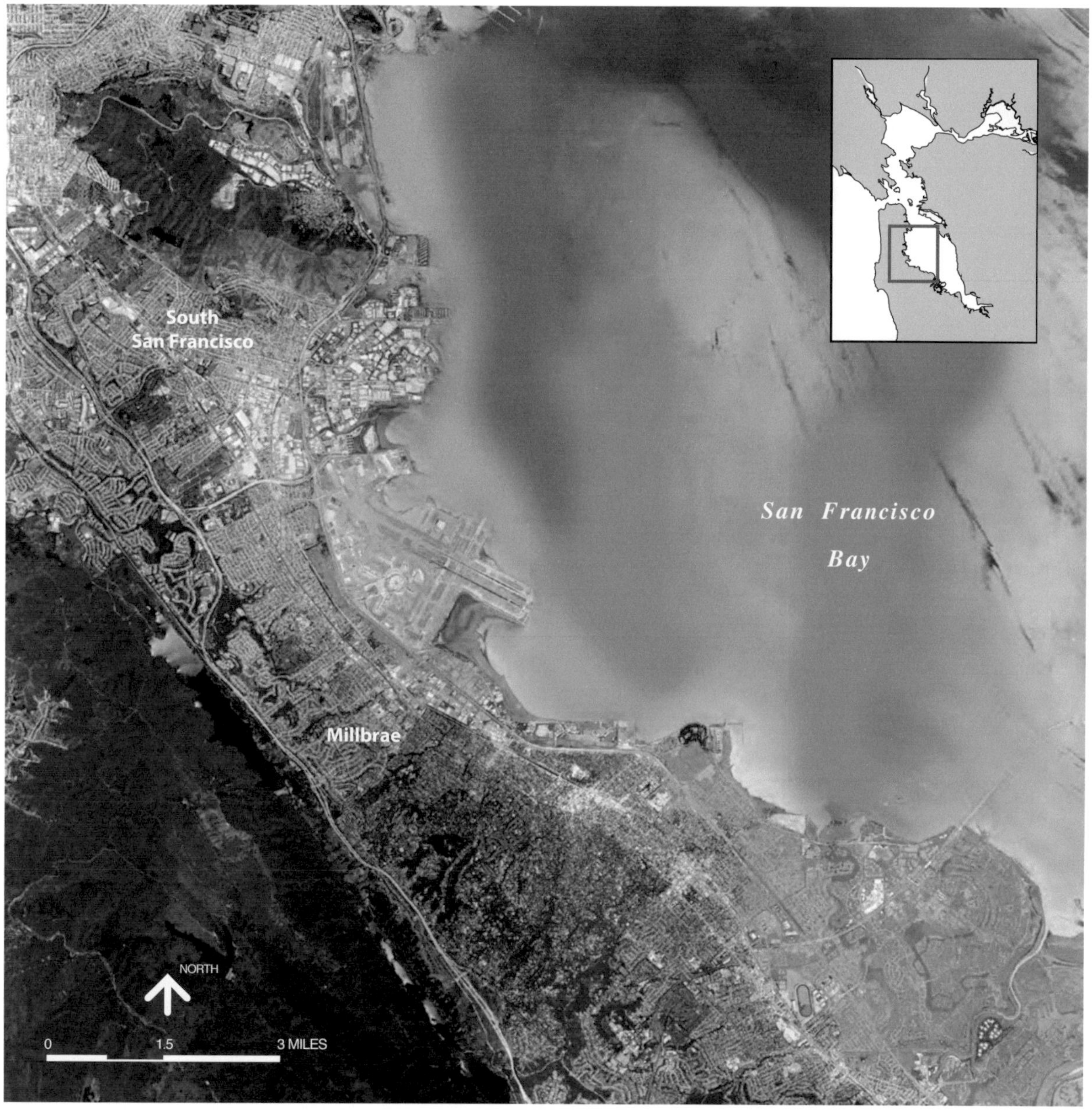

FIGURE 1.2 Expected inundation of low-lying areas, including the San Francisco International Airport (center), in the San Francisco Bay Area with a 40 cm rise in sea level (light blue shading). SOURCE: Bay Conservation and Development Commission, <http://www.bcdc.ca.gov/planning/climate_change/index_map.shtml>.

COMMITTEE APPROACH

Assessments are intended to yield a judgment on a topic, based on review and synthesis of scientific knowledge. Beginning in 1989, the primary assessments of global sea-level change have been carried out by thousands of scientists working under the auspices of the Intergovernmental Panel on Climate Change (IPCC). IPCC assessments, which are made every 5 or 6 years, evaluate observations, models, and analyses of climate change, including sea-level change. For Task 1, this report summarizes the latest IPCC (2007) findings on global sea-level rise and its major components, then updates them with more recent results.

For Task 2, the committee drew on published research on sea-level change along the west coast of the United States and also carried out its own analyses. Prior assessments of the rate of local sea-level rise have been made for Washington (Mote et al., 2008)

FIGURE 1.3 High surf during a high tide of nearly 2.7 m removed the front lawn of the Pacific Sands Resort at Neskowin, Oregon, on January 9, 2008. SOURCE: Courtesy of Armand Thibault.

BOX 1.1
Committee Charge

The committee will provide an evaluation of sea-level rise for California, Oregon, and Washington for the years 2030, 2050, and 2100. The evaluation will cover both global and local sea-level rise. In particular, the committee will

1. Evaluate each of the major contributors to global sea-level rise (e.g., ocean thermal expansion, melting of glaciers and ice sheets); combine the contributions to provide values or a range of values of global sea-level rise for the years 2030, 2050, and 2100; and evaluate the uncertainties associated with these values for each timeframe.
2. Characterize and, where possible, provide specific values for the regional and local contributions to sea-level rise (e.g., atmospheric changes influencing ocean winds, ENSO [El Niño-Southern Oscillation] effects on ocean surface height, coastal upwelling and currents, storminess, coastal land motion caused by tectonics, sediment loading, or aquifer withdrawal) for the years 2030, 2050 and 2100. Different types of coastal settings will be examined, taking into account factors such as landform (e.g., estuaries, wetlands, beaches, lagoons, cliffs), geologic substrate (e.g., unconsolidated sediments, bedrock), and rates of geologic deformation. For inputs that can be quantified, the study will also provide related uncertainties. The study will also summarize what is known about
    a. climate-induced increases in storm frequency and magnitude and related changes to regional and local sea-level rise estimations (e.g., more frequent and severe storm surges);
    b. the response of coastal habitats and geomorphic environments (including restored environments) to future sea-level rise and storminess along the west coast;
    c. the role of coastal habitats, natural environments, and restored tidal wetlands and beaches in providing protection from future inundation and waves.

and California (e.g., Cayan et al., 2009), and numerous studies have been published on individual contributors to sea-level change along the U.S. west coast. The committee also analyzed tide gage records and Global Positioning System data from California, Oregon, and Washington for their local (around the station) and regional (along the coast of one or more states) trends, and extracted regional information from satellite altimetry data and glacial isostatic adjustment models.

The most challenging aspect of the committee charge was the projections of sea level for 2030, 2050, and 2100. The numerical global climate models developed for the IPCC Fourth Assessment Report[3] project global sea-level rise to 2100. However, they do not account for rapid changes in the behavior of ice sheets and glaciers as melting occurs (ice dynamics) and thus likely underestimate future sea-level rise. The new suite of climate models for the Fifth Assessment Report was not available at the time of writing this report. Consequently, the committee projected global sea-level rise (Task 1) using model results from the IPCC Fourth Assessment Report, together with a forward extrapolation of land ice that attempts to capture an ice dynamics component. The committee also considered results from semi-empirical projections, which are based on the observed correlation between global temperature and sea-level change (e.g., Vermeer and Rahmstorf, 2009). For the projections of sea-level rise along the U.S. west coast (Task 2), the committee derived local values using regional ocean information extracted from global models, GPS data from along the coast, and ice loss rates of large or nearby glaciers.

Uncertainty

In the IPCC Fourth Assessment Report, the major components of global sea-level rise were estimated at the 90 percent confidence level (Bindoff et al., 2007). That is, values given as $x \pm e$ mean that there is a 90 percent chance that the true value is in the range $x - e$ to $x + e$. This report follows the IPCC convention unless specified otherwise.

Uncertainty in projecting climate-related sea-level changes arises from three sources: internal variability of the climate system, which fluctuates on interannual to multidecadal and longer timescales and on regional to global spatial scales; model uncertainty; and scenario uncertainty (Hawkins and Sutton, 2009). The first is particularly important for projections based on extrapolation of observations because observational records tend to be short relative to the timescale of variability in the climate system. Models have uncertainties because they are mathematical approximations that depart in important ways from the actual system. Uncertainty in models used to describe key elements of sea-level change results from uncertainties in model parameters (e.g., initial conditions, boundary conditions) as well as structural uncertainties from incomplete understanding of some climate processes or an inability to resolve the processes with available computing resources (Knutti et al., 2010). Finally, future emissions of greenhouse gasses and other factors that drive changes in the climate system depend on a collection of human decisions at local, regional, national, and international levels, as well as potential but unknown technological developments. The IPCC deals with this uncertainty by providing a range of possible futures (scenarios) based on assumptions about trends in concentrations of greenhouses gases and other influences on the climate (e.g., Moss et al., 2010).

This report uses both model and extrapolation approaches to make projections. Each approach has different uncertainties (e.g., extrapolations take no account of emission scenarios), which were combined into a single uncertainty range for the projections. Although isolating the various sources of uncertainty may have been useful for some applications (e.g., evaluating costs and risks of various mitigation strategies), it was not required in the committee charge and would have required a different analysis approach.

OVERVIEW OF SEA-LEVEL CHANGE

Sea level is neither constant nor uniform everywhere, but changes continually as a result of interacting processes that operate on timescales ranging from hours (e.g., tides) to millions of years (e.g., tectonics). Processes that affect ocean mass, the volume of ocean water, or sea-floor topography cause sea level to change on global scales. On local and regional scales, sea level is

[3] More than 20 such models from around the world were analyzed and compared through the World Climate Research Program's Coupled Model Intercomparison Project (CMIP3). See <http://www-pcmdi.llnl.gov/ipcc/about_ipcc.php>.

also affected by vertical land motions and local climate and oceanographic changes. The primary factors that contribute to global and local sea-level change are illustrated in Figure 1.4 and discussed below.

Global Sea-Level Change

Global sea level has varied significantly throughout Earth's history. Sediment and ice-core records of these changes provide a pre-anthropogenic context for understanding the nature and causes of current and future changes. Over the past 2.5 million years, large continental ice sheets grew during long intervals of cold global temperatures (glacial periods or ice ages) and retreated during intervals of warm global temperatures (interglacial periods). Traces of paleo-shorelines, found along many of the world's coastlines, provide robust evidence that global mean sea level was at least 6 m higher during the last interglacial period (~116,000–130,000 years ago) than at present (Kopp et al., 2009). During the Last Glacial Maximum (~26,000 years ago), approximately 40×10^6 km^3 of sea water was transferred to the continents and stored as ice. During that period, ice sheets covered much of North America, northern Europe, and parts of Asia, and sea levels were 125–135 m lower than present (Peltier and Fairbanks, 2006; Clark et al., 2009). The onset of deglaciation more than 20,000 years ago (Peltier and Fairbanks, 2006) caused sea level to rise at an average rate of about 10 mm yr^{-1} (Alley et al., 2005). Empirical and glacial isostatic modeling studies suggest that the rate of ice melt dropped significantly 7,000 years ago (Gehrels, 2010), then declined steadily to a value of zero change around 2,000 years ago (Fleming et al., 1998; Peltier, 2002b; Peltier et al., 2002). Geological data from salt marshes show a clear acceleration from relatively low rates of sea-level change during the past two millennia (order 0.25 mm yr^{-1}; Figure 1.5) to modern rates (order 2 mm yr^{-1}) sometime between 1840 and 1920 (Kemp et al., 2011).

Since the industrial era began, changes in global sea level have been driven in part by the accumulation of greenhouse gases in the atmosphere, which trap heat and raise global temperatures. The primary processes responsible for modern sea-level rise are thermal expansion of ocean water and melting from glaciers,

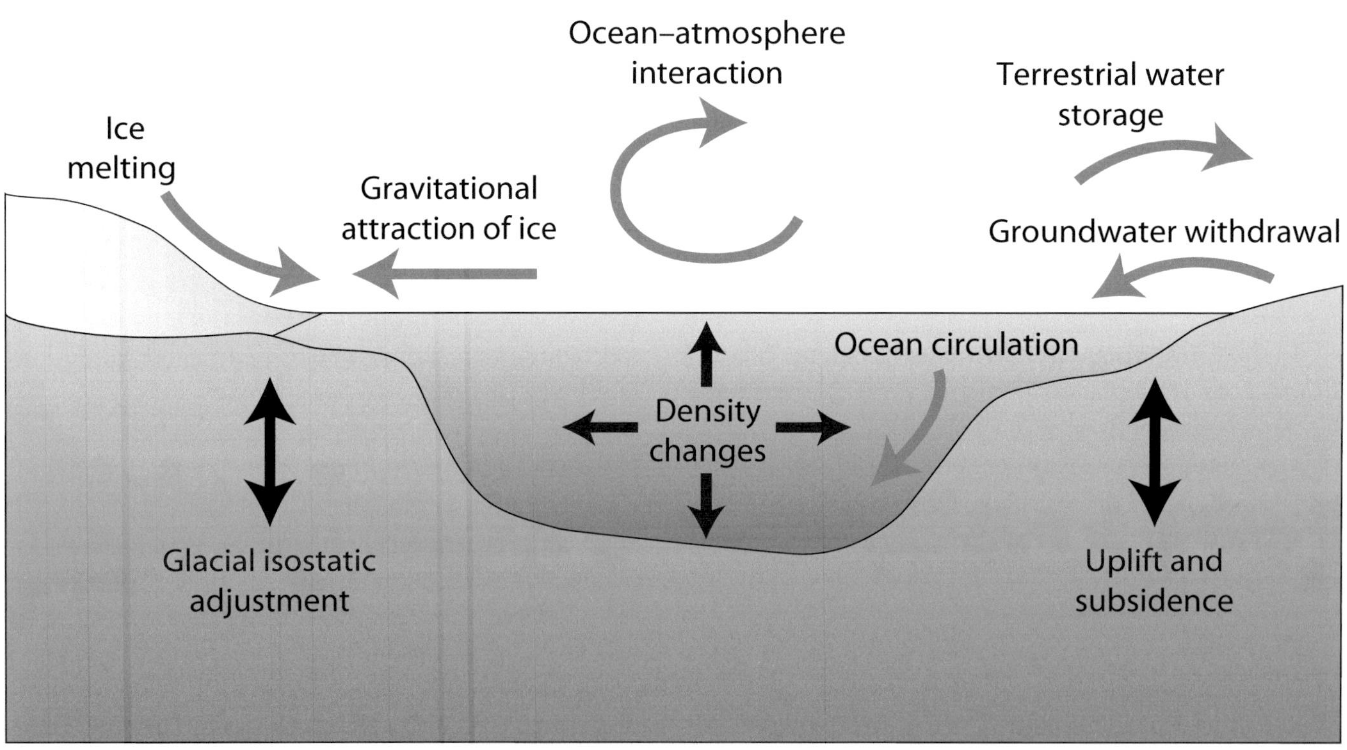

FIGURE 1.4 Processes that influence sea level on global to local scales. SOURCE: Modified from Milne et al. (2009).

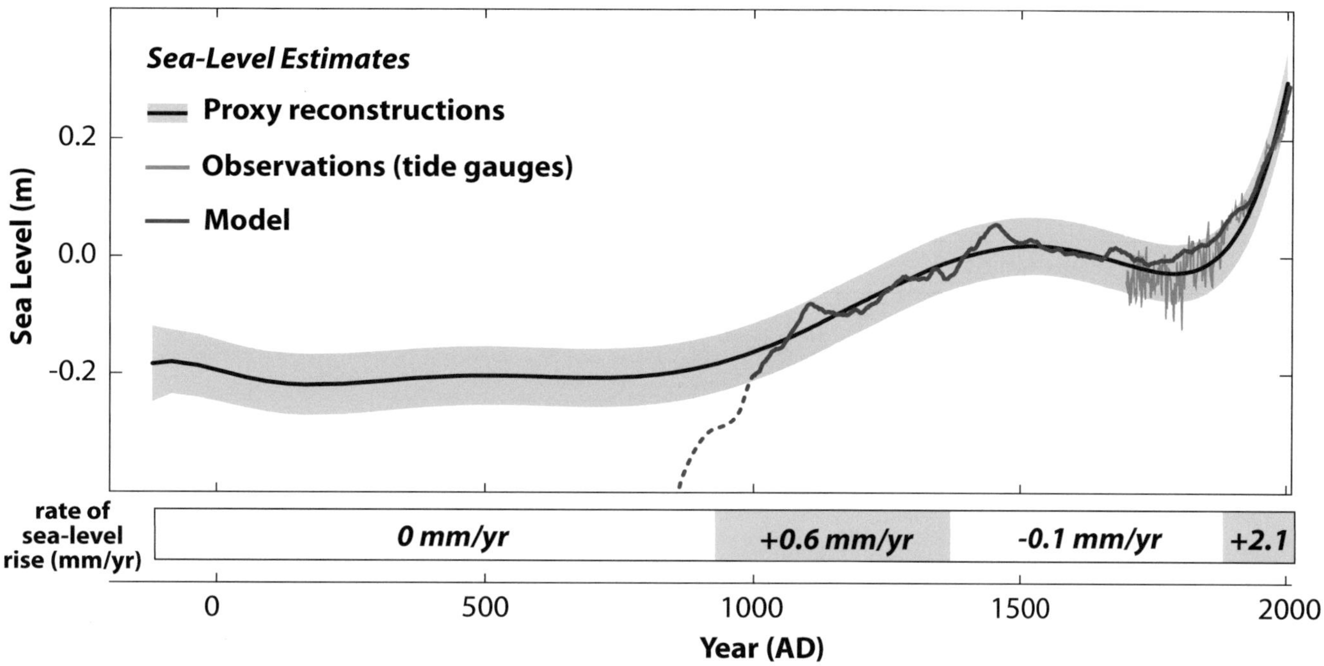

FIGURE 1.5 Sea-level estimates for the past 2000 years, adjusted for glacial isostatic effects, from proxy (geological) evidence (blue), tide gage observations (green), and modified semi-empirical model hindcasts (red). Dotted red line shows where the model hindcast deviates from the proxy record. The lower panel shows rates of sea-level change in mm yr^{-1} based on the proxy reconstructions. SOURCE: Data from Jevrejeva et al. (2008), Vermeer and Rahmstorf (2009), and Kemp et al. (2011).

ice caps, and the Greenland and Antarctic ice sheets (Figure 1.6). Changes in the amount of water stored in land reservoirs have a smaller effect on global sea level. In general, groundwater extraction transfers water to the ocean and causes sea level to rise, and filling of land reservoirs causes sea level to fall.

Local and Regional Sea-Level Change Along the U.S. West Coast

Relative (or local) sea level is the mean level of the sea with respect to the land, both of which change with time, as summarized below.

Changes in Ocean Levels

Sea level in the Pacific Ocean is affected by ocean circulation, short-term climate variations, storms, and gravitational and deformational effects of land ice changes. Changes in ocean circulation affect regional sea level on seasonal to decadal and longer timescales by redistributing ocean mass and altering seawater temperature and salinity patterns. These changes in ocean circulation are driven primarily by changes in winds and ocean surface density associated with the El Niño-Southern Oscillation (ENSO), which has a period of 2 to 7 years, and the Pacific Decadal Oscillation, which has a typical period of several decades. During a strong El Niño, a pulse of warm water in the eastern equatorial Pacific moves northward, forming a bulge in sea level along the California, Oregon, and Washington coasts. The low atmospheric pressures and west-southwest winds induced by an El Niño further elevate sea levels, which can reach 30 cm above normal levels for several months (Komar et al., 2011). Sea level is lower along the U.S. west coast during cooler La Niña conditions.

Large storms raise coastal sea level for the duration of the storm, usually several hours. The path and propagation speed of storms dictate wind direction and changes in barometric pressure, which in turn influence wind waves and high water. The strongest winds and hence the biggest waves along the west coast of the United States are typically generated during winter storms. Large waves along the California coast are also generated by tropical storms that reach the eastern Pacific in summer and early fall.

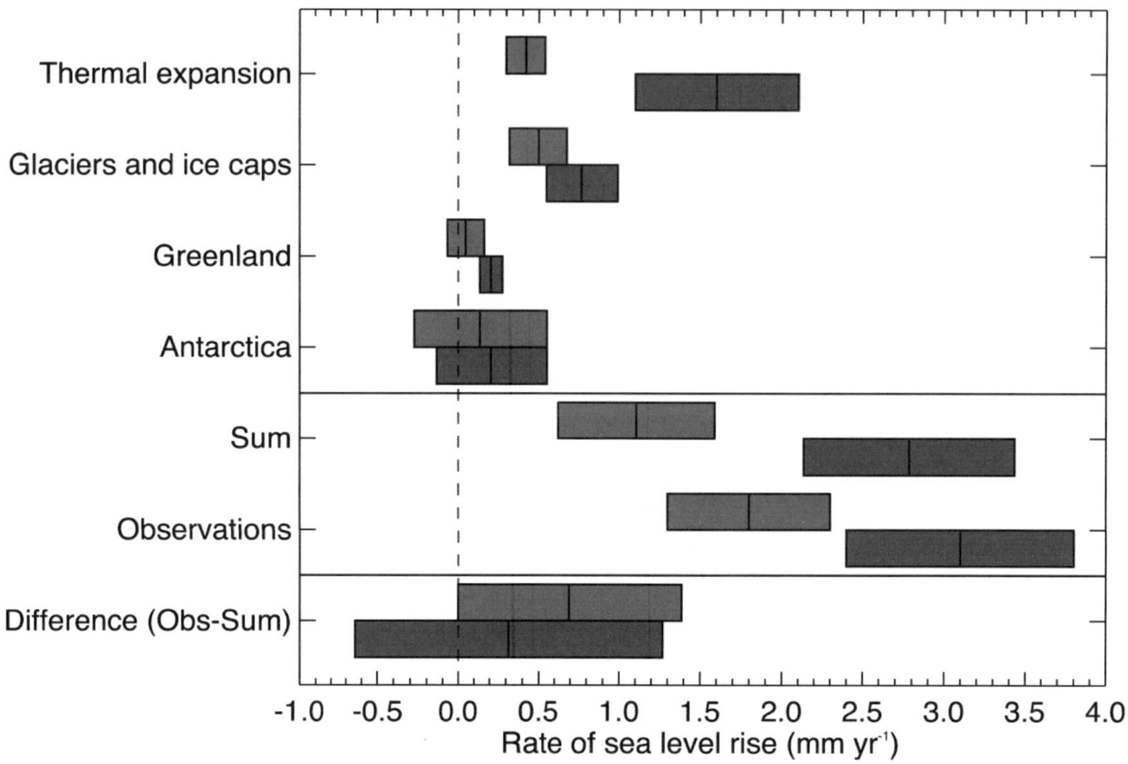

FIGURE 1.6 IPCC (2007) estimates of the primary contributions to global mean sea-level change for 1961 to 2003 (blue) and for 1993 to 2003 (brown), compared to the observed rate of global sea-level rise from tide gages and satellite altimetry. The bars represent the 90 percent error range. The relative contributions of these components has changed in recent years, as discussed in this report. SOURCE: Figure 5.21 from Bindoff et al. (2007).

Finally, the large mass of glaciers and ice sheets exerts an additional gravitational pull that draws ocean water closer. As the ice melts, the gravitational pull decreases, ice melt is transferred to the ocean, and the land and ocean basins deform in response to the loss of land ice mass. These gravitational and deformational effects create regional patterns of sea-level change. Modern melting of ice masses that are nearby (Alaska glaciers) or large (Greenland and Antarctic ice sheets) has the largest effect on sea level in the northeast Pacific Ocean, reducing the land ice contribution to local sea-level rise on the order of tens of percent. The influx of fresh melt water to the ocean also decreases seawater salinity and thus density near shore, which further contributes to regional sea-level variations.

Changes in Land Levels

Regional and local land motion along the U.S. west coast is caused by the ongoing response of the solid earth to a massive loss of ice at the end of the last ice age, tectonics, compaction of sediments, and the removal or addition of fluids from underground reservoirs. During the last glacial maximum, the weight of the ice depressed the land under the ice mass. As the ice melted, the land beneath rose at rates up to 50–100 mm yr^{-1} (e.g., Shaw et al., 2002), and the ocean floor subsided as ice melt was added to the ocean basins, exerting a considerable load (on the order of 100 t m^{-2} for a sea-level rise of 100 m; Figure 1.7). These isostatic adjustments produced a characteristic pattern of sea-level change, with land uplift and relative sea-level fall near the major ice centers, and relative sea-level rise everywhere else. Box 1.2 illustrates the effect of glacial isostatic adjustment on relative sea level along the west coast of the United States over the past 18,000 years.

The west coast of the United States is tectonically active, straddling three plate boundaries: the North American and Pacific plates, which slide past one another along the San Andreas Fault Zone in California, and the Juan de Fuca plate, which subducts under the North American plate along the Cascadia

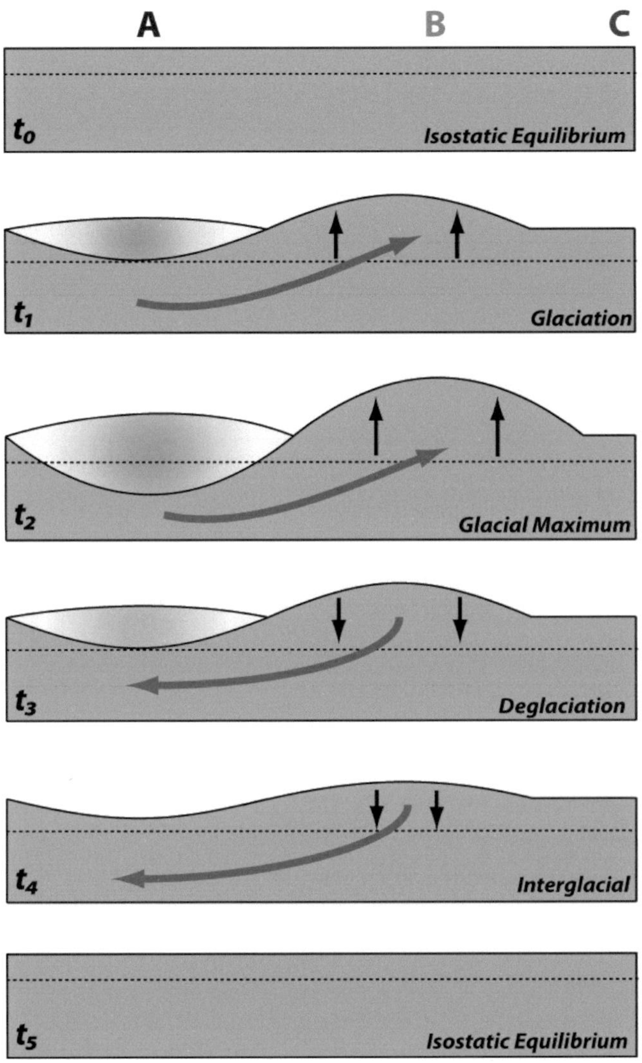

FIGURE 1.7 Response of the solid earth (brown) to the growth and melting of an ice sheet (blue) at increasing distances from the ice (A, B, and C). The addition of an ice sheet causes the land below it to subside and pushes (red arrow) a peripheral bulge outward. With deglaciation, the subsurface material flows back toward the area formerly covered by ice until equilibrium is again reached. SOURCE: Modified from Kemp et al. (2011).

Subduction Zone offshore Washington, Oregon, and northernmost California (Figure 1.8). In subduction zones, strain builds within the fault zone, causing the land to rise slowly before subsiding abruptly during a great (magnitude greater than 8) earthquake. The last great earthquake in the region occurred in 1700, causing a sudden rise in relative sea level of up to 2 m due to subsidence (Atwater et al., 2005). Since that event, much of the coastline of northern California, Oregon, and Washington has been slowly rising. Land motions along the San Andreas Fault Zone have less impact on sea level because the primary motions are horizontal and much of the fault is further inland.

Land subsidence resulting from sediment compaction and fluid (water, petroleum) withdrawal may cause relative sea level to rise. Compaction is particularly important in deltas and other coastal wetlands, where sediments have high water contents. Withdrawal of groundwater and petroleum increases the effective stresses in the surrounding sediments, resulting in consolidation and subsidence, which may be partially reversed by returning fluids to the subsurface.

GEOGRAPHIC VARIATION ALONG THE U.S. WEST COAST

How much coastal inundation can be expected with sea-level rise depends on the local geomorphology, which varies significantly along the west coast of the United States. The geomorphologic features along the coast are primarily the result of a collision between the North American and Pacific plates that began more than 100 million years ago and created steep coastal mountains, uplifted marine terraces, and sea cliffs. Over time, coastal lowlands developed, dominated by long sandy beaches, estuaries, and other wetlands. Most of the California coastline (72 percent or about 1,265 km) is characterized by steep, actively eroding sea cliffs, including about 1,040 km of relatively low-relief cliffs and bluffs, typically eroded into uplifted marine terraces (Figure 1.9), and 225 km of high-relief cliffs and coastal mountains (Figure 1.10). The remaining 28 percent of the coastline is relatively flat and comprises wide beaches, sand dunes, bays, estuaries, lagoons, and wetlands.

The coast of Oregon is dominated by resistant volcanic headlands separated by areas of lower relief. The latter are characterized by uplifted marine terraces, valleys where rivers emerge at the shoreline, and associated estuaries, sand spits, beaches, and dunes. The most extensive sand spits occur along the northern Oregon coastline. The longest continuous beach extends about 96 km, from Coos Bay to Heceta Head, near Florence. The largest coastal dune complex in the United States backs this region (Figure 1.11). Many of the estuarine wetlands have been diked, primarily to provide pasturelands.

BOX 1.2
Changes in Relative Sea Level Along the U.S. West Coast Since the Last Glacial Maximum

During the last ice age, northern Washington was covered by the Laurentide Ice Sheet. When the ice sheet retreated, coastal areas that had been depressed under the weight of the ice sheet were flooded. Relative sea level peaked in that area about 17,000 years ago, reaching values of about 90 m above present in Anacotes (#1 in the Figure) and about 40 m above present in Seattle (#2). Subsequent glacio-isostatic uplift caused relative sea level to fall to its lowest levels about 12,000 years ago (about -40 m in Anacortes and -55 m in Seattle). Relative sea level then rose as ice meltwater was transferred to the oceans and the Laurentide Ice Sheet peripheral bulge began to collapse, causing coastal subsidence.

Glacio-isostatic contributions were much lower in southern Washington, Oregon, and northern California (#3–#9 in the Figure) than for northern Washington, but they were still a dominant influence on sea level. In this area, rates of relative sea-level rise slowed as the effects of glacio-isostatic subsidence decreased. In Eureka, California (#7), for example, relative sea level rose at an average rate of about 7.5 mm yr^{-1} between 10,000 and 6,000 years before present, then rose at a decreasing rate.

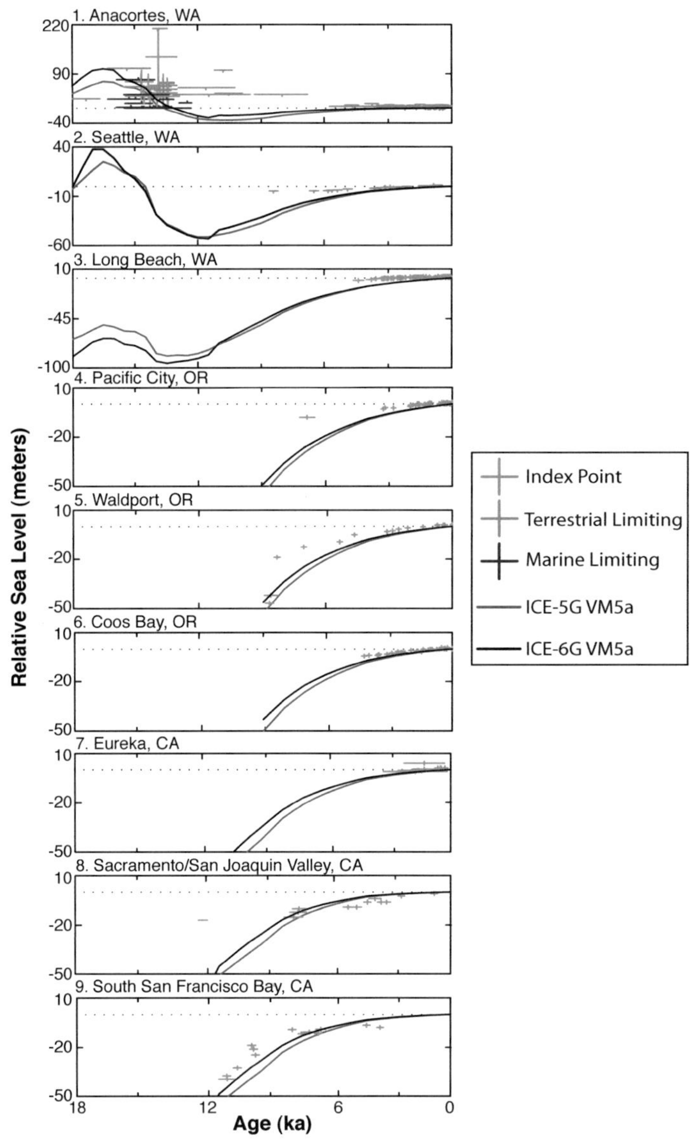

FIGURE Reconstruction of changes in relative sea level over the past 18,000 years for nine locations in Washington, Oregon, and California. Green crosses (index points) represent former sea levels inferred from dated organic sediment in salt and fresh water marshes. Limiting data are from marine shells (blue crosses) and terrestrial peat (orange crosses) that must have been laid down below and above mean sea level, respectively. Red and black lines are model predictions (Peltier and Drummond, 2008; Argus and Peltier, 2010; Peltier, 2010). SOURCE: Data provided by Richard Peltier, University of Toronto.

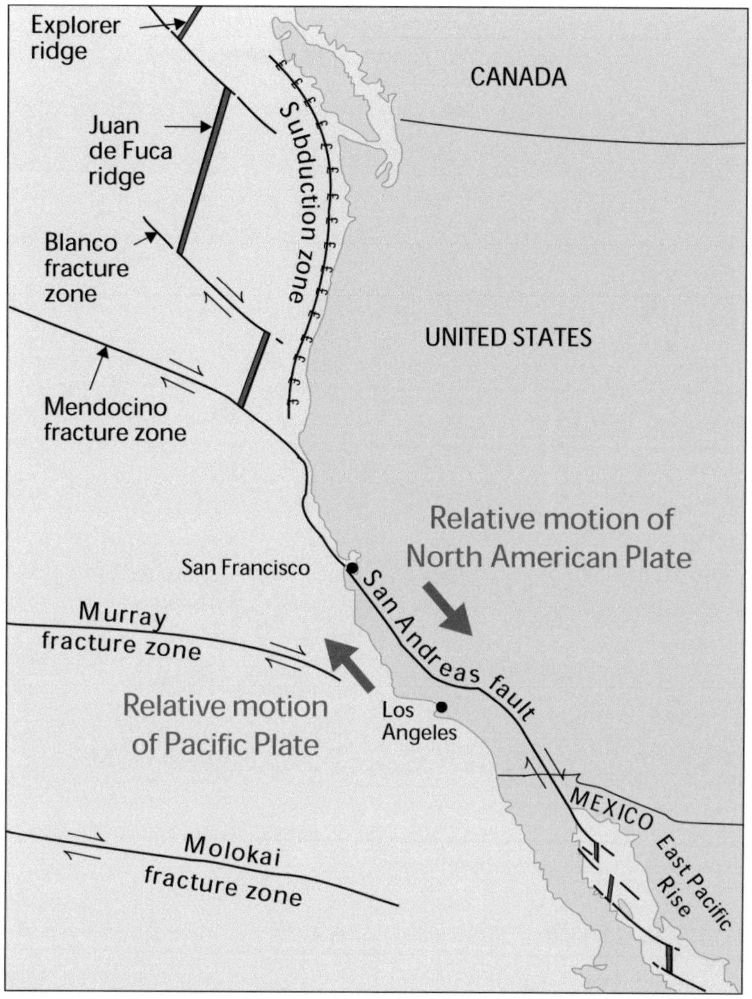

FIGURE 1.8 Major tectonic features along the western United States. Subduction of the oceanic Juan de Fuca and Gorda plates beneath the North American Plate occurs along the Cascadia Subduction Zone, which extends more than 1,000 km from Mendocino, California, to Vancouver Island. South of Cape Mendocino, the North American and Pacific plates slide past one another along the San Andreas Fault Zone. The land west of the San Andreas Fault, from San Diego to Cape Mendocino, is moving northwest relative to the rest of North America. SOURCE: U.S. Geological Survey, <http://pubs.usgs.gov/gip/dynamic/understanding.html>.

The shoreline of southern Washington is dominated by depositional landforms. Beaches, mostly backed by dunes, some developed, extend northward about 100 km from the mouth of the Columbia River to the mountainous Olympic Peninsula (Figure 1.12). The Long Beach Peninsula near the Columbia River and Grays Harbor include some of the most extensive wetlands in Washington, outside of Puget Sound. Some of these wetlands are being restored (e.g., Figure 1.13). Small coastal developments are present on portions of the peninsula and on the low-lying coastal areas to the north.

ORGANIZATION OF THE REPORT

This report evaluates changes in sea level in the global oceans and along the coasts of California, Oregon, and Washington for 2030, 2050, and 2100. Chapter 2 describes methods for measuring sea level and presents recent estimates of global sea-level rise. Chapter 3 updates the IPCC (2007) estimates of the major components of global sea-level change—thermal expansion of ocean water, melting of glaciers and ice sheets, and transfers of water between land reservoirs and the oceans. Chapter 4 assesses the factors that influence sea-level change along the U.S. west

FIGURE 1.9 Uplifted marine terraces, Santa Cruz County, California. SOURCE: Copyright 2002–2012 Kenneth & Gabrielle Adelman, California Coastal Records Project, <www.Californiacoastline.org>.

FIGURE 1.10 Steep rocky cliffs of the Marin Headlands north of San Francisco, California. SOURCE: Copyright 2002–2012 Kenneth & Gabrielle Adelman, California Coastal Records Project, <www.Californiacoastline.org>.

FIGURE 1.11 Oregon Dunes National Recreation Area. The largest coastal dune field in the United States has developed along the central Oregon coast and extends inland up to 3 km. SOURCE: Gary Griggs, University of California, Santa Cruz.

FIGURE 1.12 Long Beach Peninsula, Washington. Sandy beaches backed by dunes dominate the southern coast of Washington. SOURCE: Courtesy of Phoebe Zarnetske, Oregon State University.

FIGURE 1.13 Tidal wetlands along the mouth of the Nisqually River, Washington, are being restored following removal of a dike built a century ago to drain the area for cattle ranching. SOURCE: Courtesy of Carl Safina; photo taken for the PBS television series *Saving the Ocean*.

coast, including regional changes in ocean circulation, climate-induced changes in storms, gravitational and deformational effects of land ice change, and vertical land motions. It also summarizes the results of the committee's analysis of tide gage and GPS records from the California, Oregon, and Washington coasts, which is discussed in detail in Appendix A. Sea-level data from the northeast Pacific Ocean is presented in Appendix B. Data and uncertainties associated with the analysis of gravitational and deformational effects of land ice change are given in Appendix C. The tide gage and vertical land motion analyses draw on leveling data, and a description of leveling data compiled and analyzed for California by James Foster, University of Hawaii, appears in Appendix D. Chapter 5 summarizes recent projections of global and regional sea-level rise and presents the committee's projections for 2030, 2050, and 2100. The method used to project the cryospheric component of global sea-level rise is described in Appendix E. Chapter 5 also describes what rare, extreme events, such as a great earthquake along the Cascadia Subduction Zone, might mean for local sea-level rise. Chapter 6 summarizes the literature on natural shoreline responses to and protection from sea-level change. Biographical sketches of committee members are given in Appendix F, and a list of acronyms and abbreviations appears in Appendix G.

2

Measured Global Sea-Level Rise

Rates of global sea-level rise over the past several millennia are inferred from geological and archeological (proxy) evidence. Modern rates are estimated using tide gage measurements, which in some places date back to the 17th century, and satellite altimetry measurements of sea-surface heights, which have been available for the past two decades. Gravity Recovery and Climate Experiment (GRACE) satellite measurements, beginning in 2002, offer a possible additional estimate of global sea level.

Following a few thousand years of relative stability, global sea level began rising shortly after the beginning of the industrial era. The Intergovernmental Panel on Climate Change (IPCC) Fourth Assessment Report estimated that more modern rates of sea-level rise began sometime between the mid-19th and mid-20th centuries, based on geological and archeological observations and some of the longest tide gage records (Bindoff et al., 2007). Tide gage measurements indicate that global mean sea level rose 1.7 ± 0.5 mm yr^{-1} over the 20th century and 1.8 ± 0.5 mm yr^{-1} from 1961 to 2003. Rates from satellite altimetry and tide gages were higher from 1993 to 2003—3.1 ± 0.7 mm yr^{-1}—but the IPCC was unable to determine whether the higher rate was due to decadal variability of the oceans or to an acceleration in sea-level rise. This chapter describes how sea level is measured and summarizes rates of sea-level rise estimated since the IPCC Fourth Assessment Report was published.

PROXY MEASUREMENTS

Salt-marsh sediments, micro-atolls, and archaeological indicators are capable of capturing sub-meter-scale sea-level changes during the past 2000 years (Box 2.1). The most robust signal in these proxy records is an acceleration from relatively low rates of sea-level change during the past two millennia (order 0.1 mm yr^{-1}) to higher modern rates of sea-level rise (2–3 mm yr^{-1}; e.g., Lambeck et al., 2004; Gehrels, 2010; Kemp et al., 2011). Both the magnitude and timing of the acceleration vary among reconstructions, likely because of different assumptions about the underlying geophysical processes and uncertainties in determining height and time from proxy records. Recent reconstructions place the onset of acceleration in sea-level rise between 1840 and 1920 (Donnelly et al., 2004; Gehrels et al., 2006, 2008; Kemp et al., 2009, 2011). This late 19th or early 20th century acceleration in sea-level rise is also visible in the longest tide gage records of Brest (Wöppelmann et al., 2008), Amsterdam (Jevrejeva et al., 2008), Liverpool (Woodworth, 1999), Stockholm (Ekman, 1988), and San Francisco (Breaker and Ruzmaikin, 2010).

TIDE GAGES

Tide gages measure the water level at the location of the gage (Box 2.2). Originally designed for navigational purposes, the first gages began operating in the ports of Stockholm, Sweden, and Amsterdam, The Netherlands, in the 17th century. There are now more

BOX 2.1
Inferring Sea Level from Proxy Measurements

Sea-level "proxies" are natural archives that record rates of sea-level rise prior to the mid-19th century, when tide gage measurements became relatively common. Proxy indicators are generally calibrated against data from modern instruments and then used to reconstruct past sea levels. Three types of proxy archives can be measured with sufficient precision to be compared with the instrumental record: salt-marsh sediments, micro-atolls, and archaeological observations. Stratigraphic sequences from salt marshes record changes in the frequency and duration of tidal inundation, and thus past sea levels. The recent discovery of correlations between microfossils, such as foraminifera, and tidal elevation has significantly improved the precision of many sea-level reconstructions based on salt marshes (Horton and Edwards, 2006). Coral microatolls grow in a narrow range of sea levels. Growth at the upper surface of the coral potentially records fluctuations in relative sea level (e.g., Smithers and Woodroffe, 2001). Finally, some archaeological observations are relatable to sea level, including coastal water wells and Roman fish ponds (e.g., Lambeck et al., 2004).

Detailed proxy studies have not been done along the west coast of the United States. An example of the use of salt-marsh sediments from North Carolina to estimate rates of sea-level rise is shown in the figure below. Analysis of sediment cores suggest that the rate of sea-level rise changed three times: increasing between 853 and 1076, decreasing between 1274 and 1476, then substantially increasing between 1865 and 1892 (Kemp et al., 2011).

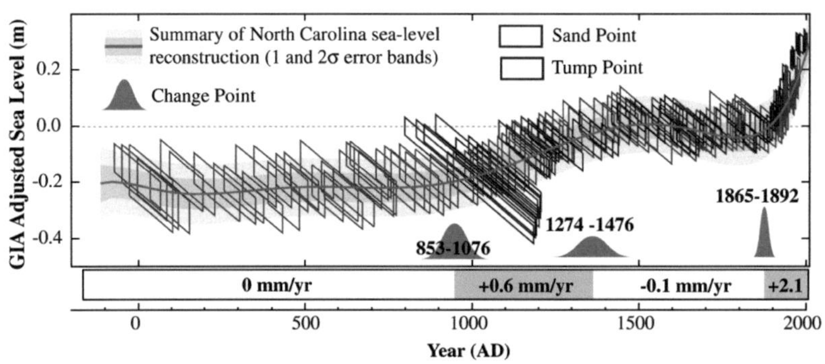

FIGURE Two thousand years of sea-level rise estimates from two North Carolina salt marshes (Sand Point and Tump Point). Errors in the data are represented by parallelograms; the correction for glacial isostatic adjustment is larger at the old end of the error box. The red line is the best fit to the sea-level data. Green shapes indicate when significant changes occurred in the rate of sea-level rise. SOURCE: Kemp et al. (2011).

than 2,000 tide gages worldwide, most of which were established since 1950 (Jevrejeva et al., 2006).

By averaging the water levels measured at the gage over a long period of time (daily, monthly), the effect of daily tides is removed, leaving only the relative sea level. This water level reflects not only the sea level, but also the effects of the weather, such as persistent wind systems and changes in atmospheric pressure; interannual to decadal climate variability; changes in oceanic currents; and vertical motions of the land on which the gage sits. These effects must be removed from the tide gage measurement to obtain the change in sea level caused by changes in ocean water volume or mass (see Appendix A).

The global mean sea level is determined by spatially averaging all of the qualified tide gage records from around the world. Spatial averaging provides a means to avoid bias due to regional climate variations. Sampling bias due to the small number of tide gages, particularly before 1950, and their concentration in the Northern Hemisphere and along coasts and islands is a major source of uncertainty in sea-level change estimates (Peltier and Tushingham, 1989; Church, 2001; Holgate and Woodworth, 2004). Long tide gage records (e.g., at least 50–60 years) are commonly used to average out decadal variability of the oceans' surface (Douglas, 1992).

The rate of sea-level change is estimated by fitting a curve through the historical tide gage readings. The curve could be a straight line or a higher order polynomial over the whole length of the record or shorter sections. More sophisticated data-dependent decompositions of the tide gage record also have been used (e.g., Peltier and Tushingham, 1989; Moore et al.,

BOX 2.2
Tide Gage Measurements

Tide gages measure the height of the water relative to a monitored geodetic benchmark on land (Figure). Tide gages originally used a float to track the water level inside a vertical tube. The bottom of the tube was closed except for a hole that permitted a small amount of water to enter the tube with time, thus serving as a temporal filter. Slow changes in the sea surface caused by tides or storm surges have sufficient time to fill the tube, while passing waves do not. Today, electronic sensors or bubbler gages have replaced tide gage floats.

Two organizations collect and preserve tide gage records from around the world: the Global Sea Level Observing System, which has established a network of 290 tide gages worldwide; and the Permanent Service for Mean Sea Level, which stores and disseminates the tidal records from more than 2,000 stations around the world.

FIGURE Examples of tide gage stations. (*a*) A float and stilling-well gage at Holyhead, UK. SOURCE: UK National Oceanography Centre. (*b*) A float gage at Vernadsky, Antarctica. SOURCE: British Antarctic Survey. (*c*) A radar tide gage at Alexandria, Egypt. SOURCE: Courtesy of T. Aarup, Intergovernmental Oceanographic Commission. (*d*) An acoustic gage at Vaca Key, Florida. Acoustic gages now form the majority of the U.S. sea-level network. SOURCE: National Oceanic and Atmospheric Administration.

2005; Jevrejeva et al., 2006). Because sea level exhibits considerable interannual and decadal variability, the calculated rate of change depends on the length and start date of the record used. For example, Church and White (2006) found that the global rate of sea-level rise was 1.7 ± 0.3 mm yr^{-1} for the 20th century, 0.71 ± 0.4 mm yr^{-1} for 1870–1935, and 1.84 ± 0.19 mm yr^{-1} for 1936–2001. Their results are shown in Figure 2.1, compared to other independent estimates of global sea-level rise from tide gages.

The time dependency of global sea level can be seen in the analysis of Church and White (2011), who calculated the sea-level rise using 16-year moving windows of data, as shown in Figure 2.2 (see also Box A.1 in Appendix A). In this example, the linear trend in global sea-level rise was 1.7 mm yr^{-1} from 1900 to 2009, with some 16-year intervals yielding rates of 2–3 mm yr^{-1} in the 1940s, 1970s, and 1990s. This variability has been attributed to natural climate variability (e.g., El Niño-Southern Oscillation [ENSO]), which causes short-term variations in global mean temperature, and to large volcanic eruptions, which briefly cool the Earth's surface and troposphere (e.g., Hegerl et al., 2007).

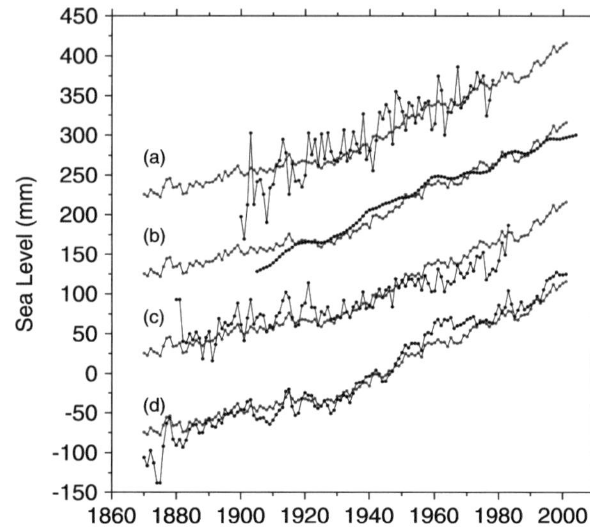

FIGURE 2.1 Global sea-level time series from Church and White (2006; red) compared with independent global sea-level time series from (*a*) Trupin and Wahr (1992), (*b*) Holgate (2007), (*c*) Gornitz and Lebedeff (1987), and (*d*) Jevrejeva et al. (2006) in black. Time series are arbitrarily shifted vertically for clarity. SOURCE: Woodworth et al. (2009).

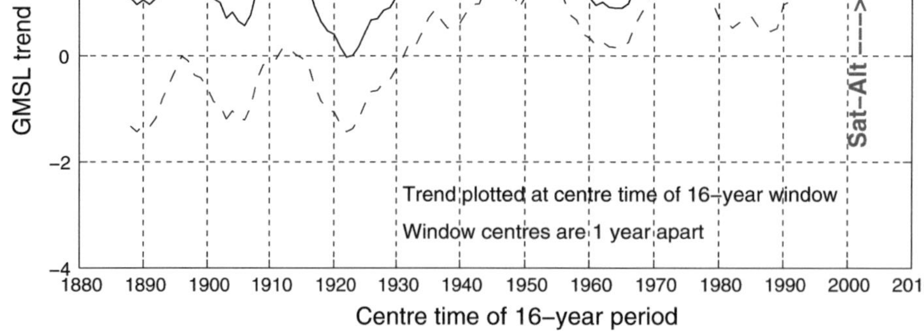

FIGURE 2.2 Sixteen-year running averages of global sea-level rise trends showing variability in rates over short timescales. SOURCE: Church and White (2011).

Recent Estimates

Recent estimates of rates of global sea-level rise are presented in Table 2.1. In general, the new estimates over the entire 20th century are similar to those reported in the IPCC Fourth Assessment Report. Rates for the last decade of the 20th century are higher and similar to IPCC (2007) rates estimated from satellite altimetry and confirmed by tide gages (see results of Jevrejeva et al., 2008; Merrifield et al., 2009; and Church and White, 2011). Because of natural temporal (e.g., Figure 2.2) and spatial variability in the sea-level signal, the meaning of the higher rates of global sea-level rise since the early 1990s is subject to interpretation. For example, Merrifield et al. (2009) attributed most of the recent rise to higher rates of sea-level rise in the Southern Hemisphere and tropical regions, which had been seen by Cabanes et al. (2001) in satellite altimetry data.

It is also possible that the recent higher rate of sea-level rise represents an acceleration in the long-term trend. The record of sea-level rise is punctuated by periods of acceleration and deceleration. Jevrejeva et al. (2008) used a Monte-Carlo-Singular Spectrum Analysis to remove the 2- to 30-year variability from more than 1,000 tide gage records from around the world. They found an acceleration of 0.01 mm yr^{-2} over the entire 300-year period, with 60- to 65-year periodicity in acceleration and deceleration for the preindustrial 18th and 19th centuries. The fastest rises in sea level occurred between 1920 and 1950 (up to 2.5 mm yr^{-1}) and between 1992 and 2002 (3.4 mm yr^{-1}; Jevrejeva et al., 2008). Many, but not all long tide gage records around the world show an acceleration in global sea-level rise around 1920–1930 and a deceleration around 1960 (Woodworth et al., 2009; see also Figure 2.1). Although Houston and Dean (2011) found a slight deceleration since 1930, Rahmstorf and Vermeer (2011) argued that this result reflects the choice of start date (1930) and the regional character of the gages used in their analysis.

Even if the higher rates since the 1990s represent a persistent acceleration in sea-level rise, significant additional acceleration would be required to reach commonly projected sea levels (e.g., Hansen, 2007; Rahmstorf, 2007; Vermeer and Rahmstorf, 2009). For example, taking a rate of 3.1 mm yr^{-1} from satellite altimetry, sea level would rise only 0.28 m over the next 89 years. To reach 1 m by 2100 would require a positive acceleration of 0.182 mm yr^{-2} for the entire time period, based on the following quadratic equation:

$$H = H_0 + (b \times t) + (c/2)\, t^2,$$

where H_0 is the current sea level, b is the linear rate of sea-level rise, and c is the acceleration in units of mm yr^{-2}. In this example, acceleration would account for more than 72 percent of the future sea-level rise. Such rapid acceleration is not seen in the 20th century tide gage record, except for short periods of time, such as the 1930s and the 1990s (Figure 2.2).

TABLE 2.1 Rates of Global Sea-Level Rise Estimated from Tide Gages

Source	Period	Sampling	Rate of Sea-Level Rise (mm yr^{-1})
IPCC (2007)	1900–2000	Not specified	1.7 ± 0.5
	1961–2003		1.8 ± 0.5
Church and White (2006)	1870–1935	400 gages, global coverage	0.71 ± 0.4
	1936–2001		1.84 ± 0.19
Holgate (2007)	1904–1953	9 gages, mostly Northern Hemisphere	2.03 ± 0.35
	1954–2003		1.45 ± 0.34
	1904–2003		1.74 ± 0.16
Shum and Kuo (2011)	1900–2006	500 gages, global coverage	1.65 ± 0.4
Domingues et al. (2008)	1961–2003	Not specified	1.6 ± 0.2
Church and White (2011)	1900–2009	400 gages, global coverage	1.7 ± 0.2
	1993–2009		2.8 ± 0.8
Jevrejeva et al. (2008)	1992–2002	1,023 gages, global coverage	3.4
Merrifield et al. (2009)	1993–2007	134 gages, global coverage	3.2 ± 0.4

SATELLITE ALTIMETRY

Satellite altimeters measure the sea-surface height with respect to the Earth's center of mass (Box 2.3). The satellite measurement also includes large-scale deformation of the ocean basins caused by glacial isostatic adjustment (GIA), which must be removed from the signal to obtain the ocean volume change. The global mean sea level is calculated by averaging measurements of sea-surface height made by the various altimeters, three of which are currently operating, which revisit a given spot on the Earth every 10 to 35 days.

Recent altimetry estimates of sea-level rise are similar to those reported in the IPCC Fourth Assessment Report, ranging from 3.2 to 3.3 mm yr^{-1} from 1992 to 2010 (Table 2.2), and 2.9 ± 0.4 mm yr^{-1} from 1985 to 2010. The latter estimate includes data from higher latitudes and has a gap in data from 1988 to 1991 (Figure 2.3). A recent analysis of the total error budget due to instrument, orbit, media propagation errors, and geophysical corrections and their drifts suggests an uncertainty of ~0.4–0.5 mm yr^{-1} (Ablain et al., 2009), in agreement with external calibration using data from island tide gages (Mitchum et al., 2010).

The regional variability in sea level seen in many tide gage analyses has been confirmed by satellite altimetry records. Figure 2.4 shows the regional variation in sea-level trends in the global oceans based on 25 years (1985–2010 with a 3-year data gap) of satellite altimetry data. The largest variations are in the western Pacific and eastern Indian oceans, where sea level has been rising much faster than the global mean (warm colors in Figure 2.4). Sea level has been dropping in other areas, including the eastern Pacific Ocean (cool colors in Figure 2.4). The IPCC concluded that these spatial patterns reflect interannual to interdecadal variability resulting from the El Niño-Southern Oscillation, the North Atlantic Oscillation, the Pacific Decadal Oscillation, and other climate patterns (Bindoff et al., 2007).

Satellite altimetry and tide gage estimates of sea-level change over the same timespan are in good agreement (e.g., Nerem et al., 2010). However, there are significant differences between long-term trends in tide gage records and the shorter satellite altimetry records. For example, Shum and Kuo (2011) estimated a tide-gage trend of 1.50 mm yr^{-1} for 1880–2008 and a satellite altimetry trend of 2.59 mm yr^{-1} for 1985–1987 and 1991–2010 (Figure 2.5). Differences in trends for the two types of measurements for other data periods have also been reported (e.g., Church and White, 2011). These differences are likely due to contamination of the altimetry trend by interannual or longer variations in the ocean (e.g., Willis et al., 2010; Shum and Kuo, 2011) and, to a smaller extent, to sampling biases. Satellite altimetry records are shorter than tide gage records but cover more of the global ocean (81.5°N–81.5°S in Figure 2.5). In addition, the sea-level signal from altimetry is dominated by the open ocean whereas the signal from tide gages is more strongly affected by the coastal ocean (e.g., Holgate and Woodworth, 2004).

GRAVITY RECOVERY AND CLIMATE EXPERIMENT (GRACE)

The GRACE mission makes detailed measurements of the Earth's gravity field and its variability over time. Among the gravity variations detected by GRACE are mass changes in the ocean and land reservoirs (e.g., land ice, groundwater) that contribute to sea-level change (Box 2.4). The land ice and water components are discussed in Chapter 3. For the ocean component, GRACE measures the ocean bottom pressure—the sum of the mass of the ocean and atmosphere above—at spatial resolutions of ~500 km. Ocean bottom pressure changes when winds move water across the ocean surface or when water is added to the oceans (e.g., through ice melt, stream runoff), increasing the ocean mass. The ocean mass change is determined by computing gravity field changes from the GRACE signal (Chambers et al., 2004; Tapley et al., 2004), then correcting for the effect of glacial isostatic adjustment and high frequency ocean responses to wind and surface pressure forcing. When combined with other observations—such as altimetry data that have been corrected for temperature and salinity effects—GRACE data offer a potential means of distinguishing how much global sea-level change is due to changes in mass and how much is due to changes in temperature and salinity.

Currently, however, there are difficulties associated with using GRACE data to infer ocean mass changes. Changes in gravity over the ocean, and thus the ocean bottom pressure signal, are small relative to

BOX 2.3
Satellite Radar Altimetry Measurements

The first altimeter mission observing the global ocean was launched in 1978 (Seasat), but routine measurements of sea level from satellites began with the launch of TOPEX/Poseidon (1992–2006) and ERS-1 (1991–2000), and continued with ERS-2 (1996–2011), Geosat Follow-on (1998–2001), Jason-1 (2001–present), Envisat (2002–present), Jason-2 (2008–present), and Cryosat-2 (2010–present). Although these satellites are sometimes maneuvered in geodetic phases or interleave orbits, they have occupied essentially the same ground tracks as 10-day, 17-day, or 35-day repeat orbits, providing a long data set of compatible observations. These satellites were equipped with radar altimeters to determine the distance between the satellite and the sea surface (see Figure). The location of the satellite, which has to be accurately known at all times, is determined using tracking data from the Satellite Laser Ranging network, the Doppler Orbitography and Radio-positioning Integrated by Satellite (DORIS) land-based beacons, and the Global Navigation Satellite System (GNSS). Using the range or range-rate information from these tracking systems, the position and velocity of the satellite are determined and the radial orbit is then calculated. The sea surface is estimated by averaging measurements taken over a 10-, 17-, or 35-day satellite track repeat cycle. The accuracy of the sea surface height measurements for TOPEX-class altimetry systems, considered to be the most accurate among the radar altimetry missions due to their optimal orbital sampling and high instrument precision, is a few cm (1 σ), after correcting for instrument and media errors and geophysical phenomena.

The TOPEX and Jason satellites measure(d) the global ocean to latitudes of 66° north and south. Satellite altimeters that extend observations into the polar ocean include Geosat (1984–1987) and Geosat Follow-on, which covered latitudes of 71° north and south; ERS-1 and -2 and Envisat, which cover latitudes of 81.5° north and south; and Cryosat-2, which covers latitudes of 88° north and south. Their repeat orbits are longer than the TOPEX and Jason satellites: 17 days for Geosat and Geosat Follow-on, 35 days for ERS-1 and -2 and Envisat, and 365 days with 30-day subcycles for Cryosat-2.

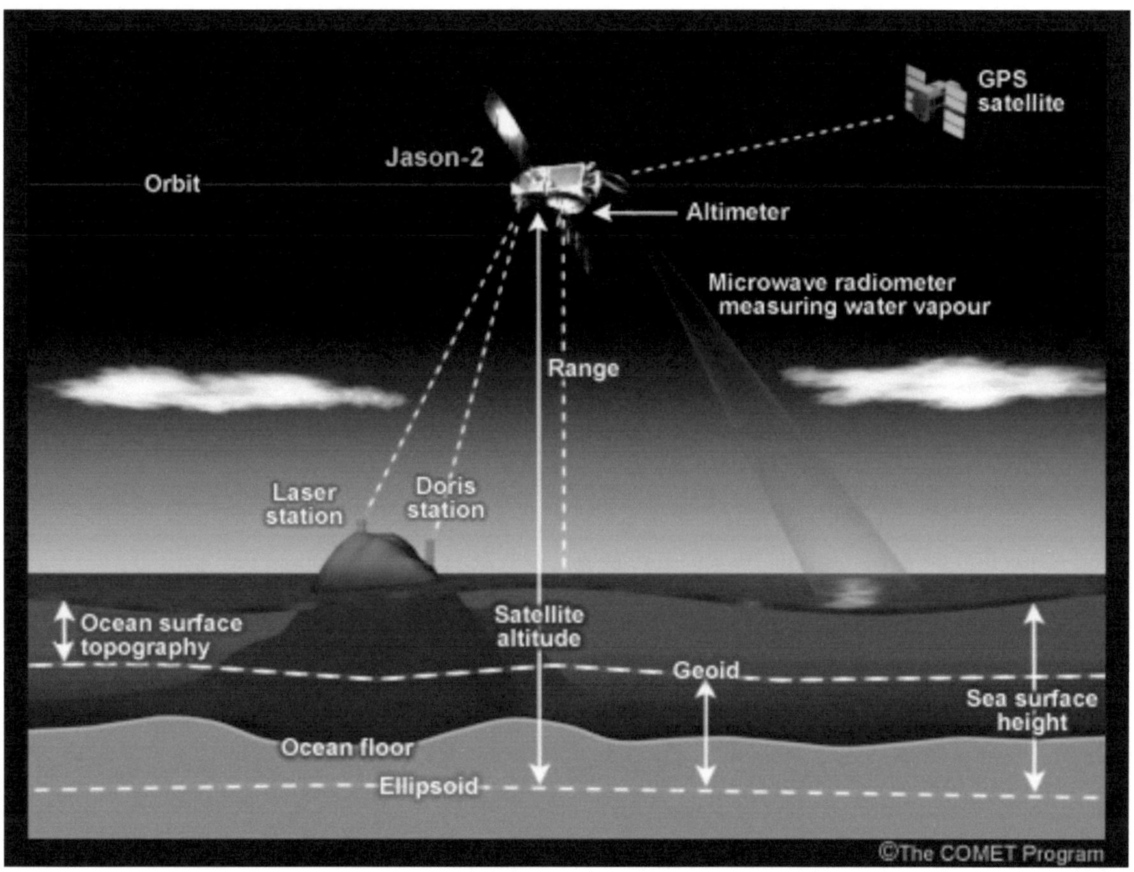

FIGURE The Jason-2 satellite uses a radar altimetry instrument to accurately measure sea-surface heights. SOURCE: COMET® Website at <http://meted.ucar.edu/> of the University Corporation for Atmospheric Research, sponsored in part through cooperative agreement(s) with the National Oceanic and Atmospheric Administration, U.S. Department of Commerce. ©1997–2011 University Corporation for Atmospheric Research. All rights reserved.

TABLE 2.2 Rates of Global Sea-Level Rise Estimated from Satellite Altimetry

Source	Period	Latitude	Instruments	Rate of Sea-Level Rise (mm yr^{-1})[a]
D. Chambers (personal communication)	1992–2010	± 66°	TOPEX and Jason-1, -2	3.3 ± 0.5
Nerem et al. (2010)	1992–2010	± 66°	TOPEX and Jason-1, -2	3.3 ± 0.5
Leuliette and Miller (2009)	1992–2010	± 66°	TOPEX and Jason-1, -2	3.2 ± 0.3
Cazenave et al. (2009)	1992–2010	± 66°	TOPEX and Jason-1, -2	3.3 ± 0.2
Church and White (2011)	1993–2009	± 66°	TOPEX and Jason-1, -2	3.2 ± 0.4
Shum and Kuo (2011)	1985–2010	± 81.5°	Geosat, Geosat Follow-on, ERS, TOPEX, Envisat, and Jason-1, -2	2.9 ± 0.5

[a] All rates were corrected for glacial isostatic adjustment using the ICE-5G (VM2) model (Peltier, 2004) and atmospheric pressure effects (see Appendix B).

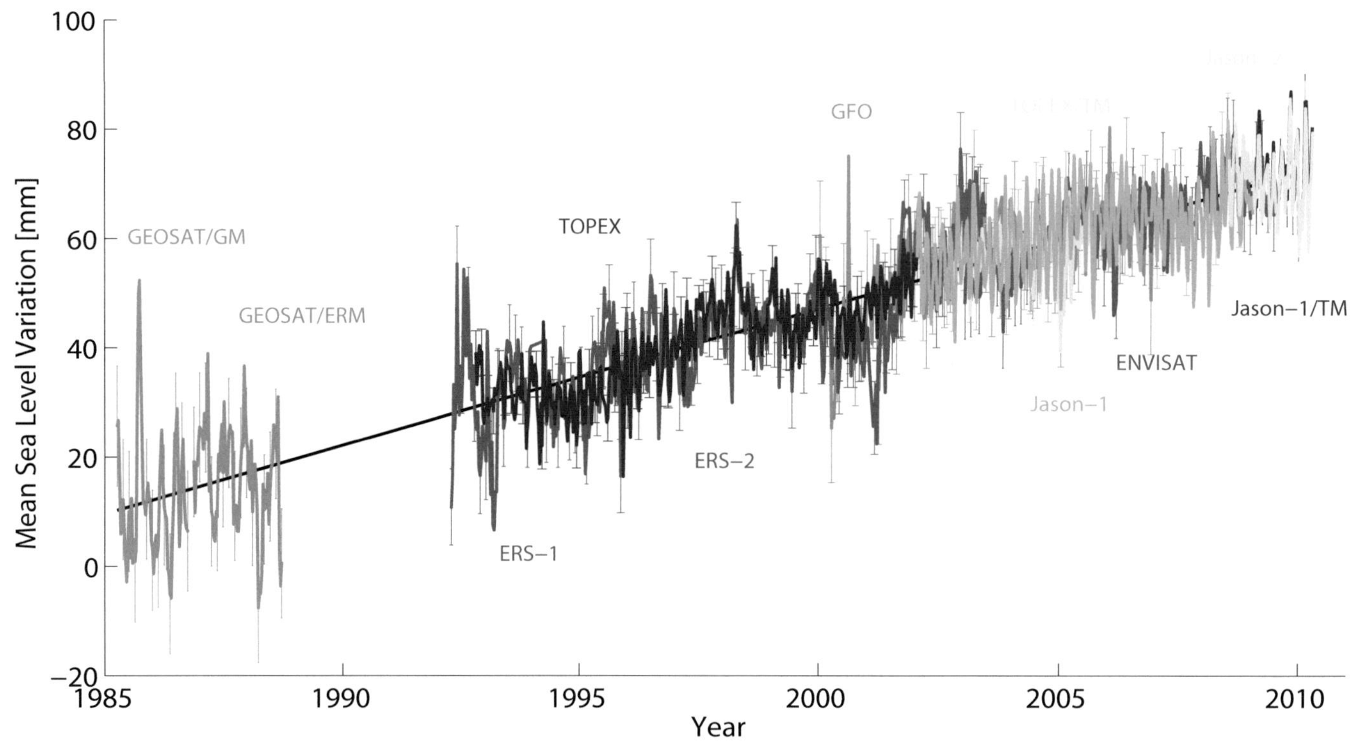

FIGURE 2.3 Global sea-level rise trends from different satellite altimeters for 1985–2010. The measured trend is 2.6 ± 0.4 mm yr^{-1}, and the trend corrected for glacial isostatic adjustment and atmospheric effects is 2.9 ± 0.4 mm yr^{-1}. Seasonal variations in the time series were not removed, but the trend was estimated simultaneously with periodicities associated with seasonal variations. SOURCE: Updated from Shum and Kuo (2011).

the GRACE accuracy limit and to the land gravity signal. Moreover, uncertainties in GIA models strongly affect the ocean mass calculated from GRACE (e.g., Cazenave et al., 2009). Finally, GRACE data must be adjusted to reduce high-frequency barotropic signals over the ocean and over land (Flechtner, 2007) and to account for motion of the geocenter (e.g., using laser ranging or Global Positioning System [GPS] data; Swenson et al., 2008). Once a consensus is reached on how to handle the processing and corrections, GRACE data may provide a valuable constraint on the ocean mass component of sea level and on the total sea-level budget.

CONCLUSIONS

Recent estimates of global sea-level rise are in close agreement with estimates in the IPCC Fourth Assess-

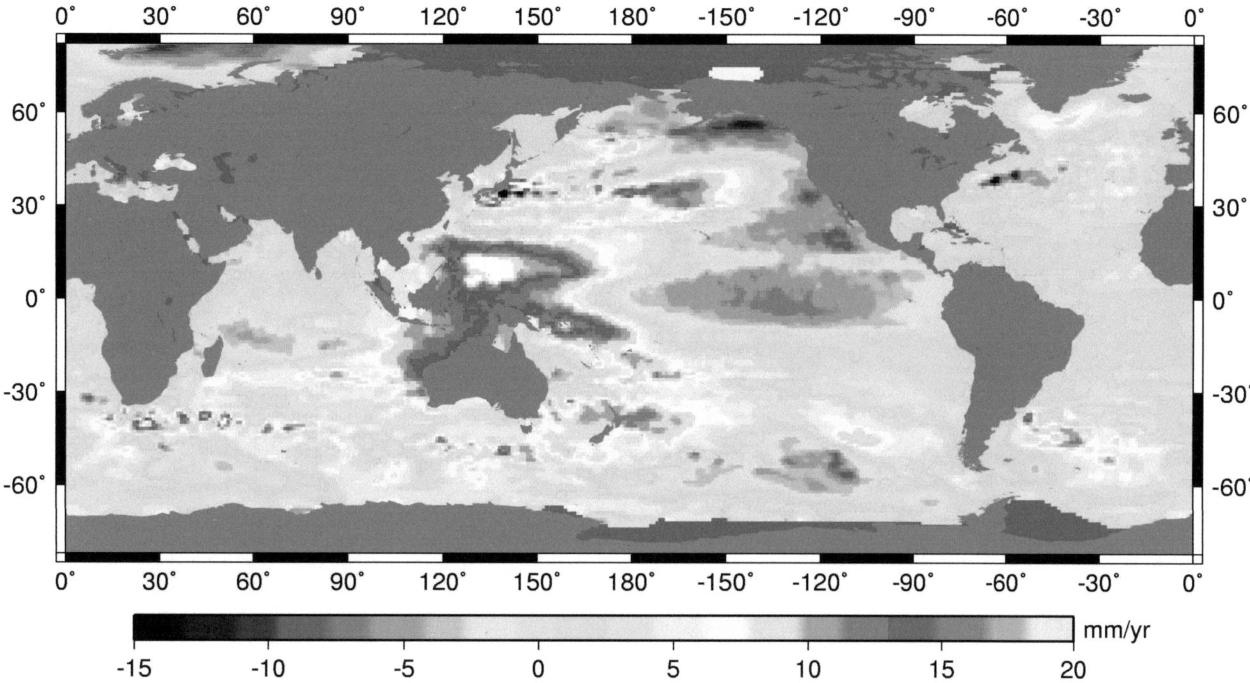

FIGURE 2.4 Regional variations in global sea-level rise based on observations from satellite altimetry from 1985 to 2010. The data were corrected for glacial isostatic adjustment, atmospheric barotropic pressure response, and various instrument, media, and geophysical effects. SOURCE: Updated from Shum and Kuo (2011).

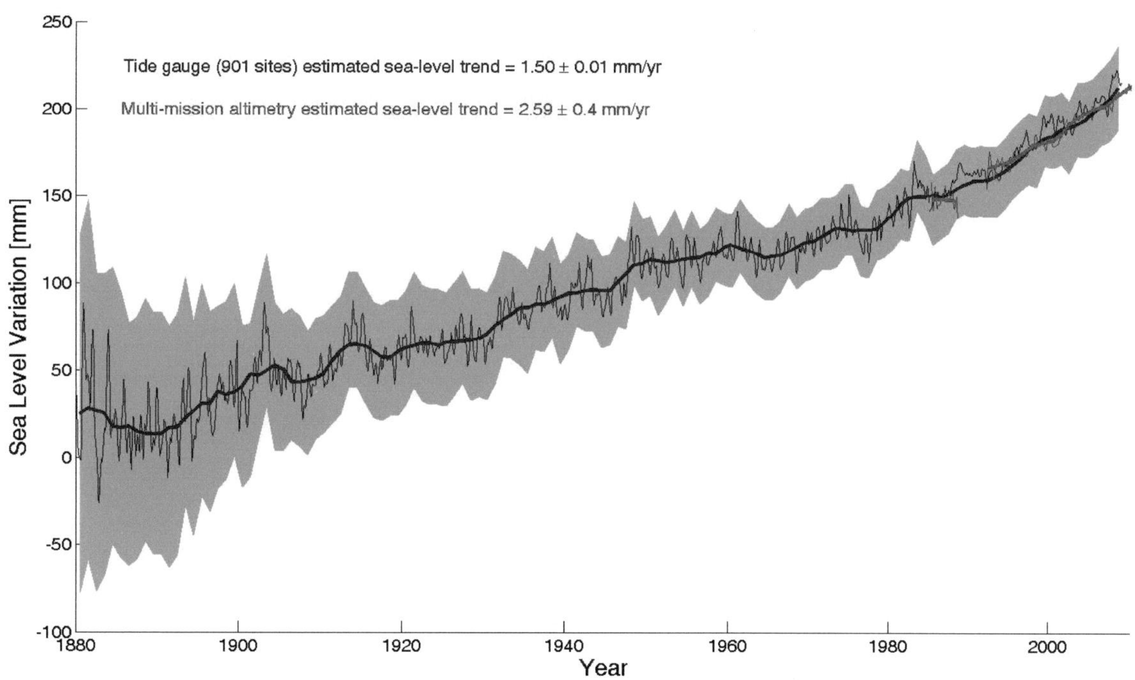

FIGURE 2.5 Comparison of sea-level time series from tide gages (1880–2008; blue lines) and from satellite altimetry (1985–1987 and 1991–2010; red lines) after corrections for atmospheric barotropic pressure effects and glacial isostatic adjustment (using the ICE-5G [VM2] model, Peltier, 2004). The thin blue line represents average monthly sea level from global tide gage data. The thick blue line represents yearly sea-level changes from a moving average of tide gage observations, and the shaded area represents the sea-level uncertainty, which reflects the number of gage sites used in the global averages, the number of data points, and the standard deviations of the fit of seasonal signals and the trend of the original gage time series. The thick red line is the yearly averaged altimetry sea-level data. SOURCE: Updated from Shum and Kuo (2011).

BOX 2.4
GRACE Measurements

The Gravity Recovery and Climate Experiment measures changes of the mass distribution on Earth. The twin satellites travel in the same polar orbit 500 km above the Earth, with one satellite leading the other by approximately 220 km (Figure). When the lead satellite passes over a region of relatively high mass, it will accelerate because of increased gravitational attraction and will increase the distance between the satellites. On the other side of the region of high mass, it will slow again. The same effect applies to the trailing satellite. By monitoring the changing distances between the satellites, and knowing their positions in space accurately via GPS and star cameras, the distribution of mass below the satellites can be determined. Mass redistributions of the Earth are manifested in temporal gravity signals with a monthly sampling and spatial resolution longer than 300–400 km (half-wavelength; Tapley et al., 2004). GRACE data can be used to measure changes in mass of the ocean and its land reservoirs (e.g., land ice and groundwater; see Chapter 3). Launched in 2002, the mission is expected to end in 2015.

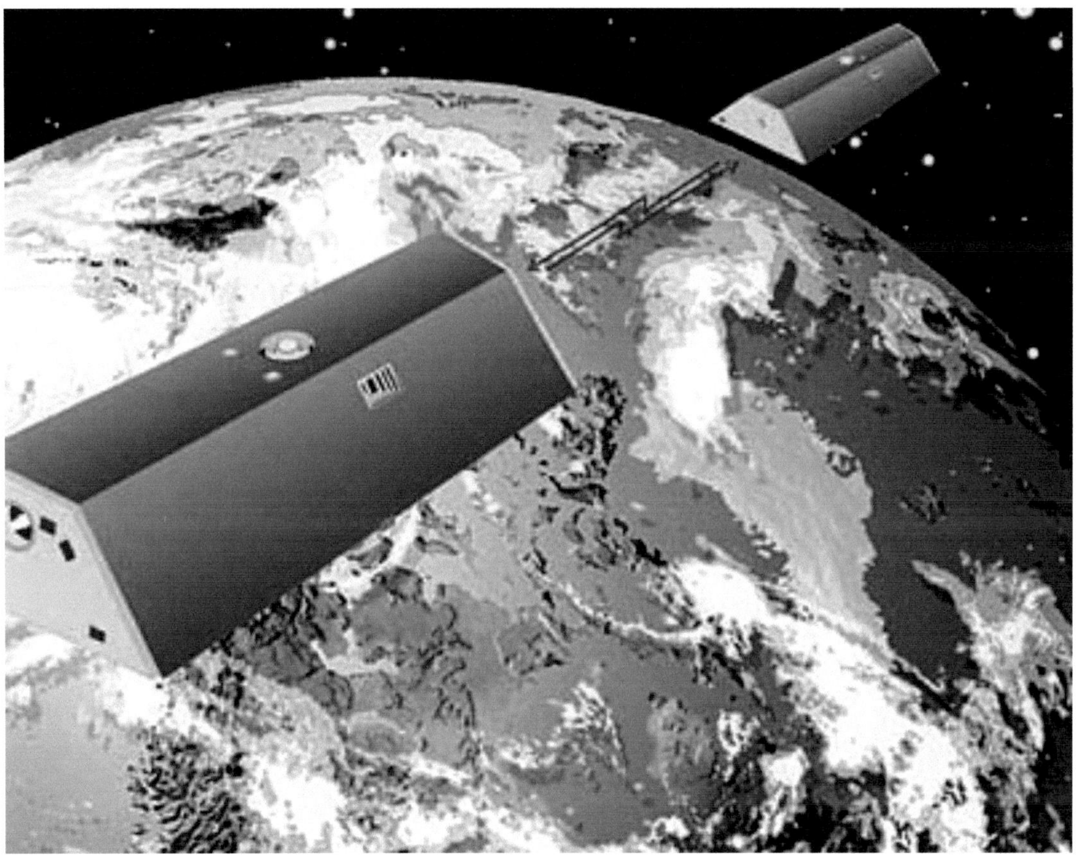

FIGURE An artist's concept of GRACE satellites with ranging link between the two craft. SOURCE: National Aeronautics and Space Administration.

ment Report, with long-term (50–100 years) rates of about 1.8 mm yr^{-1} estimated from tide gages, and recent (post-1990) rates of about 3.2 mm yr^{-1} estimated from satellite altimetry and tide gages. The higher rates of recent sea-level rise may reflect interannual and longer variations due to ENSO and other climate patterns. Increases of 3–4 times the current rate would be required to realize scenarios of 1 m sea-level rise by 2100. Such an acceleration has not yet been detected.

3

Contributions to Global Sea-Level Rise

Sea-level rise is governed by processes that alter the volume of water in the global ocean—primarily thermal expansion of sea water and transfers of water from terrestrial reservoirs, such as land ice and groundwater, to the ocean. The Intergovernmental Panel on Climate Change (IPCC) Fourth Assessment Report found that thermal expansion accounted for about one-quarter of the observed sea-level rise for 1961–2003, melting of land ice accounted for less than half, and changes in land water storage accounted for less than 10 percent (Bindoff et al., 2007). For the last 10 years of that period (1993–2003), the IPCC estimated that thermal expansion and land ice melt each contributed about half to the total sea-level rise. The improved agreement between estimates of the individual contributions and the total sea-level rise for the later time period was attributed to the availability of satellite altimetry data and other global ocean data sets and to better knowledge of the processes causing sea-level rise. Subsequent work has corrected instrument biases, reducing estimates of the thermal expansion contribution to sea-level rise, and recorded increased rates of land ice loss. In the most recent estimate, for 1993–2008, the contribution from land ice increased to 68 percent, the contribution from thermal expansion decreased to 35 percent, and land water storage contributed -3 percent (sea-level fall; Church et al., 2011).

This chapter evaluates the contributions of thermal expansion, glaciers, ice sheets, and other terrestrial sources of water to global sea-level rise. Each section begins with a summary of findings from the IPCC Fourth Assessment Report, then evaluates more recent results.

THERMAL EXPANSION

Sea level is affected by changes in the density of sea water, induced by temperature changes (thermosteric) or by salinity changes (halosteric). Freshening of the water column (halosteric expansion) has been estimated to account for about 10 percent of the global average steric sea-level rise during recent decades (e.g., Antonov et al., 2002; Munk, 2003; Ishii et al., 2006). However, only about 1 percent of the halosteric expansion contributes to the global sea-level-rise budget because ocean mixing increases the salinity and thus decreases the volume of the added freshwater (Bindoff et al., 2007). Consequently, only the thermosteric component is discussed below.

When the ocean warms, seawater becomes less dense and expands, raising sea level. Because warm water expands more than cold water with the same amount of heating, and seawater at higher pressure expands more than seawater at lower pressure, global sea-level change depends on the distribution of ocean temperature change throughout the ocean, from top to bottom. Thermosteric sea-level change is calculated from temperature and pressure measurements made from a wide variety of instruments that descend through the water column, are towed from ships, or are attached to moored and drifting buoys and profiling floats (see Johnson et al., 2006).

Estimates from the IPCC Fourth Assessment Report

The IPCC Fourth Assessment Report found that warming in all three of the major ocean basins has occurred over the past few decades (Bindoff et al., 2007). Further, thermal expansion of the global ocean (thermosteric sea-level rise) exhibits significant decadal and interannual variations. Thermosteric sea-level rise was estimated to account for approximately one-quarter of the observed rate of global sea-level rise from 1961 to 2003, contributing 0.32 ± 0.12 mm yr^{-1} down to 700 m depth and 0.42 ± 0.12 mm yr^{-1} down to 3000 m depth. For the last 10 years of that period (1993–2003), the contribution of thermal expansion was estimated to have increased to 1.5 ± 0.5 mm yr^{-1} above 700 m and 1.6 ± 0.5 mm yr^{-1} above 3,000 m, about half of the observed rate of global sea-level rise.

Recent Estimates

At about the time the IPCC Fourth Assessment Report was published, systematic efforts began to be made to correct for biases in the expendable bathythermograph (XBT) and mechanical bathythermograph (MBT) data, which constitute the majority of ocean temperature observations prior to 2002, and in Argo data (Box 3.1). These biases affected the temperature inferred from measurements and thus the calculated rate of thermosteric sea-level rise. Thermosteric sea-level trends have recently been reanalyzed using bias-corrected temperature data, and the record has been extended by new observations. In addition, a few new estimates of the thermosteric fraction of sea level have been made using data assimilation products and satellite data.

BOX 3.1
Bathythermograph and Argo Measurements

Bathythermographs are dropped from ships and transmit temperature via a thin wire as they sink through the water column. Mechanical bathythermographs (MBTs) record temperature at 5 m depth intervals down to approximately 285 m. Thus, they are useful only for studying the thermal structure of the upper ocean. The successor expendable bathythermographs (XBTs) can provide temperature profiles to depths of approximately 760 m (standard instruments) or 1,830 m (special instruments). Data from MBTs and/or XBTs are available since 1948.

Ocean profiling floats are deployed under the multi-national Argo program and by individual countries. Argo profiling floats began measuring the temperature and salinity of the upper 1,000–2,000 m of the ocean in 2000. The Argo array currently comprises more than 3,000 ocean profiling floats distributed around the world (see Figure). Data from these floats are collected via satellite.

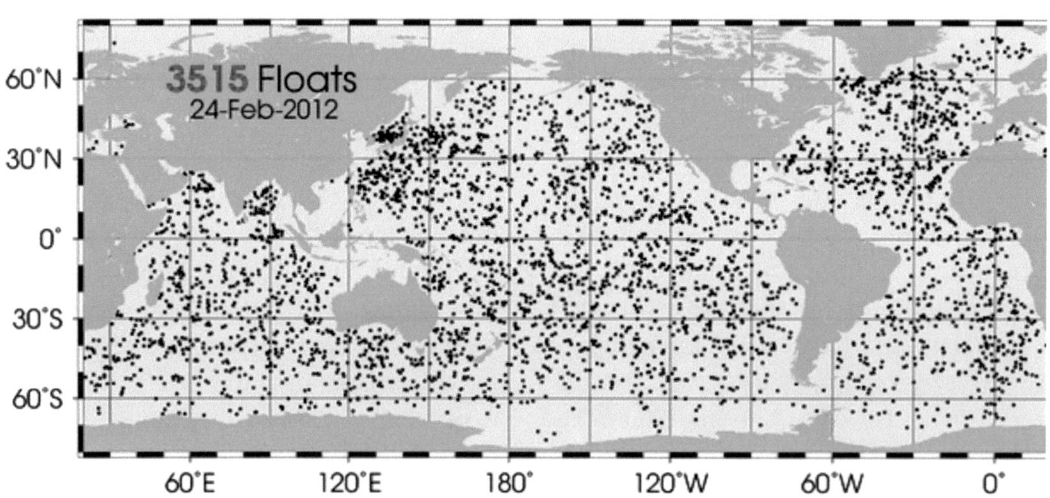

FIGURE Distribution of Argo profiling drifters on February 24, 2012. These floats measure salinity and temperature over the upper 1,000–2,000 m of the ocean. SOURCE: These data were collected and made freely available by the International Argo Program and the national programs that contribute to it (<http://www.argo.ucsd.edu>, <http://argo.jcommops.org>).

In Situ Data

A time-varying warm bias (systematically warmer temperature than the true value) has been found in the global XBT data, and a cold bias (systematically colder temperature than the true value) has been found in a small fraction of Argo float data (e.g., Gouretski and Koltermann, 2007; Wijffels et al., 2008; Ishii and Kimoto, 2009; Willis et al., 2009). XBT and MBT temperature observations are subject to instrument bias, such as depth bias. The depth of each temperature observation is calculated using a fall-rate equation and the time elapsed since the XBT entered the water. Inaccuracies in the fall rate affect the apparent depth at which the temperature profile is taken, which in turn causes a temperature bias that varies with depth. The MBT depth bias may have resulted from a delayed response by the diaphragm used to sense pressure and thus infer depth (Gouretski and Koltermann, 2007). The Argo biases were associated with a particular set of instruments deployed mainly in the Atlantic Ocean (Willis et al., 2009). The sensors on these instruments use pressure measurements to infer depth, but a flaw caused temperature and salinity values to be associated with incorrect pressure values, biasing the data.

Correcting for XBT depth bias reduced the magnitude of the interdecadal variability previously seen in the thermosteric sea-level signal during the 1970s (Domingues et al., 2008). An apparent sharp rise in thermosteric sea level during the 1970s was greatly decreased in the corrected data of Levitus et al. (2009), and essentially disappeared in the corrected data of Ishii and Kimoto (2009; compare the dotted and solid red and blue lines in Figure 3.1, top). Correcting for depth bias also changed the estimated rate of global thermosteric sea-level rise. For example, Ishii et al.'s (2006) original estimate of thermosteric sea-level rise for the upper 700 m was 0.26 ± 0.06 mm yr^{-1} from 1951 to 2005. After correcting XBT and MBT temperatures for depth bias and using an improved temperature climatology, Ishii and Kimoto (2009) found a slightly higher rate of 0.29 ± 0.06 mm yr^{-1} for the same time period (Table 3.1).

Discarding biased Argo profiles removed an apparent cooling trend from 2003 to 2006 (Willis et al., 2009). The linear trend from January 2005 to September 2011 in the newly analyzed data is 0.48 ± 0.15 mm yr^{-1} (Figure 3.2, Table 3.1).

Recent observational estimates of thermosteric sea-level rise have all been corrected for XBT and MBT depth bias (Table 3.1). The new estimates are based on updates of the Ishii and Kimoto (2009) data set (e.g., Ishii, personal communication; Kuo and Shum, personal communication), which corrects for depth bias, or the Ingleby and Huddleston (2007) data set (e.g., Domingues et al., 2008), which corrects for both XBT fall-rate bias and undersampling bias. Their estimates of the long-term thermosteric trend (beginning 1951–1961) in the upper 700 m of the ocean range from 0.29 ± 0.06 mm yr^{-1} to 0.52 ± 0.08 mm yr^{-1} (Table 3.1). The latter, by Domingues et al. (2008), is higher than the rates estimated by IPCC (2007) for the same period and by other investigators for similar periods. Estimates of the thermosteric trend in the upper ocean since 1993 range from 0.71 ± 0.31 mm yr^{-1} to 1.23 ± 0.30 mm yr^{-1} (Table 3.1). These rates are generally lower than those estimated by the IPCC (2007) for 1993 to 2003.

Observations for the deep ocean are sparse, so thermal expansion estimates for the full ocean depth are more uncertain than those for the upper ocean. The only recent estimates of the rate of thermosteric sea-level rise for the full ocean depth are by Domingues et al. (2008) and Church et al. (2011), who used a thermal expansion value of 0.2 ± 0.1 mm yr^{-1} and 0.17 mm yr^{-1}, respectively, for the deep ocean. This deep-ocean value is comparable to a recent estimate of ~0.15 ± 0.08 mm yr^{-1} based on abyssal (below 4,000 m) and deep ocean (1,000–4,000 m) observations south of the SubAntarctic Front taken in the 1990s and 2000s (Purkey and Johnson, 2010). Kouketsu et al. (2011) estimated thermosteric sea-level change of ~0.11 mm yr^{-1} for the ocean below 3,000 m from the 1990s and to the 2000s based on observed data, and 0.12 mm yr^{-1} based on an ocean model data assimilation product. The IPCC (2007) assessment, based on work by Antonov et al. (2005), was 0.1 mm yr^{-1} from 700 m to 3,000 m (Bindoff et al., 2007). Given the scarcity of data, however, it is difficult to assess the uncertainty in deep ocean warming.

Domingues et al. (2008) estimated that thermosteric sea-level rise for the full ocean depth increased from 0.72 ± 0.13 mm yr^{-1} for 1961–2003 to 1.0 ± 0.4 mm yr^{-1} for 1993–2003 (Table 3.1). The Church et al. (2011) estimates for 1993–2008 are 0.88 ± 0.33 mm yr^{-1}. For comparison, the committee calculated thermosteric sea-

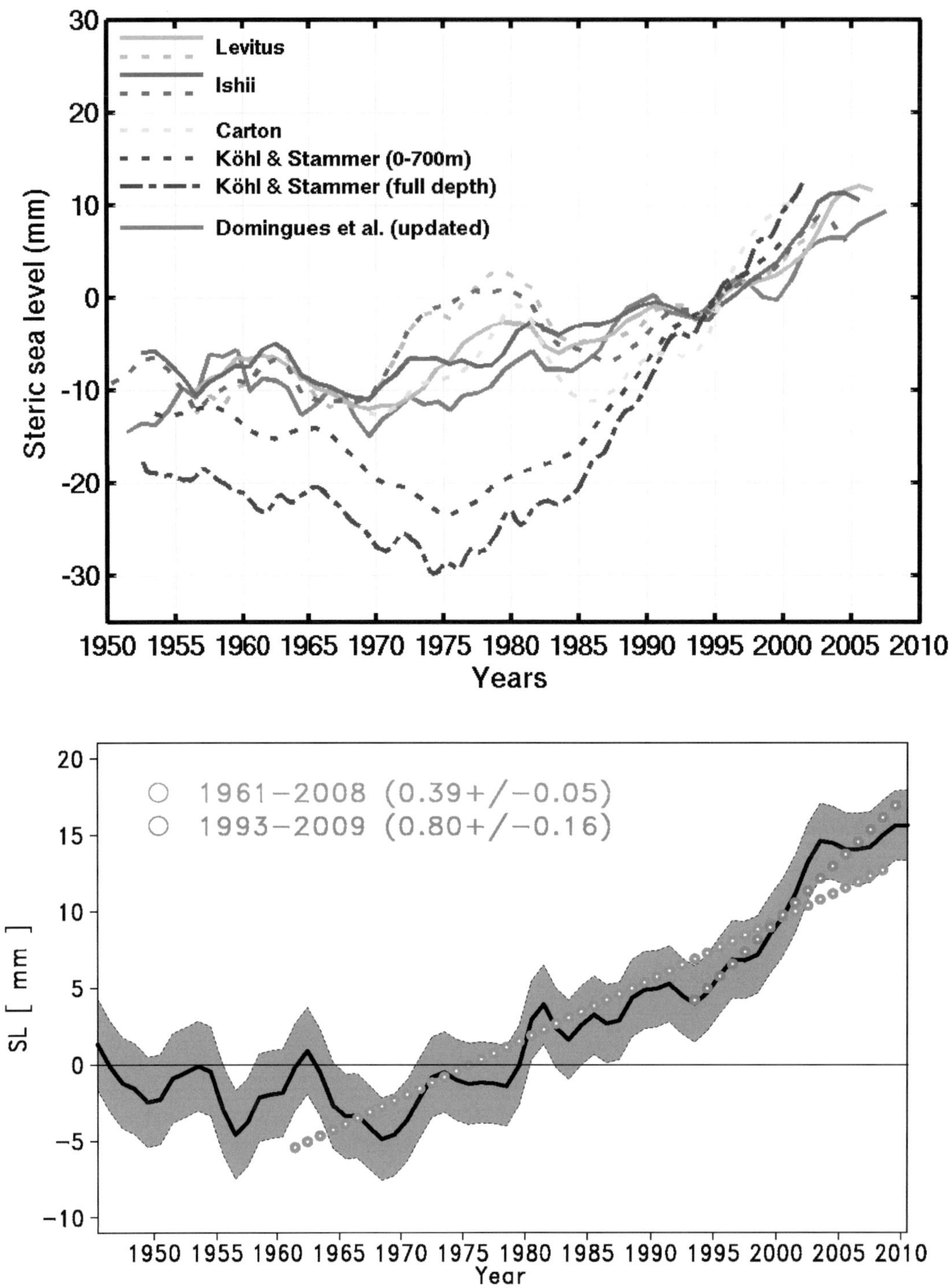

FIGURE 3.1 (*Top*) Estimates of global mean thermosteric sea level for the past six decades. The dotted blue and red lines are the IPCC (2007) estimates for the upper 700 m. The solid blue and red lines are the equivalent curves after correction for XBT biases. Also shown are a bias-corrected estimate for the upper 700 m by Domingues et al. (2008; brown line with 1 standard deviation shaded) and an uncorrected estimate down to 1,000 m from the Simple Ocean Data Assimilation model by Carton et al. (2005; green dotted line). Estimates from the ocean data assimilation model of Kohl and Stammer (2008) to 700 m (gray dotted line) and full depth (gray dash-dotted line) also are shown. SOURCE: Church et al. (2010). (*Bottom*) New estimate of global mean thermosteric sea-level rise for the upper 700 m using an updated version of bias-corrected data from Ishii and Kimoto (2009). The orange and blue symbols and values are linear thermosteric sea-level trends for different time periods. The gray shading represents 1 standard deviation. SOURCE: Ishii and Kimoto (2009).

TABLE 3.1 Recent Estimates of Global Mean Thermosteric Sea-Level Rise

Source	Period	Depth Range (m)	Instrument Bias Corrections	Thermosteric Sea-Level Rise (mm yr^{-1})
IPCC (2007)	1961–2003	0–700	None	0.32 ± 0.12
		0–3,000		0.42 ± 0.12
Domingues et al. (2008)	1961–2003	0–700	XBT fall-rate bias	0.52 ± 0.08
		Full depth		0.72 ± 0.13
Ishii and Kimoto (2009)	1951–2005	0–700	XBT and MBT depth bias	0.29 ± 0.06
Kuo and Shum (personal communication, 2011)[a]	1955–2009	0–700	XBT and MBT depth bias	0.33 ± 0.01
Ishii (personal communication, 2011)[b]	1961–2008	0–700	XBT and MBT depth bias	0.39 ± 0.05
IPCC (2007)	1993–2003	0–700	None	1.5 ± 0.5
		0–3,000		1.6 ± 0.5
Domingues et al. (2008)	1993–2003	0–700	XBT fall-rate bias	0.79 ± 0.39
		Full depth		1.0 ± 0.40
Ishii and Kimoto (2009)	1993–2005	0–700	XBT and MBT depth bias	1.23 ± 0.30
Ishii (personal communication, 2011)[b]	1993–2009	0–700	XBT and MBT depth bias	0.80 ± 0.16
Church et al. (2011)[c]	1993–2008	0–700	XBT fall-rate bias, ARGO	0.71 ± 0.31
		Full depth	pressure bias	0.88 ± 0.33
Willis (personal communication, 2011)[d]	2005–2011	0–900	Biased ARGO data removed	0.48 ± 0.15

[a] Based on the Ishii and Kimoto (2009) data set, calculated for a different time period.
[b] Updated from Ishii and Kimoto (2009) using the latest observational data.
[c] Updated from Domingues et al. (2008) and other recently updated data sets, including ARGO.
[d] Updated from Leuliette and Willis (2011) for thermosteric sea level.

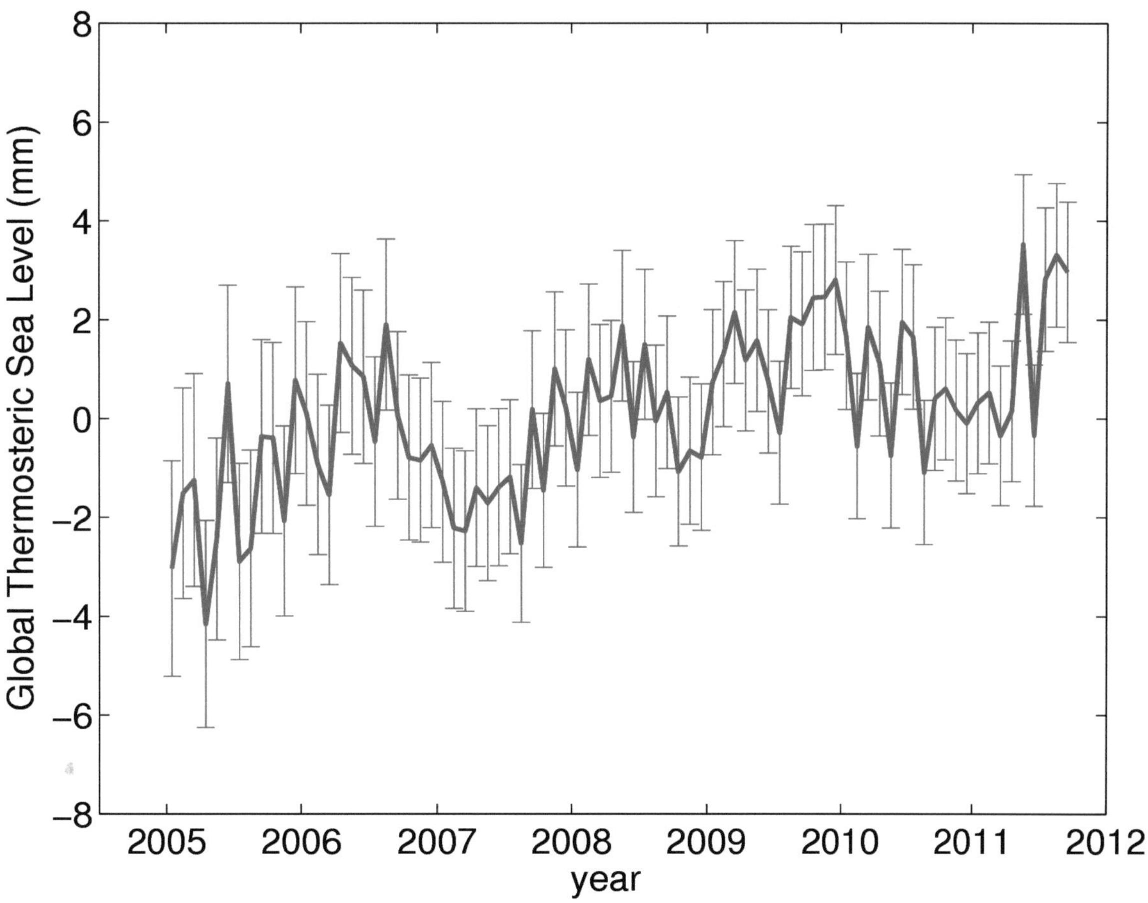

FIGURE 3.2 Thermosteric sea-level rise estimated from Argo data for the upper 900 m using updated data from Leuliette and Willis (2011). The error bars are 1 standard deviation. SOURCE: Courtesy of J.K. Willis, Jet Propulsion Laboratory, California Institute of Technology.

level rates for the full ocean depth using measured rates for the upper 700 m by other investigators (Table 3.1) and the Domingues et al. (2008) value for the deep ocean below 700 m (Table 3.2). For the longer observational period (approximately five decades), the committee calculated rates ranging from 0.5 ± 0.12 mm yr^{-1} (based on Ishii and Kimoto, 2009) to 0.59 ± 0.11 mm yr^{-1} (based on Ishii, personal communication, 2011). These rates are lower than the Domingues et al. (2008) rates, but they are comparable within errors. For the post-1993 observational period, the committee's calculated rates are 1.0 ± 0.19 mm yr^{-1} and 1.43 ± 0.31 mm yr^{-1} (Table 3.2). The most recent estimate of Ishii (personal communication, 2011) is comparable to estimates of Domingues et al. (2008) and Church et al. (2011), within their reported errors.

The above estimates of the global thermosteric sea-level trend and its variability on interannual and decadal timescales differ, sometimes substantially. For example, Domingues et al. (2008) shows a continued thermosteric sea-level rise after 2004, whereas Levitus et al. (2009) and Ishii and Kimoto (2009) show a plateau (top panel of Figure 3.1). These differences result from uncertainties in the data and the choice of instrument bias corrections, processing approach, baseline mean climatology, mapping technique, and treatment of unsampled or undersampled areas. Correcting for XBT fall-rate bias reduced the errors in the thermosteric sea-level trend (S. Levitus, personal communication, 2011). However, uncertainties in the bias corrections remain the dominant source of error, especially for recent decades (Ishii and Kimoto 2009; Willis et al., 2009; Gouretski and Reseghetti, 2010; Lyman et al., 2010).

Different data processing approaches also may account for some differences among thermosteric sea-level estimates, such as the relatively high estimates of Domingues et al. (2008) for 1961–2003 and the relatively low estimates of Ishii (personal communication, 2011) for 1961–2008 for the upper 700 m. The treatment of data in unsampled and undersampled regions of the world's oceans also can introduce uncertainties (Purkey and Johnson, 2010). Sampling problems are particularly acute in the Southern Ocean and likely result in estimates of thermosteric sea-level rise that are biased low (Gille, 2008; Church et al., 2010).

Models

The warming observed in the upper ocean also has been inferred from ocean-atmosphere climate models. For example, Pierce et al. (2006) found general consistency between models and observations for ocean warming, with the signal disappearing around 600 m depth. Climate model simulations also suggest heat uptake by the deep ocean (Katsman and van Oldenborgh, 2011; Meehl et al., 2011). Song and Colberg (2011), using an ocean general circulation model constrained by sea-surface temperature and atmospheric radiation measurements, found a strong warming signal of 1.1 mm yr^{-1} below 700 m for the 1993–2008 period. This value is much higher than observational estimates (Purkey and Johnson, 2010; Kouketsu et al., 2011; Loeb et al., 2012), for reasons that are currently under debate.

Data Assimilation

Ocean data assimilation techniques can be used to obtain estimates of deep-ocean warming and the resulting thermosteric sea-level rise by constraining the numerical models with available data. There are, however, significant differences between the various data assimilation products and direct observations, arising in part from uncertainties in direct observa-

TABLE 3.2 Committee Estimates of Thermosteric Sea-Level Rise for the Full Ocean Depth

Data Source Used in the Estimate	Period	Thermosteric Sea-Level Rise Estimates, This Report (mm yr^{-1})[a]
Ishii and Kimoto (2009)	1951–2005	0.5 ± 0.12
Kuo and Shum (personal communication, 2011)	1955–2009	0.53 ± 0.14
Ishii (personal communication, 2011)	1961–2008	0.59 ± 0.11
Ishii and Kimoto (2009)	1993–2005	1.43 ± 0.31
Ishii (personal communication, 2011)	1993–2009	1.0 ± 0.19

[a] Calculated from estimates of the upper 700 m of the ocean by various investigators and the Domingues et al. (2008) rate of 0.2 ± 0.1 mm yr^{-1} for the deep ocean below 700 m.

tions and differences in data-assimilation approaches for estimating the state of the ocean (e.g., Church et al., 2010). The Simple Ocean Data Assimilation model (Carton et al., 2005; Carton and Giese, 2008) uses a multivariate sequential approach to force the ocean model toward observed temperature and salinity data. Ocean dynamics and other properties are not preserved. Using this approach, the estimated thermosteric sea-level trend from 1968 to 2001 is similar to the observed estimates (Figure 3.1). Kohl and Stammer (2008) used a more sophisticated approach, which synthesizes the observed data into a dynamically consistent model using the adjoint assimilation technique. To ensure dynamical consistency, the model forcing fields are modified. The estimated thermosteric sea-level trend using this method shows a large decrease until 1975 and then a larger rise afterward (Figure 3.1).

Ocean data assimilation has been an active research topic only since the 1990s. Over time, it may become a more reliable source for studies of decadal sea-level variability and change (Church et al., 2010).

Satellites

A few investigators have inferred global steric sea-level rise from the Gravity Recovery and Climate Experiment (GRACE) and altimeter data (e.g., Lombard et al., 2007; Cazenave et al., 2009). Satellite altimetry measures the total sea-level change (steric plus ocean mass) and GRACE measures ocean mass change. The difference between the two measurements provides an independent estimate of the steric sea-level change. However, estimates made this way vary significantly.

Summary

The thermal expansion estimates in the IPCC Fourth Assessment Report were made before temperature biases due to the XBT and MBT depth errors were discovered. Efforts to improve the IPCC (2007) estimates have focused on using new temperature data, correcting instrument bias, and improving data processing methods. New estimates of thermosteric sea-level rise are generally higher than those estimated by the IPCC (2007) for the past four or five decades and generally lower than those estimated by the IPCC (2007) for the past 10–15 years (Figure 3.3). However, the new estimates overlap significantly with the IPCC (2007) estimates, within errors.

Estimates of thermosteric sea-level rise for the upper 700 m of the ocean have lower uncertainties than

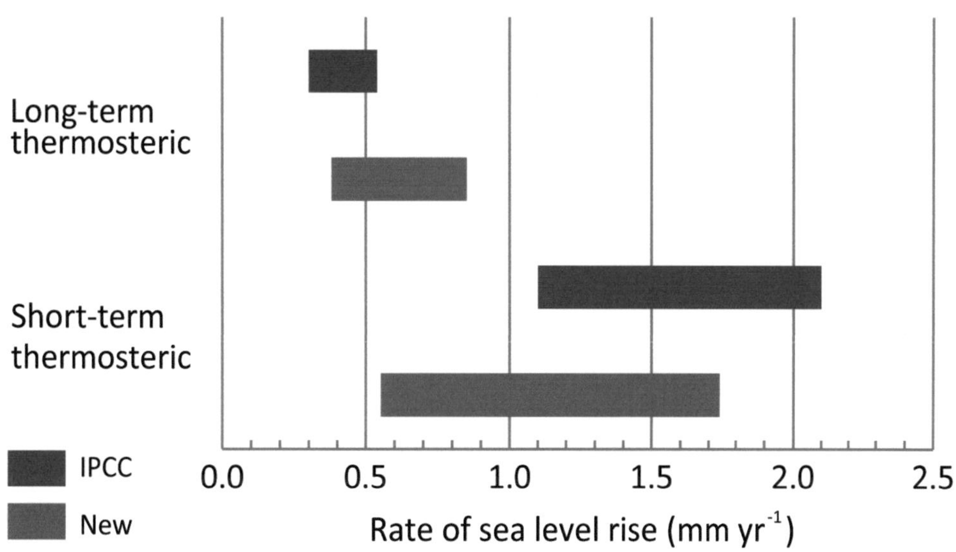

FIGURE 3.3 Comparison of thermosteric sea-level estimates for the full ocean depth from IPCC (2007; blue) and subsequent estimates (red). The bars represent the highest and lowest estimates. Long-term trends are for 1961–2003 (IPCC) and 1951–2005 (new estimates); short-term trends are for 1993–2003 (IPCC) and 1993–2008 (new estimates). SOURCE: IPCC estimates from Bindoff et al. (2007); new estimates are from Tables 3.1 and 3.2 based on data from Domingues et al. (2008), Ishii and Kimoto (2009), and Church et al. (2011).

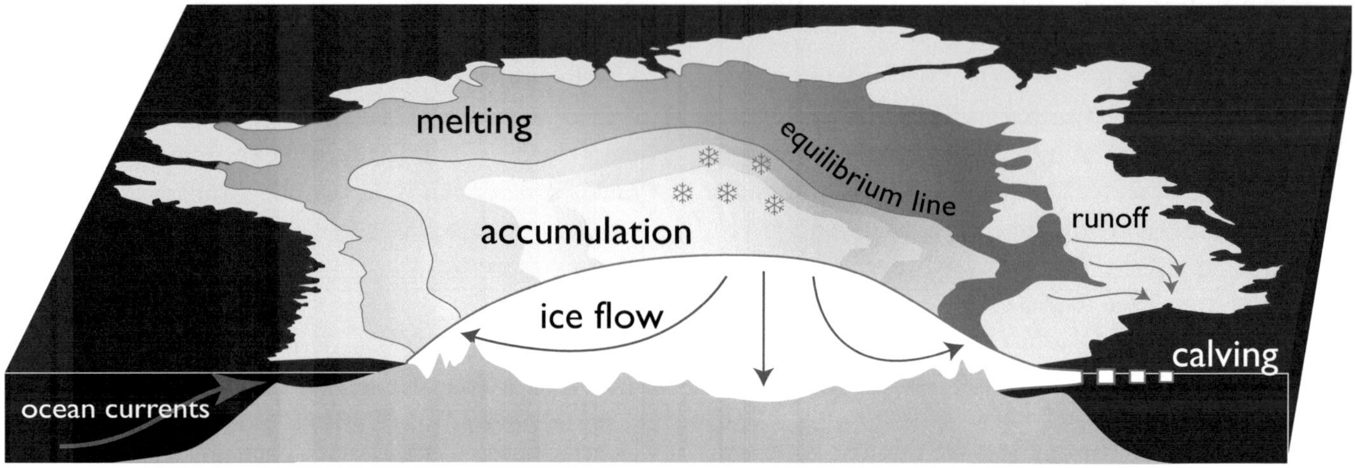

FIGURE 3.4 Glacier and ice sheet mass balance components. Ice accumulates at high elevations and is lost at lower elevations through melting, sublimation, or iceberg calving. The boundary between areas of net gain and loss is called the equilibrium line.

estimates for the full ocean depth due to the paucity of deep-ocean measurements. Studies suggest that sampling problems cause a low bias in upper-ocean thermosteric sea-level rise estimates, and also make it difficult to assess the uncertainty in the deep-ocean thermosteric sea-level rise. Data assimilation and model results are not yet robust enough to be used to fill in missing data.

GLACIERS, ICE CAPS, AND ICE SHEETS

Loss of land-based ice is a major contributor to global sea-level rise, equal to or exceeding the contribution of thermal expansion. The equivalent of at least 65 m of sea level is stored in glaciers, ice caps, and ice sheets. The Greenland and Antarctic ice sheets store the equivalent of about 7 m and 57 m of sea level, respectively (Bamber et al., 2001; Zhang et al., 2003),[1] and glaciers and ice caps store the equivalent of 0.6 ± 0.07 m, about one-third of which is around the periphery of Greenland and Antarctica (Radic and Hock, 2010).

The response of glaciers and ice sheets to climate change depends on processes acting at the upper surface; at the base, where glacial meltwater and the properties of the bedrock affect the rate of ice flow; and, in some locations, at the marine margin, where iceberg calving and melting occur (Figure 3.4). Glaciers, ice caps, and ice sheets typically gain mass through snow accumulation and lose mass through melting and runoff (ablation), iceberg calving, and, to a lesser extent, sublimation and wind erosion and transport. Calving can be the dominant mechanism of mass loss, accounting for 50–100 percent of the loss on the Antarctic Ice Sheet, about 50 percent of the loss on the Greenland Ice Sheet (Rignot and Jacobs, 2002; van den Broeke et al., 2010), and, where it has been measured, about 50 percent of loss from ocean-terminating ice cap complexes (Blaszczyk et al., 2009). In general, mass is gained at higher elevations and on the upper surface of a glacier or ice sheet, and mass is lost at lower elevations and at the base. The difference between accumulation and ablation is called the mass balance, and it is determined through a combination of in situ and satellite measurements (Box 3.2), often combined with models.

To determine the contributions of land ice to sea-level rise, mass balance estimates are converted to sea-level equivalent (SLE), the change in global average sea level that would occur if a given amount of water or ice were added to or removed from the oceans. SLE is computed by dividing the observed mass change of the ice by the surface area of the world's oceans (362×10^6 km^2). When working with changing ice volume (e.g., rates of iceberg flux), the volume is converted to mass using the density of ice (900 kg m^3). Using these values, 1.11 km^3 ice = 1 km^3 of water = 10^9 kg water = 1 GT water, and 362 GT water = 1 mm SLE. For glacier ice resting on bedrock below sea level, a

[1] See also data compiled for the Sea-level Response to Ice Sheet Evolution (SeaRISE) assessment project, <http://websrv.cs.umt.edu/isis/index.php/SeaRISE_Assessment>.

BOX 3.2
Measuring the Earth's Ice

Monitoring the world's land ice masses is a challenging task complicated by the size, wide distribution, and generally remote and hostile environments in which most glaciers are located. Changes in glacier or ice sheet volume can be calculated by mass budget methods (balancing input and output fluxes), by repeated geodetic measurements, by combinations of the two, and by measurements of mass change through gravity surveys using the GRACE satellite system. Quantitative determination of glacier and ice sheet mass balance requires a variety of data sets, including ice surface elevation and ice thickness, the rate of ice flow, and the rate of ice (snow) accumulation and ablation. Measurements are made both in situ (ideal for individual glaciers and process studies) and remotely (ideal for covering large regions). Satellite remote sensing instruments collect data at visible, near-infrared, and microwave wavelengths and may image the surface in blocks or along the ground track below the satellite (see the review in Quincey and Luckman, 2009).

Ice Thickness. The thickness of glaciers and ice sheets is generally measured using radar sounding from aircraft or at the ice surface. The 25–400 MHz radar signal penetrates to the bedrock below the ice, and the difference between returns from the upper and lower surfaces is used to calculate the ice thickness. Radar sounding works best in cold, clean ice. Ice with substantial fractions of liquid water or crevasses scatter radar energy, creating complications for radar soundings of fast-moving outlet glaciers, especially those in warmer environments.

Surface Topography. Topography is measured using aerial photogrammetry, airborne and satellite laser altimetry, and satellite radar altimetry. Radar altimeters on satellites (ERS-1, -2, Envisat, and CryoSat-2) are used to measure ice sheet surface elevation with decimeter accuracy, but the footprint of the sensor (the area on the surface within the field of view of the antenna) is relatively large (a few kilometers) and varies with surface roughness and slope. Laser altimeters have much smaller footprints (tens of meters; about 70 m for NASA's Geoscience Laser Altimeter) and may be mounted on aircraft or on satellites. Laser altimeters mounted on small aircraft are used for repeat surveys of glaciers in Alaska, where optimal flight lines are poorly suited for satellite orbital paths (Larsen et al., 2007). Repeat mapping of surface topography can be used to derive volume change, as long as neither the density nor the bed topography change between surveys.

Ice Velocity. The rate of ice flow can be calculated using repeated measurements of the locations of features, either a survey monument or a natural feature (e.g., a crevasse intersection), on the ice surface. Locations can be determined using ground-based optical survey measurements, Global Positioning System surveys, photogrammetry, or satellite image processing. Ice flow also can be determined from radar interferometry (e.g., Figure), which uses the change between observations in the phase of the returning radar wave to make a high-precision measurement of ground displacement relative to the spaceborne radar.

Gravity. Ice sheet mass changes since 2002 can be determined from the GRACE satellite system (see Box 2.4). The ice sheet changes must be separated from other mass change signals such as those caused by glacial isostatic adjustment. In some instances the modeled corrections are robust, but in others the uncertainties can be large. The spatial resolution of the measurements depends on details of the processing and the latitude of interest (Wahr et al., 2004).

Mass Balance. Accumulation and ablation are traditionally determined using in situ measurements made at least twice yearly, at the end of the accumulation season and ablation season. This technique remains the only way to make direct observations of the components of the mass budget, but it is too time consuming and expensive to be used as an operational tool on the ice sheets. Accordingly, remote sensing methods are used extensively, with reliance on limited point climate and meteorological observations and on meteorological and surface energy balance models. Snow accumulation can be estimated from atmospheric models coupled with satellite observations or by analyzing annual layers in ice cores and interpolating between core sites using radar sounding of the ice layers. Melt can be estimated using energy balance models driven by atmospheric models. Runoff cannot be measured remotely, and in most cases is determined solely by modeling.

continued

BOX 3.2 Continued

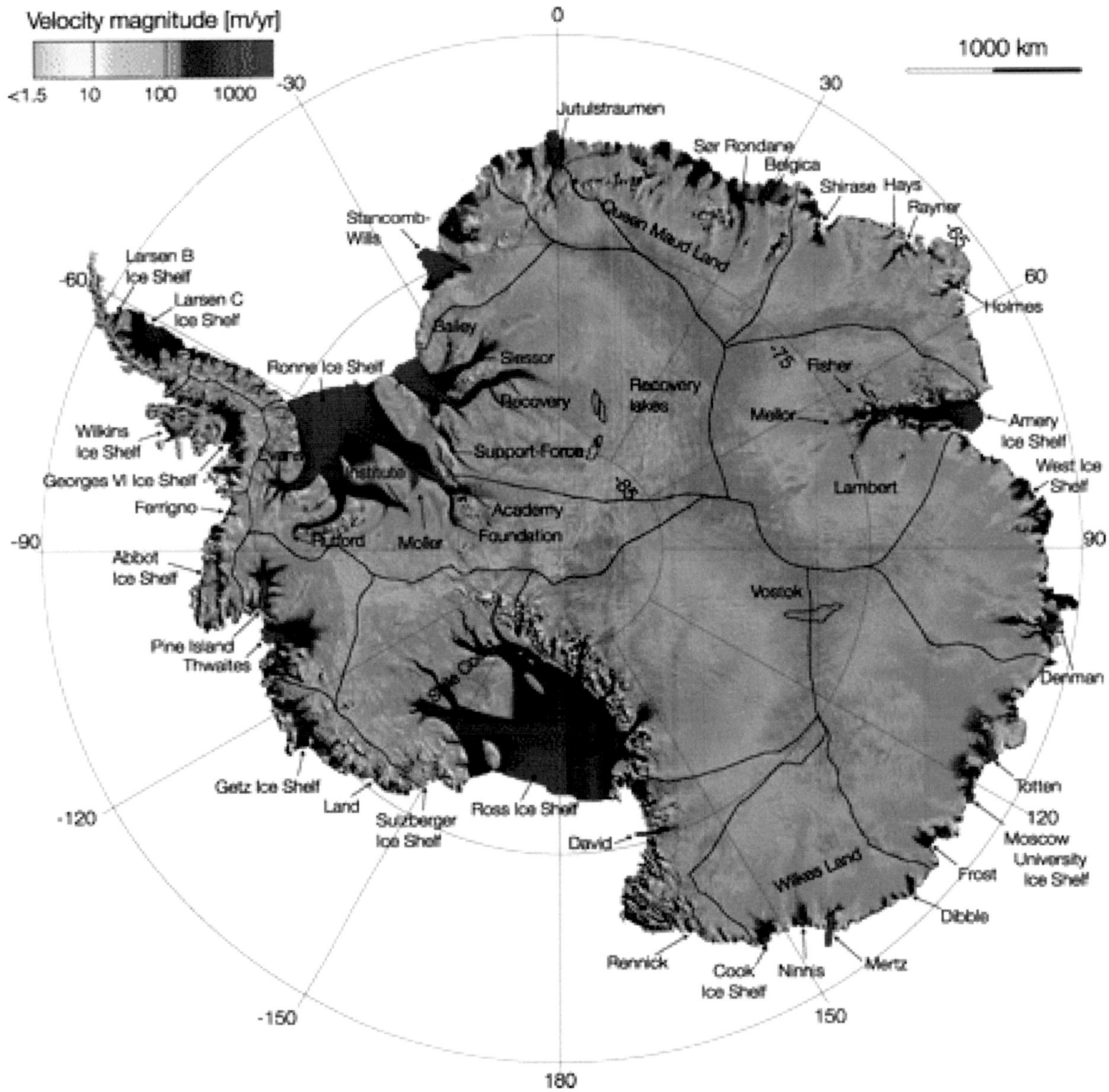

FIGURE Antarctic glacier velocity (in m yr^{-1}) derived from radar interferometry. Black lines delineate major ice divides. Velocities can reach a few km yr^{-1} on fast-moving glaciers (e.g., Pine Island) and floating ice shelves. SOURCE: Rignot et al. (2011b).

correction term is added to account for the ice volume below the water line that has already affected sea level by its presence. On sufficiently long timescales, a correction for glacial isostatic adjustment of the underlying bedrock, based on forward models, also may be made.

The conversion of ice mass loss to SLE assumes that all land ice melt enters the ocean. Land storage of ice melt may be significant for land-terminating glaciers in continental interiors (e.g., high mountains in Asia), but its occurrence is unconfirmed. SLE is a globally uniform value and thus may be higher or lower than the sea-level value in any particular region.

Estimates from the IPCC Fourth Assessment Report

The IPCC Fourth Assessment Report estimated that losses from glaciers and ice caps contributed 0.58 ± 0.18 mm yr^{-1} to sea-level rise from 1961 to 2003 and 0.77 ± 0.22 mm yr^{-1} from 1993 to 2003 (Bindoff et al., 2007), with the most rapid ice losses occurring in Patagonia, Alaska, northwest United States, and southwest Canada (Lemke et al., 2007). Uncertainties in the net loss rate were significant, however, because of sparse point observations and incomplete knowledge of global glacier area and volume distribution for upscaling point observations. On the Greenland Ice Sheet, the IPCC (2007) found that mass was gained at high elevations because of increasing snowfall, and mass was lost near the coast because of increases in melting and in the flow speed of outlet glaciers. The IPCC estimated that the Greenland Ice Sheet contributed 0.05 ± 0.12 mm yr^{-1} to sea-level rise from 1961 to 2003 and 0.21 ± 0.07 mm yr^{-1} from 1993 to 2003. Changes in Antarctica were more challenging to interpret because of the relatively small changes in snow accumulation rates (Monaghan et al., 2006) and to different trends in the flow of individual West Antarctic outlet streams. The IPCC estimated that the Antarctic Ice Sheet contribution was between -0.28 and +0.55 mm yr^{-1} from 1961 to 2003 and between -0.14 and +0.55 mm yr^{-1} from 1993 to 2003, allowing for the possibility that the Antarctic mass change may have reduced sea-level rise, especially prior to 1993 (Bindoff et al., 2007; Lemke et al., 2007). The rate of ice loss appears to have increased since 1993 because of increasing surface melt on the Greenland Ice Sheet and faster flow of some outlet glaciers in both Greenland and Antarctica.

Recent Results

Since the IPCC Fourth Assessment Report was published, more observations are available, and rapid flow changes at marine margins of ice sheets and glaciers, which were recognized but not included in the IPCC (2007) projections, are now represented in some projections. Ice sheet velocity and mass balance distributions are now better mapped, but the potential for rapid future increases in calving losses from ocean-terminating outlet glaciers is still poorly understood, in part because of inadequate knowledge of the underlying physics.

Glacier and Ice Cap Assessments

Owing to the delay in assimilation of new observations and the incomplete but evolving glacier inventory, most post-IPCC (2007) assessments of glacier and ice cap change (Table 3.3) are based on data collected prior to 2007. The various analyses span different periods and use different methods to average sparse data and to scale up regionally heterogeneous trends to estimate the global total, resulting in significant uncertainties. Estimated rates of ice loss, expressed as SLE, vary in time and space. For example, gravity observations indicate that the rate of mass loss in the Gulf of Alaska decreased from 2004 to 2008 (Luthcke et al., 2008; Table 3.3), but increased in the Canadian Arctic over a similar interval (Gardner et al., 2011). These patterns reflect the influence of rapid changes in the rate of ice flow (rapid dynamical response) associated with ice-ocean interaction in coastal regions.

The most recent published compilation (Cogley, 2012), and the only one to use data from after 2007, shows a substantial decrease in glacier and ice cap loss rates from 1.41 mm yr^{-1} SLE for 2001–2005 to 0.92 mm yr^{-1} for 2005–2010. The cause of this decrease is unclear, but suggests the potential for significant variability on 5- to 10-year timescales and highlights the difficulty of extracting meaningful trends from short-term observations. Jacob et al. (2012) determined an overall loss rate for global glaciers and ice caps of 0.41 ± 0.08 mm yr^{-1} for 2003–2010, but this value does not include the glaciers and ice caps on the periphery of the Greenland and Antarctic ice sheets. They estimated the contribution of these peripheral glaciers and ice

TABLE 3.3 Estimates of Glacier and Ice Cap Sea-Level Equivalent

Source	Period	Region	Method	Sea-Level Equivalent (mm yr^{-1})
Global Estimates				
IPCC (2007)	1993–2003	Global	Combination of various estimates	0.77 ± 0.22
	1961–2003			0.58 ± 0.18
Leclercq et al. (2011)	1850–2005	Global	Glacier length	0.06 ± 0.01
Kaser et al. (2006)	2001–2004	Global	Combination of three independent methods: Cogley (2009), Dyurgerov (2010), and Ohmura (2004)	0.98 ± 0.19
Cogley (2009)	2001–2005	Global	Spatial polynomial interpolation	1.41 ± 0.20
Dyurgerov (2010)	2002–2006	Global	Area weighting	0.95 ± 0.05
Cazenave and Llovel (2010)	2003–2007	Global	Uncertainty-weighted average of available estimates	1.03 ± 0.06
Cogley (2012)	2005–2009	Global	Spatial polynomial interpolation	0.92 ± 0.05[a]
Jacob et al. (2012)	2003–2010	Global	GRACE	0.41 ± 0.08[b]
Regional Estimates				
Matsuo and Heki (2010)	2003–2009	High mountain Asia	GRACE	0.13 ± 0.04
Jacob et al. (2012)	2003–2010	High mountain Asia	GRACE	0.01 ± 0.05
Luthcke et al. (2008)	2004	Gulf of Alaska	GRACE	0.39 ± 0.06
	2007			0.13 ± 0.06
Gardner et al. (2011)	2004	Canadian Arctic	GRACE	0.09 ± 0.02
	2006			0.25 ± 0.03

[a] Representative of 2005/6–2009/10, but reports for the 2009/10 balance year are still incomplete. Value updated from Cogley (2012) and upscaled to all glaciers, including peripheral glaciers surrounding the ice sheets, using the method of Kaser et al. (2006).
[b] Value excludes peripheral glaciers surrounding the Greenland and Antarctic ice sheets. Estimated loss rate (SLE), including peripheral glaciers given in Jacob et al. (2012), is 0.63 ± 0.23 mm yr^{-1}.

caps using a simple adjustment factor and arrived at a global total of 0.63 ± 0.23 mm yr^{-1}, assigning substantially less confidence in this rate than in the rate without peripheral glaciers and ice caps.

Glaciers in high mountain Asia (more than 110,000 km^2 of glacier area, including the Himalayas, Karakoram, Pamirs, Caucasus, and Tien Shan regions) have experienced losses in recent decades, but the region is sparsely observed and uncertainties are generally large. Shortly after the publication of the IPCC Fourth Assessment Report, which contained an error concerning the disappearance of Himalayan glaciers, several other erroneous reports were published, all based in part on gray literature and media stories. These created considerable confusion about the state of glaciers in the Himalayas and their near-term fate (see the summary in Cogley et al., 2010). Subsequent analyses continue to show substantial uncertainties, however. Matsuo and Heki (2010) used GRACE gravity methods to determine ice losses from high mountain Asia and estimated that the sea-level contribution of the entire region was 0.13 ± 0.04 mm yr^{-1} SLE for 2003–2009. This value was somewhat higher than the loss rate of 0.10 mm yr^{-1} determined by Dyurgerov and Meier (2005) for 1993–2003 and 1998–2003, but Matsuo and Heki (2010) arrived at their value by assigning 0.027 mm yr^{-1} SLE (10 GT yr^{-1}) to groundwater extraction. This may be an underestimate of groundwater extraction, given that the region includes the plains south of the Himalayas and part of the region where Tiwari et al. (2009) saw losses of ~54 GT yr^{-1} for 2002–2008. If a larger groundwater extraction signal were used, the GRACE data used by Matsuo and Heki (2010) would indicate a smaller high mountain Asia glacier loss rate. The most recent and detailed analysis of high mountain Asia is presented in Jacob et al (2012), who found a much lower total loss rate of 4 ± 20 GT yr^{-1} for 2003–2010, corresponding to 0.01 ± 0.05 mm yr^{-1} SLE. The authors ascribe the difference between their totals and other GRACE analyses (e.g., Matsuo and Heki, 2010) to better treatment of mass concentration (mascon) calculations in the GRACE processing and improved removal of the terrestrial groundwater signal through modeling.

All glacier and ice cap loss rates reported to date are based on a global glacier and ice cap inventory that represents only ca. 48 percent of the world's 704 ± 56 × 10^3 km^2 of glacier-covered area exclusive of the ice sheets (Figure 3.5). The Randolph Glacier

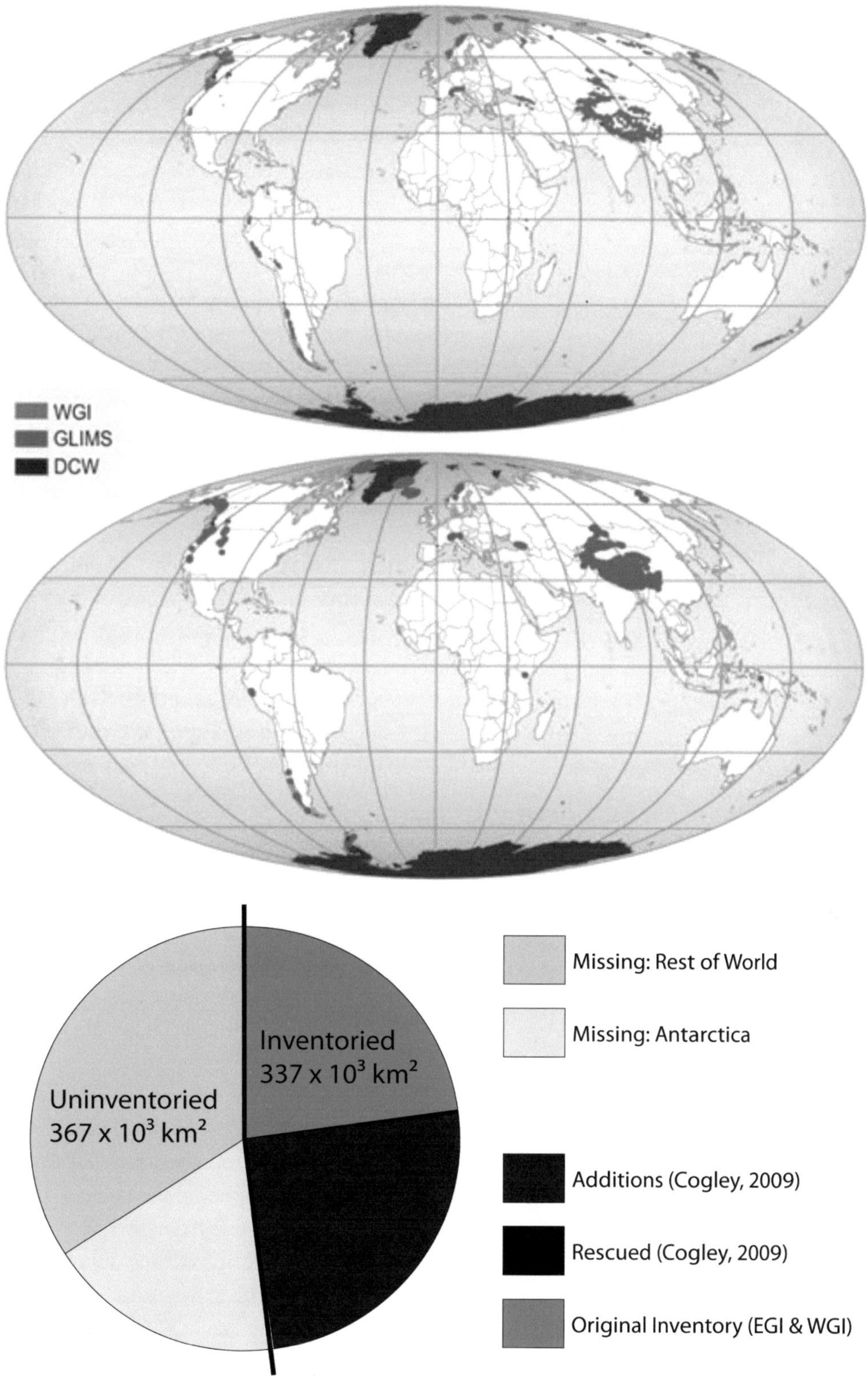

FIGURE 3.5 (*Top*) Map coverage of global glacier inventories—including the World Glacier Inventory (WGI), Global Land Ice Measurements from Space (GLIMS), and Digital Chart of the World (DCW)—used in published assessments. The figure shows the status prior to publication of the Randolph Glacier Inventory. SOURCE: United Nations Environment Programme, <http://www.grid.unep.ch/glaciers/>. (*Bottom*) Completeness of global glacier and ice cap inventories used in published assessments. Approximately one-third of the uninventoried area is in the peripheral glaciers and ice caps surrounding the two ice sheets, and most of the rest is in North America. "Missing" indicates that data are absent in the Cogley (2009) extended inventory. SOURCE: Cogley (2009).

Inventory, a new, complete inventory providing 100 percent coverage of glaciers and ice caps, including those on the peripheries of ice sheets, has recently been completed.[2] Several groups are working to update present-day analyses and projections using the new inventory. The glacier and ice cap loss rates presented here are likely to change once the new inventory is fully incorporated into assessments.

In addition to deficiencies in the global glacier and ice cap inventory, measurements of mass balance terms are sparse (Dyurgerov, 2002; Kaser et al., 2006). Observations of glacier variations extend back into the 18th century, but in situ mass balance measurements that reveal climatic patterns do not begin until the early- to mid-20th century, and most records are less than a few decades long (Zemp et al., 2009). As a result, scaling methods have been developed to translate local measurements to a global estimate (Bahr et al., 1997; Dyurgerov, 2002; Kaser et al., 2006; Cogley, 2009). The incomplete inventory and the small number of long-term observational mass balance records worldwide are the largest (and hardest to quantify) sources of uncertainty in present-day rates of glacier and ice cap mass loss.

Differences in methodology and in error reporting make quantitative comparison of the various mass balance estimates difficult. Slangen and van de Wal (2011) found that projections of future change in these systems were about equally sensitive to uncertainty in the glacier inventory as to the scaling factor used to relate temperature change to mass imbalance. Cazenave and Llovel (2010) combined all available estimates to arrive at an uncertainty-weighted average of 1.03 ± 0.06 mm yr^{-1} SLE from glaciers and ice caps, or approximately 41 percent of the total observed sea-level rise for the 2003–2007 period.

Computing mean SLE rates using the published literature requires time series data and knowledge of the uncertainties associated with the various estimates. Such information is not always available or presented in a useful way. In this sense, the best mass balance compilation available is Cogley's (2009) glacier and ice cap data set (updated in Cogley, 2012, but released as this report was being completed). For the most recent period (2005–2009), the loss rates reported for glaciers and ice caps are 0.92 ± 0.05 mm yr^{-1}.

[2] See <http://www.glims.org/RGI/randolph.html>.

Ice Sheet Assessments

Systematic assessments of ice sheets began in the mid 1980s (e.g., Bindschadler, 1985; Oerlemans, 1989). With each assessment, the mass balance has become increasingly negative (i.e., net mass loss) in both Greenland and Antarctica. A number of ice sheet assessments have been published since the IPCC Fourth Assessment Report (Table 3.4). Methods for measuring ice sheet mass balance are comparable to those used for glacier mass balance. Since 2002, however, detection of mass change using the GRACE satellite system has become a widely used tool for ice sheet mass balance owing to the operational difficulties of other measurement methods over large areas. Interpretation of GRACE data is complicated by its intrinsic mixing of gravity signals (Box 3.2). Glacial isostatic adjustment must be corrected by modeling the lithospheric response to loading changes (Velicogna and Wahr, 2006a,b; Tregoning et al., 2009), but other mass change terms (e.g., changes in terrestrial water storage) are smaller on the ice sheets than elsewhere.

As shown in Table 3.4, the reported rates of mass loss vary substantially, in part because of different uncertainties among measurement methods and improvements in the analysis of GRACE data. In addition, the ice sheet loss rates appear to experience not only a long-term trend toward faster losses but also significant interannual and multi-annual variability, so measurements made over different time intervals can be difficult to compare. The brevity of the record and differences in the spatial coverage, the quantities used to infer mass change, and the treatment of data gaps further complicate comparisons and trend assessment. The committee estimated ice sheet loss rates for the most recent period reported (2002–2009) by making a weighted average of the values in Table 3.4.[3] The average loss rates for 2002–2009 were 0.56 ± 0.13 mm yr^{-1} for the Greenland Ice Sheet and 0.37 ± 0.14 mm yr^{-1} for the Antarctic Ice Sheet.

[3] For each year, all available published values are weighted according to assessed confidence in the quality of the particular estimate, and then averaged. Some studies provide yearly values for their respective reporting periods; others provide only average values over a multi-year period, and in these cases, the average rate was assumed to apply in each year in the interval. For a multi-year interval, the weighted average is obtained through a simple linear average of the annual averages in that interval.

TABLE 3.4 Estimates of Sea-Level Equivalent from Ice Sheet Mass Loss

Source	Period	Method	Sea-Level Equivalent (mm yr^{-1})
Greenland Ice Sheet			
IPCC (2007)	1993–2003	Combination of various estimates	0.21 ± 0.07
	1961–2003		0.05 ± 0.12
Rignot et al. (2011a)	1992–2000	Mass balance method + GRACE	0.14 ± 0.14
	2000–2003		0.47 ± 0.14
	2003–2007		0.68 ± 0.14
Velicogna (2009)	2002–2003	GRACE	0.38 ± 0.09
	2007–2009		0.79 ± 0.09
Schrama and Wouters (2011)	2003–2010	GRACE	0.55 ± 0.05
Cazenave et al. (2009)	2003–2008	GRACE	0.38 ± 0.05
Luthcke et al. (2008)	2004–2009	GRACE	0.52 ± 0.20
Zwally et al. (2011)	1992–2002	ICESAT	0.02 ± 0.01
	2003–2007		0.47 ± 0.01
Sørensen et al. (2011)	2003–2008	ICESAT	0.58 ± 0.06
Wu et al. (2010)	2002–2008	GRACE + GPS	0.29 ± 0.06
Rignot et al. (2008)	1960s	Mass balance method	0.30 ± 0.19
	1970s–1980s		0.08 ± 0.14
van den Broeke et al. (2009)	2000–2008	Mass balance method	0.46
	2003–2008		0.66
Antarctic Ice Sheet			
IPCC (2007)	1993–2003	Combination of various estimates	0.21 ± 0.35
	1961–2003		0.14 ± 0.41
Rignot et al. (2011a)	1992–2000	Mass balance method + GRACE	0.18 ± 0.25
	2000–2003		0.46 ± 0.25
	2003–2007		0.56 ± 0.25
	2007–2010		0.71 ± 0.25
Velicogna (2009)	2002–2003	GRACE	0.29 ± 0.20
Chen et al. (2009)	2002–2003	GRACE	0.52 ± 0.21
Cazenave et al. (2009)	2003–2008	GRACE	0.55 ± 0.06
Wu et al. (2010)	2002–2008	GRACE	0.23 ± 0.12
Wingham et al. (2006)	1992–2003	Radar altimetry	0.07 ± 0.08

Rapid Dynamic Change

The possibility of rapid dynamic response to environmental change as a mechanism of rapid sea-level rise is a long-standing idea in glaciology (Mercer, 1978; Thomas and Bentley, 1978). Rapid flow processes have been observed on ice sheets (e.g., Bentley, 1987) and at marine-terminating glaciers for many years (Meier and Post, 1987). Increases in the rate of rapid transfer of ice from land to the ocean by glacier flow and iceberg calving were observed in Greenland between ca. 1995 and 2005 (e.g., Rignot and Kanagaratnam, 2006) and in Antarctica. These observations were published late in the compilation of results for the IPCC Fourth Assessment, so the report included the observations, but not an extensive analysis or interpretation.

A variety of observational studies are now available which, together with process studies, suggest a small set of underlying causes for changes in outlet glacier flow around the Greenland Ice Sheet, the Antarctic Peninsula, and the Amundsen Sea sector of the West Antarctic Ice Sheet. Warming ocean water appears to be increasing the rates of calving and melting (e.g., Holland et al., 2008; Nick et al., 2009; Straneo et al., 2010; Motyka et al., 2011), which in turn changes the coupling between glacier ice and the adjacent bedrock, increasing the rate of ice flow. In some extreme cases, the discharge speed increased by an order of magnitude at glacier termini, although the rate of change varied from year to year (e.g., Joughin et al., 2004; Howat et al., 2007). Climate-driven changes in sea ice in the coastal fjord environment may also be important (Amundson et al., 2010). Rapid changes at the outlet glacier terminus propagate into the interior over timescales and with magnitudes that depend on both the climate and glacier dynamics (Pfeffer, 2007). Ice sheet mass balance over the next century depends in part on how far and how rapidly that propagation proceeds (see "Recent Global Sea-Level Projections" in Chapter 5).

The position of the grounding line—the transition at which ice resting on bedrock goes afloat—depends on the ice thickness and varies with the ice flux through the transition zone. Regions where the base of the ice rests below sea level and the grounding line is relatively unprotected by adjacent floating ice are the most vulnerable to rapid acceleration and thinning (Thomas et al., 1979; Scambos et al., 2004; Schoof, 2007). Rapid retreat is possible where the bed is below sea level and slopes down toward the interior because both the thickness of the ice, and thus ice flux, and the thickness required to overcome buoyancy increase in the inland direction (Pfeffer, 2007; Schoof, 2007).

Despite rapid changes along the margins of the Greenland and Antarctic ice sheets, it is unlikely that the ice sheets will disappear over the next millennium. The ice sheets are so thick (Figures 3.6 and 3.7) that much of the surface is in higher, relatively cooler parts of the atmosphere, allowing a positive mass balance to be maintained even as the climate warms. However, if dynamic thinning reduced the Greenland Ice Sheet, for example, below some threshold size, winter snow would not compensate for the loss and the ice sheet would not re-grow under current climate conditions (Toniazzo et al., 2004; Ridley et al., 2010). Studies of such thresholds suggest that widespread denudation of Greenland and West Antarctica is possible in some warming scenarios, such as four times the preindustrial carbon dioxide (Ridley et al., 2010) or 5°C ocean warming (Pollard and DeConto, 2009), but requires thousands of years (e.g., Marshall and Cuffey, 2000; Pollard and DeConto, 2009; Ridley et al., 2010).

The rapid dynamic response from glaciers outside the ice sheets is less important than ice sheet dynamics over the long term because glaciers do not contain significant volumes of marine-grounded ice. However, the potential for significant short-term contributions is large. Between 1996 and 2007, Columbia Glacier, on Alaska's south coast, lost mass at an average rate of 6.80 GT yr^{-1}, or 0.019 mm yr^{-1} SLE, approximately 0.7 percent of the rate of global sea-level rise during this period (Rasmussen et al., 2011, corrected here for ice already grounded below sea level). The volume of Columbia Glacier, approximately 150 km^3, is too small to contribute to sea level at such a rate for long, but marine-terminating glaciers of this size can be significant factors on decadal scales.

Summary

Most post-IPCC (2007) assessments of glacier and ice cap change have been made using data collected

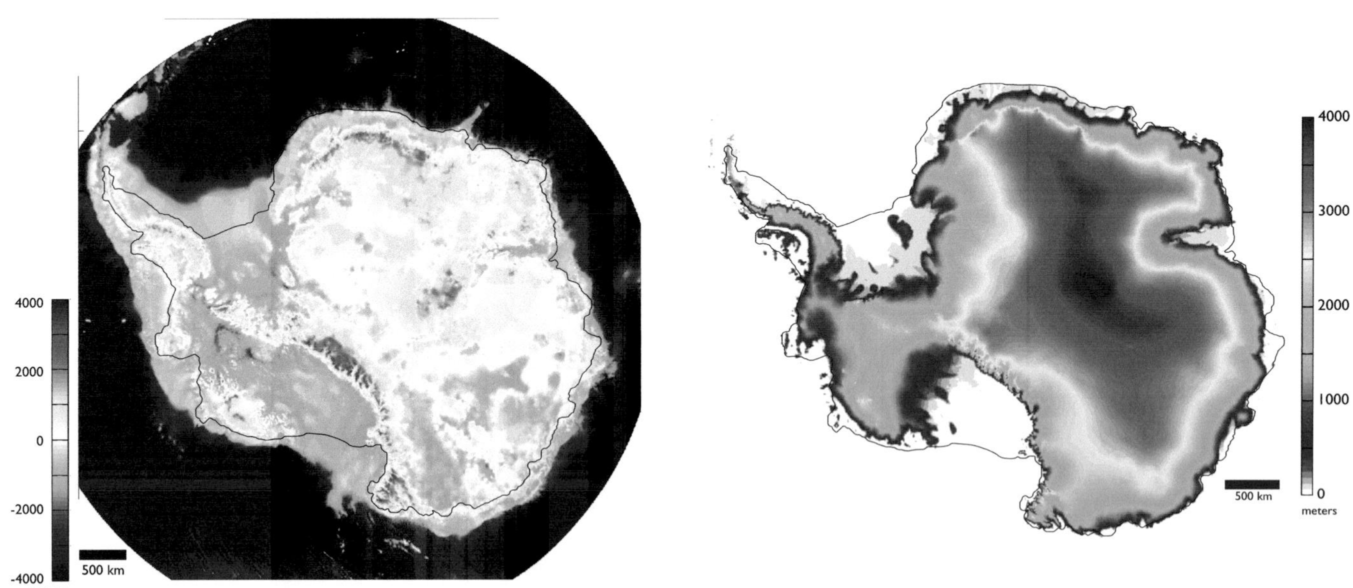

FIGURE 3.6 (*Left*) Antarctic bedrock elevations. Transition from light blue to dark blue marks the edge of the continental shelf. (*Right*) Antarctic surface elevations. Black line marks the approximate edge of the present-day ice (floating and grounded). Areas where the bed of the ice sheet is below sea level (e.g., West Antarctic Ice Sheet) are expected to be more vulnerable to rapid change than regions where the bed is above sea level. SOURCE: Data from Le Brocq et al. (2010).

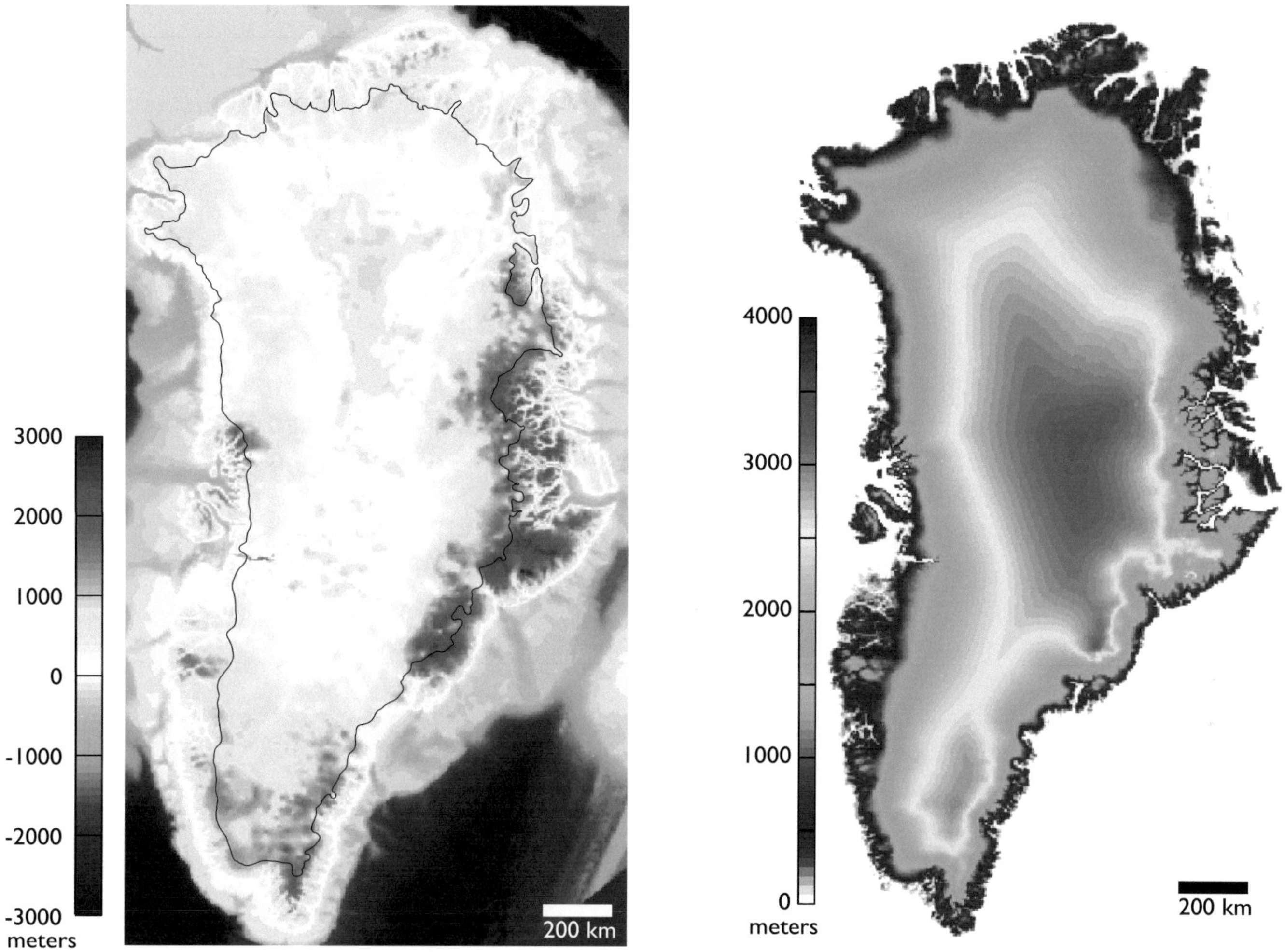

FIGURE 3.7 Greenland bedrock elevation (*left*) and surface elevation (*right*). Black line marks the approximate edge of the present-day ice (floating and grounded). SOURCE: Original bedrock elevation from Bamber et al. (2001), modified to include data in the Jakobshavn Isbrae region from the Center for Remote Sensing of Ice Sheets, <http://websrv.cs.umt.edu/isis/index.php/Present_Day_Greenland>.

prior to 2007. The new estimates of the glacier and ice cap contribution to sea-level rise tend to be at the high end of the estimates provided in the IPCC Fourth Assessment Report (Table 3.3). Most new assessments of ice sheet change are based on GRACE data, which have been available since 2002, although a few long-term assessments have been made using mass balance methods. Different methods for estimating ice-sheet mass balance yield substantially different results. Estimates made using more recent data (Table 3.4) show that the contribution of Greenland to sea-level rise is significantly higher than the IPCC (2007) estimate and the contribution of Antarctica has shifted toward the positive side of the range (raising sea level).

Since about 2006, the rate of ice loss in Greenland has increased substantially and the rate of change in Antarctica, while more difficult to quantify, appears to have shifted from negative to positive (e.g., Vaughan, 2006; Rignot and Kanagaratnam, 2006; Rignot et al., 2008; van den Broeke et al., 2009; Cazenave and Llovel, 2010; see also Table 3.4). This growing contribution arises from increases in both the amount of surface melting and the rate of ice discharge through coastal outlet glaciers. Calculated loss rates from glaciers and ice caps have decreased since about 2005 (Cogley, 2012), due to significant short-term variability in the global glacier loss rate signal and, to a lesser extent, to improvements in the global glacier inventory. Short-term (pentads to decades) glacier loss rates are strongly negative but with no clear pattern of variability, whereas the longer term trend (decade to century) is consistently negative and accelerating. In the most recent periods reported, the loss rates are 0.56 ± 0.13 mm yr^{-1} from 2002 to 2009 for the Greenland Ice Sheet, 0.37 ± 0.14 mm yr^{-1} from 2002 to 2009 for the Antarctic Ice Sheet, and 0.92 ± 0.05 mm yr^{-1} from 2005 to 2009 for glaciers and ice caps.

TERRESTRIAL WATER STORAGE

Water lost or gained by the continents generally results in a corresponding gain or loss of water by the oceans. Terrestrial water is stored in soils and the subsurface (groundwater, aquifers), in snowpack and permafrost, in surface water bodies (e.g., rivers, lakes, reservoirs, wetlands), and in biomass. Some of the water withdrawn from these sources as a result of human activities such as groundwater pumping, wetland drainage, diversion of surface water for irrigation, and deforestation eventually reaches the ocean, raising sea level at global, regional, and local scales (e.g., Bindoff et al., 2007; Milly et al., 2010). Conversely, some water that would normally reach the ocean is diverted through processes such as impoundment of water behind dams, subsurface infiltration beneath dams, and infiltration of irrigation water to depths beneath the root zone, thus lowering sea level or reducing the rate of sea-level rise.

Some changes in terrestrial water storage can be evaluated with reasonable precision at local scales, including changes caused by groundwater withdrawal, deforestation, agriculture, wetland drainage, and reservoir construction. On global scales, however, the terrestrial water balance is far more difficult to estimate. Not only must all hydrological fluxes be evaluated, but also geographic coverage of in situ measurements, such as river and stream gage records, is spotty. In some parts of the world, instrument coverage is even declining.[4] For example, the number of stream gages monitoring freshwater discharge into the Arctic Basin declined by 38 percent between 1985 and 2004 (Corell, 2005).

Terrestrial hydrologic models can be used to close observational gaps and, when coupled with global climate models, to estimate surface boundary conditions such as temperature and precipitation. Because of the complexity of hydrological processes and the wide range of spatial and temporal scales involved, fully deterministic models are generally not used. A variety of nondeterministic approaches have been developed (Eagleson, 1994; Famiglietti et al., 2009), and efforts to develop deterministic, quasi-deterministic, and hybrid models are being pursued (e.g., Kollet and Maxwell, 2008; Wood et al., 2011). These models are strongly dependent on observations, which are coming increasingly from remote sensing (Box 3.3).

The GRACE satellite system (Boxes 2.4 and 3.3) provide a sensitive means of detecting changes in land water mass, provided that other confounding mass change signals can be independently assessed and removed. Changes in groundwater mass and biomass can be observed at a precision necessary for detecting, for example, seasonal changes in soil moisture content. The limited spatial resolution of GRACE is a minor impediment to its utility in groundwater investigations, given the distributed character of most

BOX 3.3
Terrestrial Water Measurements

Prior to the launch of the GRACE gravity experiment, changes in terrestrial water storage were nearly impossible to measure directly, and the terrestrial component of the water budget was estimated largely by modeling. Reservoir impoundment was estimated by tallying the construction of reservoirs. Groundwater mining was estimated, for example, by balancing population-based estimates of well water extraction with well recharge modeled using groundwater hydrological methods.

The launch of the GRACE satellite system in 2002 provided scientists with the first means to directly measure changes in the mass of water on the Earth's surface and in the ground. Water mass can be determined at resolutions ranging from approximately 8 mm of water equivalent within a 750 km radius sample near the poles to approximately 25 mm of water equivalent near the equator (Wahr et al., 2006). The principal difficulty in interpreting GRACE data for hydrological studies lies in separating out undesired signals, including those arising from glacial isostatic adjustment (corrected using measurements or models) and from adjacent mass changes such as glacier and ice sheet changes (addressed using processing techniques that mask signals outside of the desired region; Luthcke et al., 2008).

[4] See <http://water.usgs.gov/nsip/history1.html>; <http://www.bafg.de/cln_031/nn_266918/GRDC/EN/02__Services/services__node.html?__nnn=true>.

groundwater storage, except in areas where confounding mass change signals are immediately adjacent. Distinguishing mass losses from Himalayan glaciers from groundwater losses in adjacent agricultural land to the south, for example, requires careful processing and interpretation of GRACE data.

Estimates from the IPCC Fourth Assessment Report

The IPCC Fourth Assessment Report and other previous assessments found large interannual and decadal fluctuations in the storage of water on land, likely associated with changes in precipitation, but no significant trend in land water storage due to climate change (e.g., Bindoff et al., 2007). Because land hydrology records are short, sparse, and poorly distributed for global calculations, the magnitude of changes in water storage is highly uncertain. However, the average magnitude of change over annual and longer time-scales during the reporting period (1961–2003) must have been small, given that the combined contributions of land ice and thermal expansion alone nearly match observed changes in sea level since 1993. The IPCC Fourth Assessment Report estimated that the contribution of land hydrology to sea-level change was <0.5 mm yr^{-1} (Bindoff et al., 2007).

Recent Advances

The terrestrial hydrologic processes contributing to sea-level change remain poorly constrained, although the importance of water storage in artificial reservoirs has become increasingly clear. Apart from changes in precipitation patterns and land ice volume, the primary terrestrial water fluxes are now thought to be reservoir construction, which lowers sea level, and groundwater depletion, which raises sea level. The continual development of processing techniques for analyzing data from the GRACE satellites (e.g., Ramillien et al., 2008) as well as methods for modeling global groundwater transport (e.g., Oleson et al., 2010) have made it possible to more precisely determine changes in land water storage. Several new data sets have been published since the IPCC Fourth Assessment Report, but many do not specify analysis periods, making it difficult to compare estimates or analyze trends.

Groundwater Depletion

In arid regions with significant populations and/or agricultural or industrial activity (e.g., portions of the United States, Mexico, Australia, China, Spain, and North Africa; see Shiklomanov, 1997), the rate of groundwater extraction often exceeds the rate of recharge. Huntington (2008) compiled published estimates of groundwater depletion, which ranged from 0.21 mm yr^{-1} to 0.98 mm yr^{-1} SLE (Table 3.5), but the time period to which this rate applies was not specified and the estimates are geographically incomplete. Based on hydrological modeling, Wada et al. (2010) estimated global groundwater depletion of 0.35 ± 0.1 mm yr^{-1} SLE for 1960, increasing to 0.8 ± 0.1 mm yr^{-1} SLE for 2000. Milly et al. (2010), also using modeling methods, estimated lower values of 0.12 mm yr^{-1} SLE for 1981–1998, 0.25 mm yr^{-1} SLE for 1993–1998, and 0.2–0.3 mm yr^{-1} SLE for "recent years." Milly et al. (2010) acknowledged, but did not quantify, considerable uncertainty in their estimates. Konikow (2011) estimated global groundwater depletion from 1900 to 2008, and found it increased significantly to 0.4 mm yr^{-1} during 2001–2008, double the rate of the 1990s. Most recently, Wada et al. (2012a) made an extensive assessment of groundwater extraction and depletion, arriving at a value of 0.54 ± 0.09 mm yr^{-1} SLE for 1993–2008.

Reservoir Storage

Until recently, additions to sea level from groundwater extraction were thought to be largely offset by increasing reservoir storage, although few studies estimated uncertainties in reservoir storage. Chao et al. (2008) estimated the water volume stored in 29,484 reservoirs constructed since about 1900 using the International Commission on Large Dams' World Register of Dams. Summing their stated water impoundment as the reservoirs were constructed provided the volume of water impounded as a function of time. Converting to SLE yielded a reservoir storage rate of -0.55 mm yr^{-1} for the 20th century (Chao et al., 2008). Lettenmaier and Milly (2009) found the equivalent impoundment to be -0.35 mm yr^{-1} SLE for 1940–1950 (Table 3.5). Milly et al. (2010) used results from Gornitz (2001) and others to estimate that the impoundment rate of global reser-

TABLE 3.5 Estimates of Groundwater Extraction and Reservoir Impoundment

Source	Period	Method	Sea-Level Equivalent (mm yr^{-1})
Net terrestrial depletion			
IPCC (2007)	1910–1990	Synthesis of reports	-1.5
	1990		-1.1–+1.0
Church et al. (2011)	1993–2008	Synthesis	-0.27–+0.11
Groundwater extraction			
IPCC (2007)	None given	Synthesis of reports	<0.5
Huntington (2008)	None given	Synthesis	0.21–0.98
Wada et al. (2010)	1960–2000	Hydrologic models	0.35–0.8
Milly et al. (2010)	1981–1998	Synthesis, models	0.12
	1993–1998		0.25
	"Recent years"		0.2–0.3
Konikow (2011)	2001–2008	Synthesis, models	0.4
Wada et al. (2012a)	1993–2008	Synthesis, models	0.45–0.63
Reservoir impoundment			
Chao et al. (2008)	Past half-century	Model	-0.55
Lettenmaier and Milly (2009)	1940–1950	Model	-0.35
Milly et al. (2010)	Before 1978	Synthesis, models	-0.5
	After 1978		-0.28
Wada et al. (2012b)	1993–2008	Update of Chao et al. (2008) with seepage correction	~-0.20– -0.40

voir storage declined from approximately -0.5 mm yr^{-1} SLE before 1978 to approximately -0.25 mm yr^{-1} SLE after 1978. They attributed this decrease to slowing in the rate of reservoir construction and to sedimentation, which slightly offset the storage capacity of existing reservoirs (Figure 3.8). How much sedimentation in reservoirs affects sea-level rise is a matter of debate. Huntington (2008) found that sedimentation results in a time-dependent loss of reservoir volume, which affects the rate of sea-level rise. On the other hand, Chao et al. (2008) argued that sedimentation displaces water behind dams and thus should have no effect on the contribution of reservoir storage to sea-level rise. Regardless, the effect of sedimentation is likely to be small compared with the decline in the number of reservoirs constructed. Wada et al. (2012b) estimated that decreased dam building lowered the contribution of reservoir storage to about 0.3 ± 0.1 mm yr^{-1} for 1993–2008.

Other Contributors

Snow accumulation and loss dominate seasonal variations in the terrestrial water contribution to global mean sea level but do not contribute to a long-term trend (Milly et al., 2003; Biancamaria et al., 2011). The effects of changes in permafrost on sea level are currently unknown, although the secondary hydrological effects (e.g., changes in soil hydraulic conductivity) of thawing the global permafrost area of $22 \pm 3 \times 10^6$ km^2 (Gruber, 2011) may become significant in the near future. Changes in global lake storage contributed about +0.11 mm yr^{-1} to sea level during the 1992–2008 period (Milly et al., 2010), but paleoclimatic records show that lake levels exhibit strong interannual and interdecadal variability, so this rate is not a good indicator of future trends. The magnitude and sometimes even the sign of other land water sources to sea level, including irrigation, wetland drainage, urbanization, and deforestation, are unknown (Milly et al., 2010).

Summary

Transfers of water (excluding ice melt) between the land and oceans are dominated by groundwater depletion, which raises sea level, and reservoir impoundment, which lowers sea level. Although more data (e.g., GRACE) and model results are available now than were for the IPCC Fourth Assessment Report, it remains difficult to constrain the contributions of terrestrial water to sea-level rise and uncertainties are large. Recent estimates for groundwater depletion and reservoir impoundment are in line with the IPCC (2007) estimates, on the order of 0.5 mm yr^{-1}. The two terms sum to near zero, within stated uncertainties. As

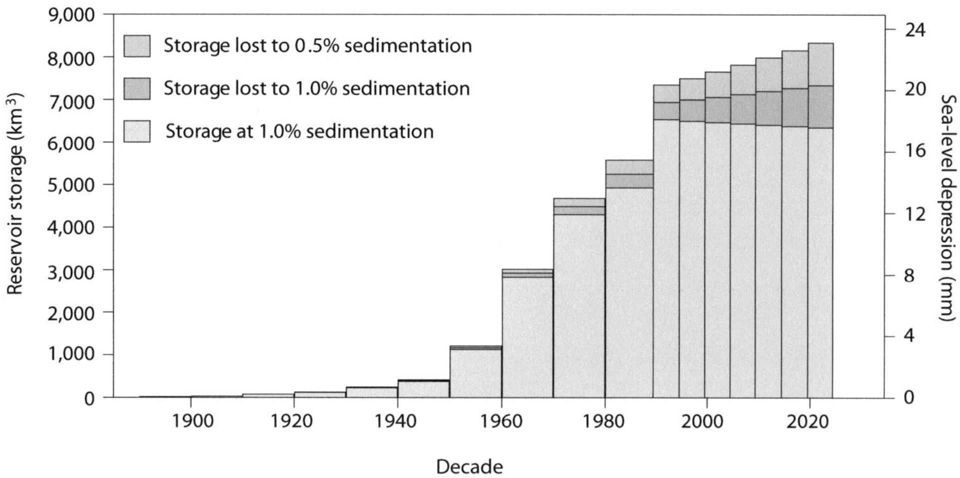

FIGURE 3.8 Accumulated global reservoir water storage in dams from 1900 to 2025 (yellow bars), based on observations and projections. The rate of water storage in dams higher than 15 m or with a capacity of more than 3 million m^3 has begun declining over the past decade because of sedimentation (blue and gray bars), potentially reducing the rate of sea-level rise. SOURCE: Lettenmaier and Milly (2009).

this report was nearing completion, a new evaluation by Wada et al. (2012b) found a net positive contribution to global sea-level rise of 0.25 ± 0.09 mm yr^{-1} during the 1990–2000 period as a result of a decrease in reservoir construction and an increase in groundwater depletion. If this result holds, terrestrial water storage could become a significant contributor to future sea-level rise.

CONCLUSIONS

The most comprehensive recent assessments of global sea-level rise is given in the IPCC Fourth Assessment Report, which evaluated data and research results published up to about mid-2006, and Church et al. (2011), which provided updated data on the components of sea-level rise. The IPCC (2007) found that the relative contributions to global sea-level rise varied over time, with thermal expansion contributing significantly more to sea-level rise for 1993–2003 than for 1961–2003. Since then, thermal expansion estimates have been corrected for instrument biases, which gave systematically warmer temperatures than the true value globally and cooler temperatures than the true value in a portion of the Atlantic Ocean. The corrected rates of thermosteric sea-level rise for the two IPCC (2007) periods are more similar, with a higher thermal expansion contribution for 1961–2003 and a lower thermal expansion contribution for the 1993–2003 period.

In addition, new types of measurements, notably the GRACE satellite system, and expanded data sets have become available since the IPCC Fourth Assessment Report was published. Estimates incorporating the new data suggest a faster growing contribution of land ice to sea-level change than was seen in IPCC (2007) for the two periods. Since 2006, ice loss rates have accelerated in the ice sheets and declined in glaciers and ice caps, likely reflecting interannual to multi-annual variability and possibly uncertainties in data processing or interpretation of short records. The most recent published estimate is that land ice melt accounted for about 65 percent of global sea-level rise for 1993–2008 (Church et al., 2011). The prospect of increased ice sheet melting is important to future sea-level rise because the Greenland and Antarctic ice sheets store the equivalent of at least 65 m of sea level.

New data and models also are available for estimating the contribution of terrestrial water (besides ice melt) to global sea-level rise. Although the contributions of the two largest terms—groundwater depletion, which transfers water to the ocean and raises sea level, and reservoir impoundment, which prevents water from reaching the ocean and lowers sea level—are significant, they are difficult to measure. As a result, most recent assessments have not assigned a rate to terrestrial storage or assigned a rate of zero, within the limits of uncertainty.

4

Sea-Level Variability and Change off the California, Oregon, and Washington Coasts

The waters of the world's oceans are subject to a variety of forces that create regional and local variations in sea level. Winds and currents move water laterally in the ocean, creating anomalous spatial patterns of sea level that can persist for a decade or longer. The high winds and low atmospheric pressures associated with El Niños and other climate patterns can significantly elevate sea level along the west coast of the United States for intervals of several months, as well as generate damaging high waves and storm surges. Melting of glaciers and ice sheets adds new water to the oceans and the associated gravitational and deformational effects distribute it nonuniformly, raising sea level in some areas and lowering it in other areas. Geologic processes (e.g., tectonics, compaction) and human activities (e.g., withdrawal of groundwater) also raise or lower the coastal land surface, increasing variability in relative (or local) sea-level rise.

This chapter evaluates the current contributions of ocean circulation, short-term climate patterns and storms, modern land ice change, and vertical land motion to sea-level rise in California, Oregon, and Washington. The discussion draws largely from published studies on the variability of sea level in this region, although the Intergovernmental Panel on Climate Change (IPCC) Fourth Assessment Report also summarizes research results on ocean circulation and short-period climate changes in the northeast Pacific Ocean. This chapter concludes with the results of the committee's analysis of tide gage records along the west coast of the United States.

CHANGES IN OCEAN CIRCULATION

Satellite altimetry data provide unambiguous evidence of significant regional differences in sea-level change in the oceans (Bindoff et al., 2007; Milne et al., 2009; Appendix B). Spatial variability in the North Pacific Ocean is associated with climate patterns—primarily the El Niño-Southern Oscillation (ENSO) but also the Pacific Decadal Oscillation (PDO; Box 4.1)—which affect ocean surface heating, surface air pressure, and wind patterns, and thus change ocean circulation (e.g., Mantua and Hare, 2002; Bond et al., 2003; Cummins and Freeland, 2007). Changes in ocean circulation change sea levels on seasonal to multidecadal timescales by redistributing mass and altering temperature and salinity in the upper ocean.

Estimates from the IPCC Fourth Assessment Report

Satellite altimetry records assessed by the IPCC showed that sea level fell about 0–6 mm yr^{-1} from 1993 to 2003 along the U.S. west coast and rose by 6 mm yr^{-1} to ~12 mm yr^{-1} in the tropical western Pacific Ocean (Bindoff et al., 2007). Temperature data from the upper 700 m of the ocean showed a similar sea-level pattern for the same period, indicating that regional sea level is influenced by changes in the thermal structure of the upper ocean, which are associated with changes in ocean circulation and surface heating. The IPCC (2007) suggested that the largest fraction of this short-term variation was caused by ENSO. Over longer periods, however, the thermosteric sea-level pattern along the U.S. west coast was different, showing a rise

BOX 4.1
Pacific Ocean Climate Patterns

ENSO. The El Niño-Southern Oscillation is a quasi-periodic climate pattern that occurs across the tropical Pacific Ocean about every 2 to 7 years. It is characterized by variations in the sea-surface temperature of the tropical eastern Pacific Ocean. In the warm El Niño phase, warm ocean temperatures in the tropical eastern Pacific are accompanied by high air surface pressures in the tropical western Pacific (Figure). In the cool La Niña phase, the pattern is reversed. The reversal in surface air pressure between the eastern and western tropical Pacific is known as the Southern Oscillation.

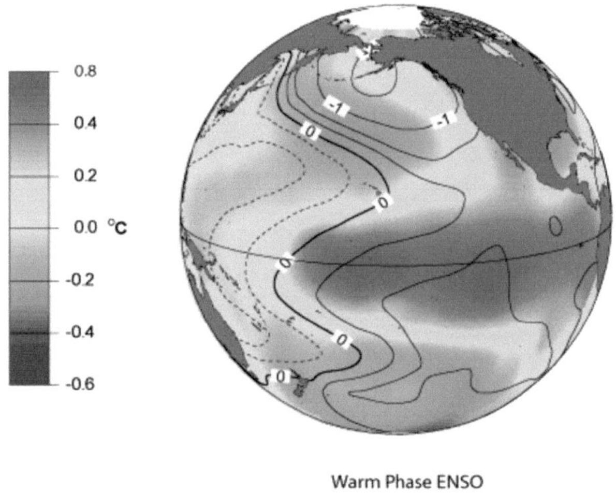

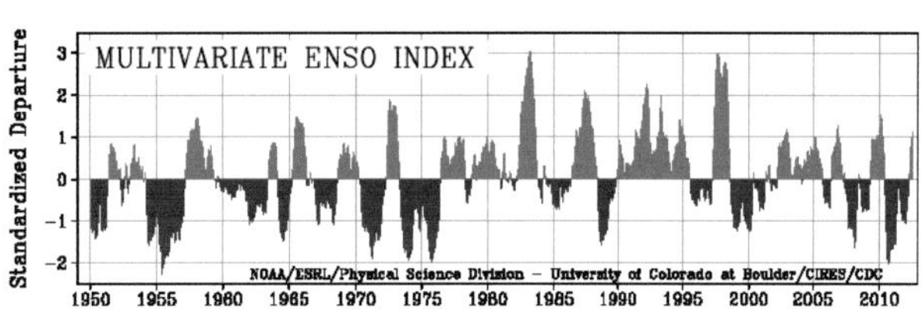

FIGURE (*Top*) Sea-surface temperature anomalies (shading) and sea-level pressure (contours) associated with the warm phase of ENSO (i.e., El Niño) for the 1900–1992 period. Positive contours are dashed and negative contours are solid. (*Bottom*) Multivariate ENSO index for 1950–2009. The index is based on variables observed over the tropical Pacific, including sea-level pressure, surface wind, sea surface temperature, surface air temperature, and cloudiness. Positive (red) index values indicate El Niño events and negative (blue) values indicate La Niña events. SOURCE: Figure and details on how the index is computed are given in <http://www.esrl.noaa.gov/psd/enso/mei/>.

PDO. The Pacific Decadal Oscillation is often described as a long-lived (i.e., decadal) El Niño-like pattern of Pacific climate variability. Like ENSO, the PDO has warm and cool phases, as defined by patterns of ocean temperatures in the northeast and tropical Pacific Ocean (Figure).

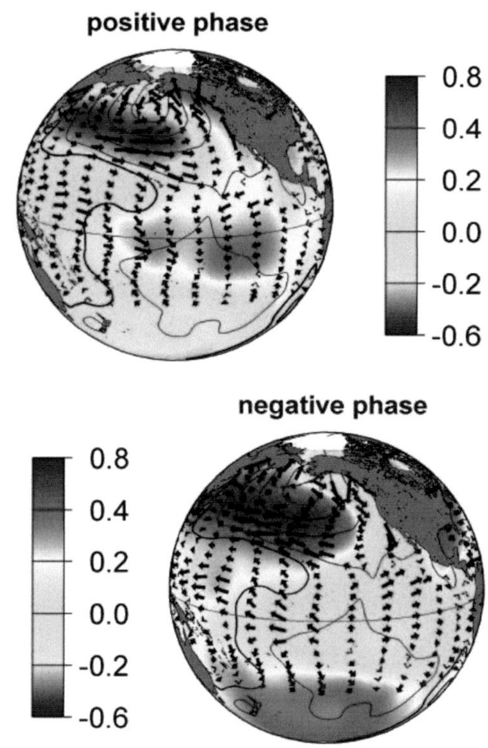

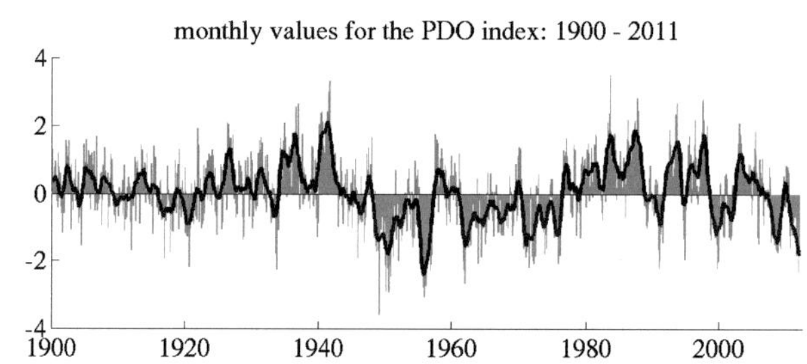

FIGURE The Pacific Decadal Oscillation. (*Top*) Typical winter patterns of sea surface temperature (colors), sea-level pressure (contours), and surface wind stress (arrows) during positive (warm) and negative (cool) phases of PDO. Temperature anomalies are in degrees Celsius. (*Bottom*) History of the PDO index (the principal component of monthly sea surface temperature anomalies in the North Pacific Ocean poleward of 20°N) from 1900 to 2010. SOURCE: Figure obtained with permission granted by Nate Mantua at the University of Washington's Joint Institute for the Study of Atmosphere and Ocean.

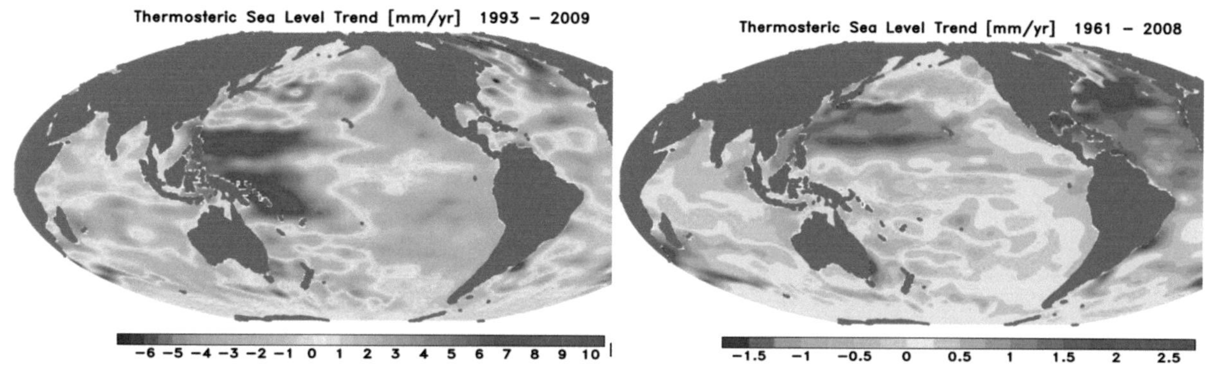

FIGURE 4.1 Trend of thermosteric sea level (mm yr^{-1}) for 1993–2009 (*left*) and 1961–2008 (*right*), based on an updated version of data from Ishii and Kimoto (2009). SOURCE: Courtesy of Masayoshi Ishii, Japan Meteorological Research Institute.

in sea level of about 0–0.8 mm yr^{-1} from 1955 to 2003, rather than a fall (Bindoff et al., 2007). This difference suggests that the spatial pattern of sea level varies on decadal and longer timescales.

Recent Advances

Changes in wind-driven ocean circulation can play an important role in determining patterns of sea-level change in the northeast Pacific Ocean on seasonal to decadal and longer timescales (e.g., Timmermann et al., 2010; Bromirski et al., 2011; Merrifield, 2011; Sturges and Douglas, 2011). Recent studies show a decrease in the rate of sea-level rise along the west coast of the United States since 1993, which is consistent with IPCC (2007) findings, but no statistically significant trends appear in tide gage records (Bromirski et al., 2011), satellite altimetry data, or in situ temperature observations since 1980. For example, thermosteric sea-level calculations show falling sea level off the U.S. west coast from 1993 to 2009 (Figure 4.1, left) and rising sea level from 1961 to 2008 (Figure 4.1, right). Bromirski et al. (2011) suggested that the flat sea-level trend since 1980 and the decrease since 1993 are associated with PDO phase changes.

Seasonal and Interannual Variability

Among all the climate modes, ENSO is the dominant cause of sea-level variability in the northeast Pacific Ocean on interannual timescales (e.g., Zervas, 2009; Bromirski et al., 2011). Sea level rises off the west coast of the United States during El Niño events and falls during La Niña events. El Niños differ in magnitude and large-scale form (Barnard et al., 2011) but commonly produce an active winter storm season in the northeast Pacific. The associated winds and ocean circulation changes may elevate sea level by 10–30 cm for several months along the west coast (Chelton and Davis, 1982; Flick, 1998; Bromirski et al., 2003; Allan and Komar, 2006; Komar et al., 2011). In fact, the highest sea levels recorded along the west coast were usually associated with El Niño events (e.g., Figure 4.2). For example, on January 27, 1983, during one of the largest El Niños in half a century, seven tide gages along the west coast (San Diego, Los Angeles, Monterey, Crescent City, Charleston, Astoria, and Seattle) recorded their highest water levels.[1] Peak sea level was 24 cm above predicted in San Diego (104 years of record), 31 cm above predicted in Los Angeles (87 years of record), and 76 cm above predicted in Seattle (112 years of record).

Large El Niño and La Niña events also can be seen in satellite altimetry data. The top panels of Figure 4.3 show the sea-level rise observed during the El Niño of 1997–1998 and the sea-level fall observed during the 1999 La Niña. The ENSO signal is strongly seasonal and reaches a peak amplitude in the Northern Hemisphere winter. Figure 4.3c shows the ocean seasonal cycle, which is occasionally magnified by ENSO.

[1] See <http://tidesandcurrents.noaa.gov/station_retrieve.shtml?type=Historic+Tide+Data>.

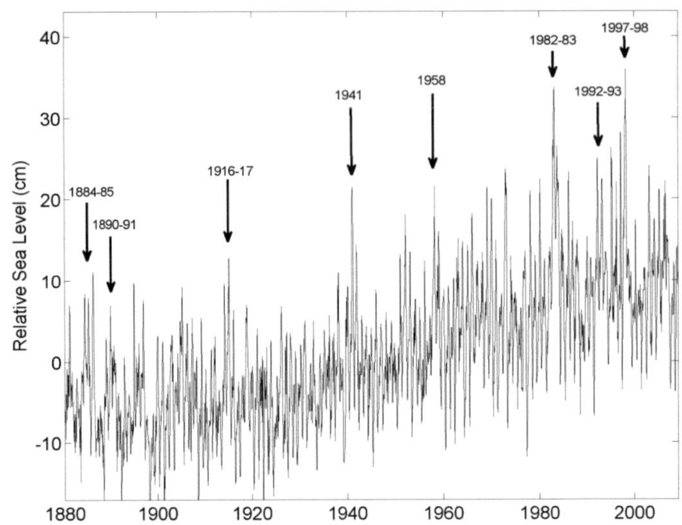

FIGURE 4.2 San Francisco tide gage record showing relative sea-level increases during major El Niño events. SOURCE: Tide gage data from the Permanent Service for Mean Sea Level.

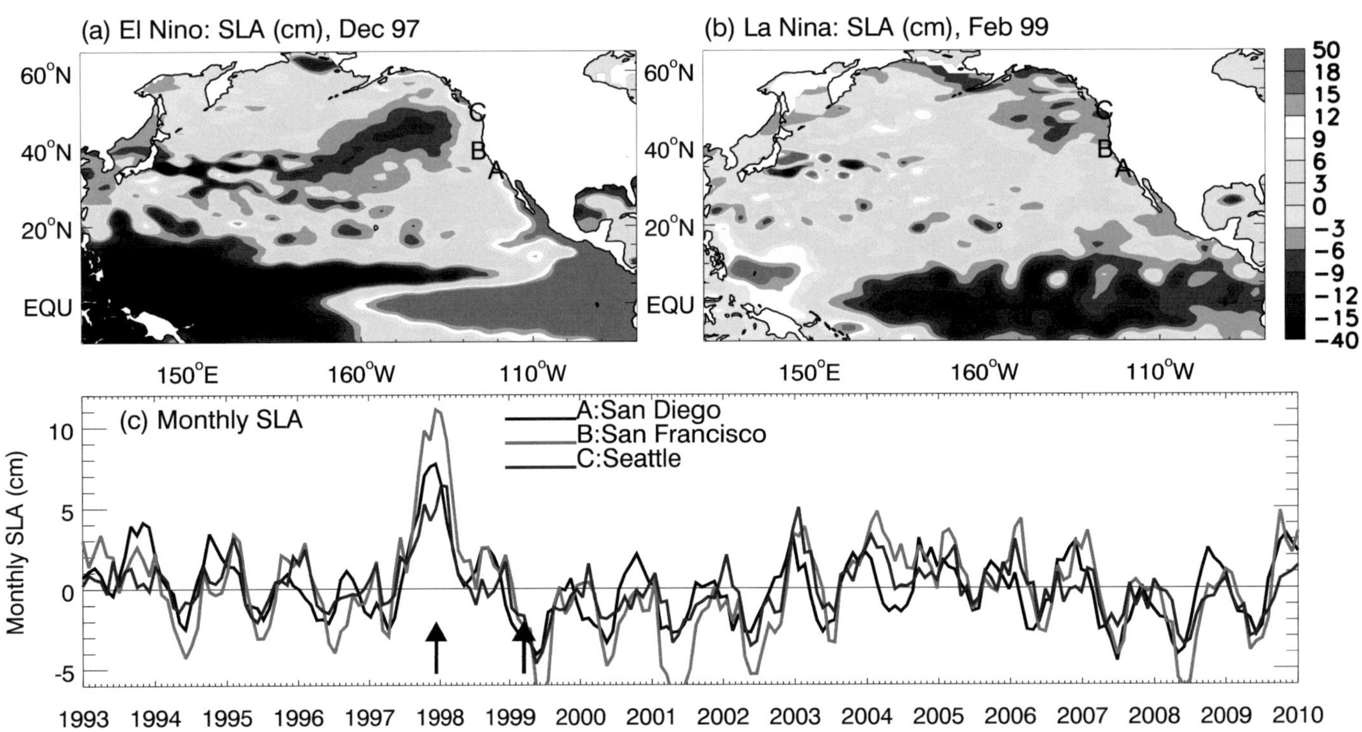

FIGURE 4.3 (*a*) Sea-level anomaly (SLA), the difference between mean sea level for 1993–2009 and sea level during the December 1997 El Niño. (*b*) Same as (a) but for a La Niña event in February 1999. Color scale on right is in cm. (*c*) Time series of monthly SLA offshore San Diego, San Francisco, and Seattle. The two black arrows correspond to the dates shown in the upper figures. SOURCE: AVISO satellite altimetry data from <http://www.aviso.oceanobs.com/>.

Decadal and Longer Variability

The low-frequency (decadal and longer) variability in sea level off the U.S. west coast often corresponds to forcing by regional and basin-scale winds associated with climate patterns such as the PDO and the North Pacific Gyre Mode (e.g., Lagerloef, 1995; Fu and Qiu, 2002; Jevrejeva et al., 2006; Cummins and Freeland, 2007; Miller and Douglas, 2007; Di Lorenzo et al., 2008, 2010; Bromirski et al., 2011; Sturges and Douglas, 2011; Merrifield, 2011). For example, ocean modeling by Bromirski et al. (2011) found that surface heating alone produced falling sea level—the opposite to that observed—whereas forcing by winds explained the rise in sea level along the U.S. west coast since 1950. They suggest that the lack of a significant trend in sea level observed in tide gages since 1980 reflects forcing by winds associated with phase changes of the PDO. Sea level rose when the PDO changed from negative (cool) to positive (warm) around 1976–1977, and it fell when the PDO changed from positive to negative at the end of the 1990s (see lower figure in Box 4.1). The PDO has largely been in a positive phase since 1977, although negative phases have occurred almost a half-a-dozen times since the 1990s.

ENSO may also play a significant role in decadal and longer sea-level variability (Newman et al., 2003). Indeed, ENSO and the PDO are not independent. ENSO can influence the PDO (Newman et al., 2003; Schneider and Cornuelle, 2005), and the PDO can modulate tropical Pacific circulation and ENSO (e.g., Vimont et al., 2009; Alexander et al., 2010).

Summary

The spatial variability of sea level in the Pacific Ocean is driven primarily by ENSO, which affects sea level on seasonal to decadal timescales, and is also associated with phase changes in the PDO, which affects sea level on decadal and longer timescales. Satellite altimetry, tide gage, and ocean temperature measurements all indicate a long-term increase in sea level off the U.S. west coast, with large amplitude seasonal to multidecadal variability. The measurements show no statistically significant sea-level trend since 1980, consistent with the PDO phase changes.

SHORT-TERM SEA-LEVEL RISE, STORM SURGES, AND SURFACE WAVES

Any climate-induced increase in storm frequency and magnitude will induce short-term changes in sea level. This issue is critical to coastal planners because storm surges and wind-driven waves are responsible for most of the flooding and erosion damage along the west coast of the United States (Armstrong and Flick, 1989; Domurat and Shak, 1989; Allan and Komar, 2006). The most severe coastal impacts tend to occur when a storm surge coincides with high tides and/or during periods of anomalously high sea level, such as those caused by El Niños. For example, the simultaneous occurrence of anomalously high sea level, high waves in late January and early March, and high astronomical tides caused significant damage along the California coast during the El Niño winter of 1983 (Figure 4.4). The amplitude of local sea-level rise from storm and wave events can greatly exceed the projected amplitude of global and regional sea-level rise, even beyond 2100, so understanding their additive effects is crucial for coastal planning. This section describes the contributions of these factors to short-term sea-level rise and the extent to which they may be changing with climate change (Task 2b).

Contributions of Tides, Storms, and El Niños to Local Sea Level

High tides along the U.S. west coast occur twice daily, often of uneven amplitude, caused predominately by the gravitational attraction of the Moon and the Sun on the Earth. The Earth-Moon-Sun orbital geometry also results in heightened high tides twice monthly (spring tides, near the times of the full and new moon) and every 4.4 years and 18.6 years (Zetler and Flick, 1985). The largest tidal amplitudes of the year along the coasts of California, Oregon, and Washington usually occur in winter and in summer (Zetler and Flick 1985). Tides in the highest winter and summer months are often more than 20 cm higher than tides in the spring and fall months.[2] The peaks in the 4.4-year and 18.6-year cycles produce monthly high tides that are about 15 cm and 8 cm, respectively, higher than they are in the intervening years (Flick, 2000). Flick et al.

[2] See data compiled at <http://tidesandcurrents.noaa.gov>.

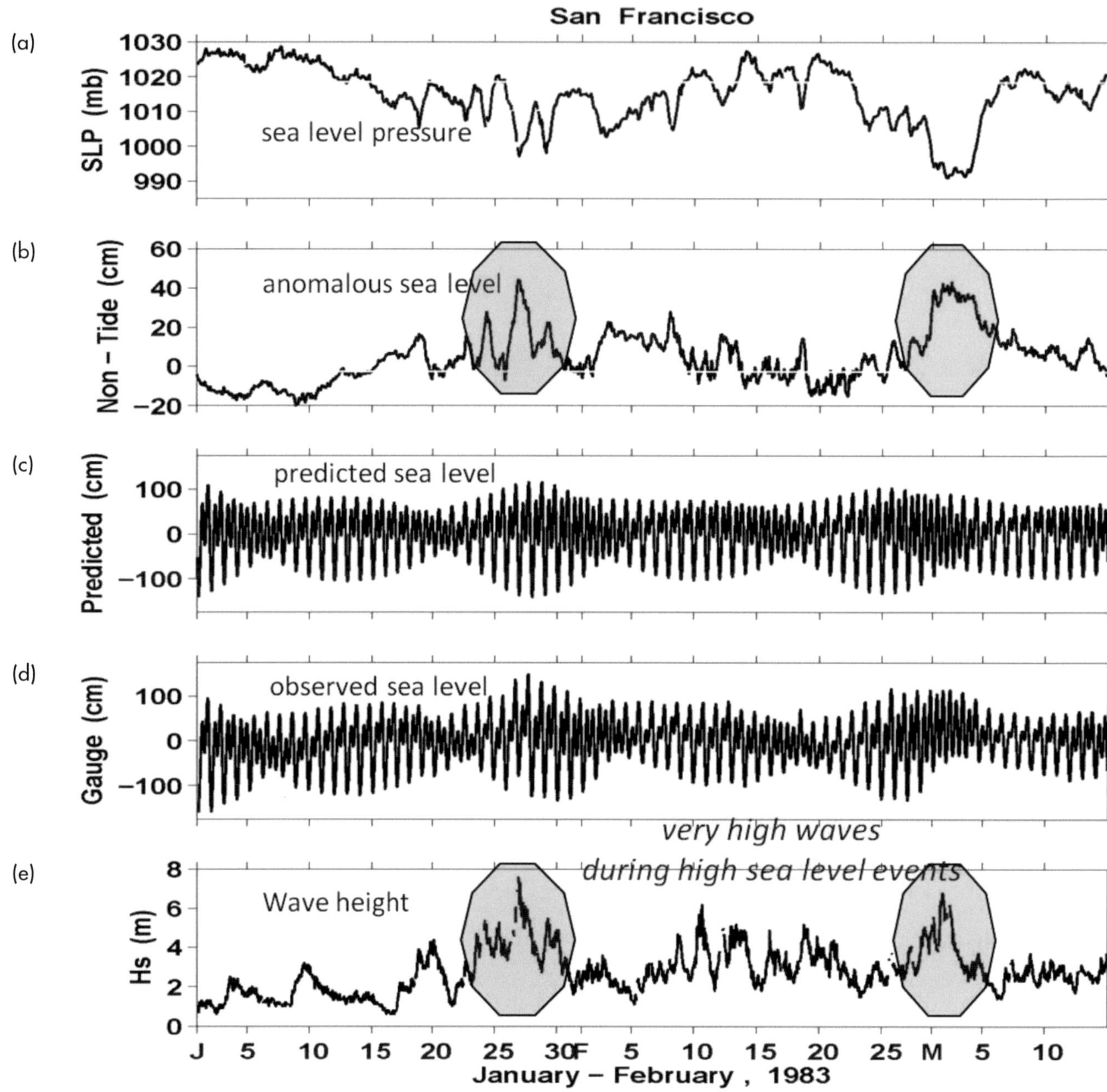

FIGURE 4.4 (*a*) Hourly sea-level pressure (SLP; mb), (*b*) sea-level anomaly (cm) above tide-predicted levels, (*c*) predicted and (*d*) observed sea level (cm) relative to a mean sea-level datum, and (*e*) significant wave height (Hs, the average height of the highest one-third of waves [m]) from a buoy sensor near San Francisco during the El Niño winter of 1983. SOURCE: Adapted from Flick (1998).

(2003) reported increases in the range from high to low astronomical tide over multiple decades at some, but not all, U.S. west coast tide gages.

Storm surges are created when high winds, the Coriolis force, and low barometric pressures from coastal storms force sea water onto the shore. During the most severe winter storms, surface atmospheric pressure along the west coast drops by 20 mb or more from long-term average levels, typically with greater pressure drops in Washington and Oregon than in California. The drop in atmospheric surface pressure raises sea level by approximately 1 cm for every 1 mb decrease in atmospheric pressure. The resulting increase in sea level is usually regional, according to the regional scale of winter cyclones, and typically lasts only a few days at most (Flick, 1998). Woodworth and

Blackman (2004) investigated high-water levels from tide gages around the world since 1975 and found that the magnitude of sea-level extremes has risen in many locations, including some parts of the U.S. west coast, and that these extremes closely followed increases in the median sea level.

Strong ocean winds also produce surface gravity or wind waves. The most extreme such waves are of two types: sustained intervals of large waves (measured by the significant wave height, the average height of the largest one-third of the waves) and rogue waves, which have individual crests that are much larger than the significant wave height. Sustained intervals of large waves occur during strong storms. These storm waves can propagate over a long distance to the shoreline. Rogue waves are produced by interactions among waves and perhaps currents, and they have the greatest impact when they arise during a sustained interval of large waves. By definition, they are expected but relatively uncommon events (Baschek and Imai, 2011).

El Niños can significantly elevate sea level along the west coast during winter months (see "Changes in Ocean Circulation" above), especially along the California coast because the North Pacific storm track is displaced toward the equator during El Niño events (Seager et al., 2010). The wind and pressure patterns that elevate sea level above climatological normals along the west coast also may occur in winters when El Niño is not present. Winters with high sea-level anomalies have usually had a few large North Pacific storms with strong westerly, southwesterly, or northwesterly winds offshore, which generate storm surges and high waves along the coast of California and sometimes the coasts of Oregon and Washington.

The path and propagation speed of storms controls the wind direction and barometric pressure, which, in turn, affects the generation of wind waves and high water (e.g., O'Reilly and Guza, 1991). The highest winds, and hence waves, along the west coast of the United States nearly always occur during strong winter extra-tropical cyclones (Wang and Swail, 2001; Bromirski et al., 2003; Caires et al., 2004; Ruggiero et al., 2010; Barnard et al., 2011; Seymour, 2011). Tropical cyclones rarely travel as far north as California, although two cases have been recorded historically (Hurd, 1939; Chenoweth and Landsea, 2004). Significant wave heights recorded by offshore coastal buoys during extra-tropical events can exceed 10 m (Figure 4.5; Ruggiero et al., 2010; Seymour, 2011), although they are usually smaller as they approach the shoreline. Significant wave heights at the shoreline vary considerably depending on incident wave direction and nearshore bathymetry.

Wave swells generated by storms propagate long distances (e.g., from the central North Pacific to the U.S. west coast) over several days. Swells generated far from the west coast tend to peak at relatively long periods (12 seconds or more), whereas more locally generated wave swells tend to peak at periods of 10 seconds or less. The largest swells are generated by winter cyclones that produce high winds with a long fetch (the total distance that wind blows over the sea surface during the storm) directed toward the west coast. A broad, deep low-pressure system over the North Pacific favors these conditions (Figure 4.6; Bromirski et al., 2005). Synoptic timescale patterns like this tend to occur during El Niño winters, but not exclusively (Seymour et al., 1984; Bromirski et al., 2005; Allan and Komar, 2006). Larger than normal waves have occurred during El Niño winters along the California coast and some parts of the Oregon and Washington coasts (Bromirski et al., 2005; Allan and Komar, 2006). La Niñas have been shown to produce smaller than normal winter wave heights at some California locations, but not everywhere along the west coast (Allan and Komar, 2006). Overall, the occurrence of large storms and high waves is clustered in time, with particular years and groups of years having many large storms, and other years having few or no large storms.

Peaks in wind waves are generally much higher than sea-level anomalies (Seymour et al., 1984; Seymour, 1998; Storlazzi and Griggs, 1998; Ruggiero et al., 2010). High breakers induce a change in mean water level at the beach (set-up), which can be about 20 percent of the breaking wave height (Dean and Dalrymple, 1991). High wave events sometimes, but not always, coincide with high sea levels (Cayan et al., 2008; Ruggiero et al., 2010).

Changes in Storminess and Extreme Wave Heights

Evidence of changes in storminess (wind intensity) in the North Pacific Ocean is mixed. Bromirski et al. (2003) examined nontidal sea-level fluctuations from 1858 to 2000 in the San Francisco tide gage record

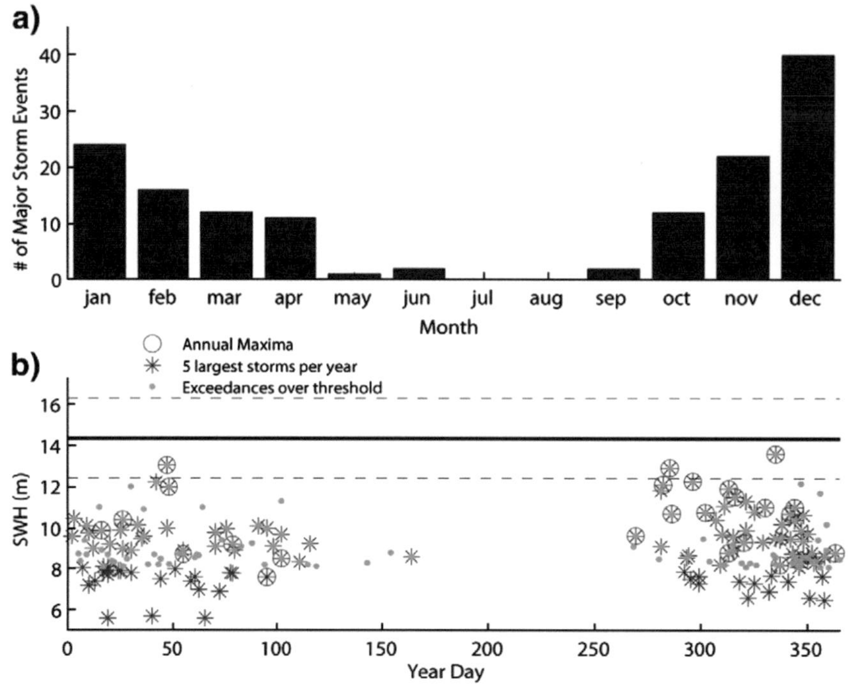

FIGURE 4.5 (*a*) Number of storm events per month off Oregon and Washington between 1976 and 2007, when the significant wave height (SWH) exceeded a threshold of 8.1 m at two deep-water wave buoys. (*b*) Days when the threshold of 8.1 m was exceeded (dots), annual maxima (circles), and the five largest storms per year (asterisks) for 1976–2007, illustrating the seasonality of the extreme wave climate. The 100-year significant wave height is shown by the solid horizontal line and its associated uncertainty is the dashed horizontal lines. SOURCE: Ruggiero et al. (2010).

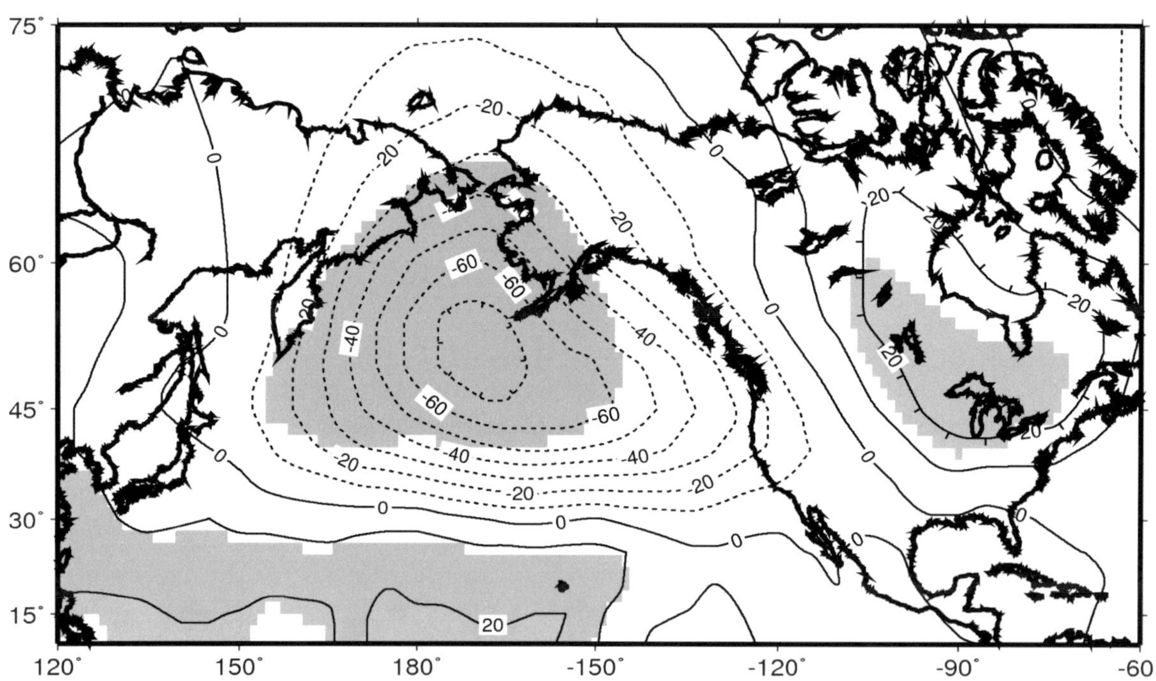

FIGURE 4.6 Atmospheric circulation during periods of high waves along the central California coast exhibits broad-scale low pressure over the North Pacific. This map shows anomalies of 700 hPa height in meters during the 15 winter months (November through March) from 1981 to 2003 when wave energy offshore San Francisco was greatest. The region of anomalously low 700 hPa indicates a low-pressure trough and increased storminess in the central and eastern North Pacific. Significant negative and positive anomalies are blue and red, respectively. SOURCE: Adapted from Bromirski et al. (2005).

and found significant decadal variability. Although the record showed an increase in storminess from 1950 to 2000, the storm intensity in recent decades did not significantly exceed that in the decades prior to 1950 (Bromirski et al., 2003). On the other hand, the IPCC Fourth Assessment Report cited several studies that reported increases in the strength of the winter westerly wind circulation across the North Pacific during the past few decades (Trenberth et al., 2007).

Lowe et al. (2010) described climate change effects on storm intensity as inconclusive, with no consensus among different model simulations on local changes in storm frequency. A simulation of San Francisco sea-level anomalies forced by 21st century climate change simulations (Cayan et al., 2008) found considerable interannual and decadal variability, driven partly by storm characteristics, superimposed on an assumed long-term rise in mean sea level. Several climate models discussed in the IPCC Fourth Assessment Report project that the mid-latitude storm tracks in both the southern and northern hemispheres will migrate poleward over the 21st century (Meehl et al., 2007). A subsequent projection by Salathé (2006) also showed a northward shift in the North Pacific winter storm track over the next several decades. The storm tracks and Pacific wind fields in some global climate model projections suggest that future wave heights might diminish somewhat over the open ocean and along the coast from southern and central California to Oregon (Salathé, 2006; Cayan et al., 2009).

If frequency or intensity of storminess changes as a result of climate change, the frequency of high sea-level extremes also would likely change. Even if the storminess regime does not change, sea-level rise will increase the exposure of the coast to storm-driven surge and high waves, magnifying their impact on the coast.

Analyses of marine weather reports discussed in the IPCC Fourth Assessment Report showed an increase in significant wave height of 8–10 cm per decade over the central and eastern North Pacific from 1950 to 2002 (Trenberth et al., 2007). Gulev and Grigorieva (2006) attributed these increases to longer period, longer distance sources of swell as well as to more locally generated wind waves. The tendency for an increase in wave energy over the eastern North Pacific is also indicated by wave hindcasts (Graham and Diaz, 2001), buoy observations (e.g., Allan and Komar, 2006), some wave buoy records (Ruggiero et al., 2010), and satellite altimeter observations (Young et al., 2011a).

A study of North Pacific wind variability on 2- to 10-day timescales from the National Centers for Environmental Prediction (NCEP) Reanalysis (Kalnay et al., 1996) indicated that wind speed trends are variable, owing to the occurrence of relatively infrequent large events. From the 1950s through the 1990s, wave model reanalyses over the North Pacific (Graham and Diaz, 2001; Caires et al., 2004) indicate a trend toward increasing wave height. From a series of buoy observations beginning in the late 1970s, Storlazzi and Wingfield (2005), Allan and Komar (2006), Ruggiero et al. (2010), and Seymour (2011) found that the largest waves along the coast from California to Washington state were larger in the period after 1990 than in the period before (Figure 4.7). This change was associated with a deepening of the winter low pressure system over the North Pacific Basin and partly to the incidence of some relatively strong El Niño years since 1995.

Increases in wind speed and wave heights in the northeastern Pacific Ocean have been reported recently,

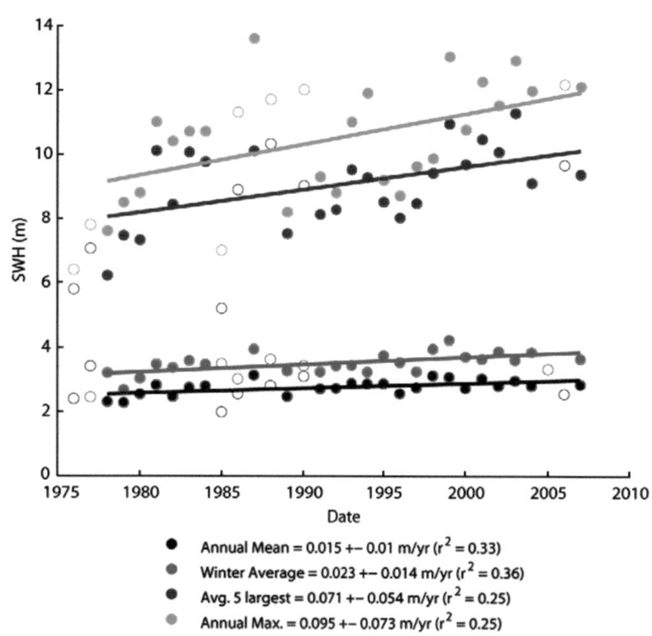

FIGURE 4.7 Increases in the annual maximum wave height (green; m), average of the five largest wave events per year (blue), winter average height (red), and annual average height (black) from northeast Pacific wave buoy sensors. Open circles represent years with too much missing data (i.e., winter months missing more than 60 percent of data). SOURCE: Ruggiero et al. (2010), after Allan and Komar (2006).

but the interpretation of these changes is controversial. Analyses of global ocean winds from ship observations (Tokinaga and Xie, 2011), satellite microwave sensors (Wentz et al., 2007), and satellite altimeters (Young et al., 2011a) indicate that wind speeds have risen over the global oceans, although the trends found by Young et al. (2011a) are greater than those derived from Tokinaga and Xie (2011) and Wentz et al. (2007) by approximately a factor of two (Wentz and Ricciardulli, 2011; Young et al., 2011b). The Young et al. (2011a) analysis also found that wind speeds within the highest 1 percent of events have risen over much of the extra-tropical oceans over the past two decades, including an increase of about 1 percent per year in the northeast Pacific, and that this increase is accompanied by increases in the extreme wave heights. The latter occurs in particular in the northeast Pacific Ocean, which is consistent with increasing extreme wave heights (by as much as 2 m over the record period) during big storms recorded in near coastal deep-water buoy records from northern California to Washington (Allan and Komar, 2006; Menéndez et al., 2008; Ruggiero et al., 2010). However, further analysis by Gemmrich et al. (2011) suggests that much of this change is spurious, caused by changes in buoy hardware and data processing. All of these estimates were made from records that are only a few decades long, and thus partly reflect changes in wind forcing associated with natural climate variability such as the Pacific Decadal Oscillation and other interannual-interdecadal fluctuations. However, the global extra-tropical pattern of extreme wave increase found by Young et al. (2011a) is atypically widespread for most decadal natural variability, and thus might indicate a longer trend. As yet there is no good explanation for why such a trend would occur.

Summary

Periods of anomalously high sea levels and wave heights along the west coast of the United States exhibit considerable variability on synoptic, interannual, and decadal timescales, in association with ENSO and other climate patterns. Some evidence suggests that wave heights have increased along the west coast from northern California to Washington during the past few decades. However, it is likely that much of this increase is associated with interannual- to decadal-scale natural variability of the Pacific atmosphere-ocean system. Some global climate models predict that the North Pacific storm track will shift northward as global climate warms during the next several decades, which would generate extreme wave heights and storm surges along the Oregon and Washington coasts. However, a northward shift in the North Pacific storm track has not yet been confirmed.

All climate models project ample winter storm activity in the North Pacific in future decades, suggesting that periods of anomalously high sea level and high waves will continue to occur along the west coast. Storm-generated bursts of high sea levels and waves are expected to vary from year to year and decade to decade. Over the next few decades, these anomalies will likely eclipse the secular rise in sea level (few to several mm per year). Short-period fluctuations of sea level may sometimes exceed 20 cm, and storm-driven wave heights of 1 m or even higher amplitudes than are seen in the historical record could easily occur. These variations will have greatest impact when they occur on days with high tides.

SEA-LEVEL FINGERPRINTS OF MODERN LAND ICE CHANGE

As glaciers and ice sheets melt and lose mass and the melt water is transferred from the continents to the ocean, the solid earth deforms and the gravitational field of the planet is perturbed. The addition of new water to the ocean basins and the associated gravitational and deformational effects create regional patterns of sea level change. Both modern melting and deglaciation of the ancient ice sheets affect sea-level change along the west coast of the United States. Melting of the ancient ice sheets caused the solid earth to rebound (glacial isostatic adjustment), resulting in significant vertical land motions in the vicinity of the California, Oregon, and Washington coasts. In contrast, modern melting affects land motions at the ice masses, which are far from the U.S. west coast, but the gravitational effect influences the height of the sea surface in the northeast Pacific Ocean. This section describes the effects of modern land ice melt on sea-level rise off the coasts of California, Oregon, and Washington. The effects of ancient ice melt are discussed in the following section (see "Glacial Isostatic Adjustment" below).

Modern melting of land ice affects sea level along the west coast of the United States in two ways. First, the large mass of glaciers and ice sheets generates an additional gravitational pull that draws ocean water closer, raising relative sea level near the ice masses. As the ice melts, the amount of ice mass on land declines, decreasing its gravitational pull on the ocean water. The loss of mass also results in uplift of the land mass under the ice. The combination of these effects causes relative sea level to fall in the vicinity of the ice mass. The fall extends, at decreasing rates, in the region within a few thousand km of the melting ice. Second, ice melt enters the ocean, raising global mean sea level. Because of gravitational and deformational effects, however, the distribution of new ice melt is nonuniform over the globe. Relative sea level falls near the shrinking ice mass and rises everywhere else. This effect is shown schematically in Figure 4.8. The combined effect of new water mass entering the ocean and altered gravitational attraction results in a spatial pattern of sea-level rise that is unique for each ice sheet or glacier (Mitrovica et al., 2001; Tamisiea et al., 2003). As a consequence, these sea-surface geometries have come to be known as sea-level fingerprints.

Only a few studies have attempted to map the sea-level fingerprints of melting land ice along the west coast of the United States (e.g., Tamisiea et al., 2003, 2005). Figure 4.9A shows the sea-level fingerprints of the three largest sources of land ice that are most likely to have significant effects on west coast sea level: Alaska, Greenland, and Antarctica. The figure shows that melting of Alaska glaciers creates a strong north-south gradient in relative sea-level change along the west coast. The gradient from uniform melting of the Greenland Ice Sheet is much smaller (Figure 4.9B). Uniform melting of either the Antarctic Ice Sheet or the West Antarctic Ice Sheet leads to a uniform change in relative sea level along the entire west coast (Figure 4.9C).

To estimate the effect of fingerprinting from these three ice masses on relative sea level, it is necessary only to multiply the global sea-level equivalent of the mass loss from each source by the appropriate scale factor (colored contours) indicated in the figure and then add the contributions from all three sources. Scale factors greater than 0 indicate that the sea-level fingerprint increases relative sea-level rise at that location, and scale factors greater than 1 indicate that the rise is higher than the global sea-level equivalent value. Scale factors less than 0 mean that the effect of mass loss from a source causes the relative sea level to fall. Scale factors for other ice sources (e.g., European Alps, northeastern Canadian Arctic, Patagonia) are not available at the resolution shown in Figure 4.9, but these sources are likely too small and/or too distant to affect the gradient in sea-level change along the U.S. west coast.

The scale factors and ice loss rates used to calculate the adjusted rates of relative sea-level rise are given in Table 4.1. Modeling or estimating individual regional land ice losses is beyond the scope of this study, so

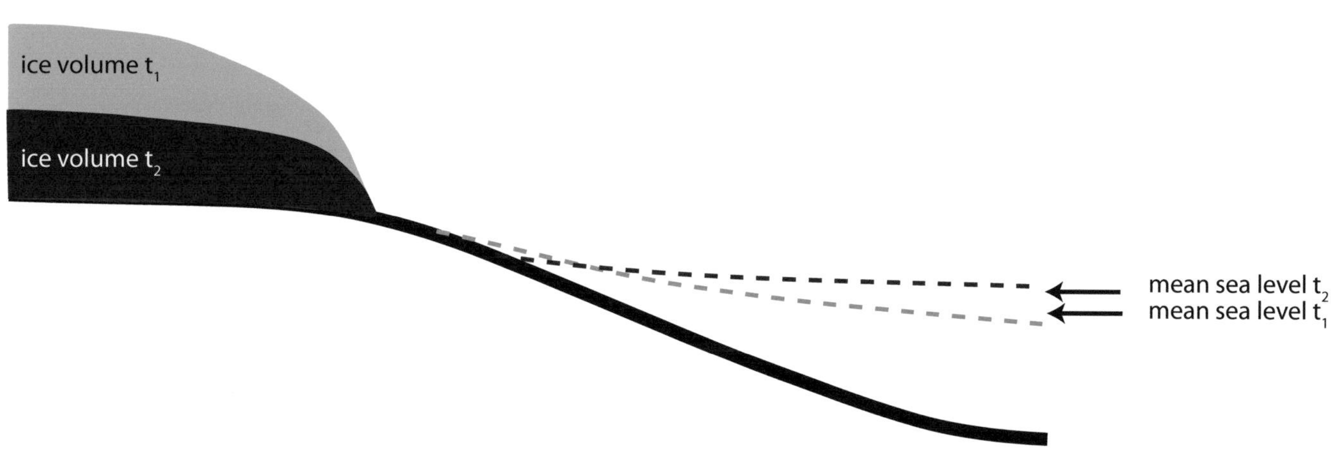

FIGURE 4.8 Schematic view of the changing sea level caused by a shrinking land ice mass. Relative sea level at time t_1 exceeds the mean sea level near the ice mass and is less than the mean at some distance beyond the mass. As the land ice mass decreases (time t_2), the local gravitational attraction decreases and the land in the vicinity of the ice rises, causing the relative sea level to fall, even though the mean sea level increases. SOURCE: Adapted from Tamisiea et al. (2003).

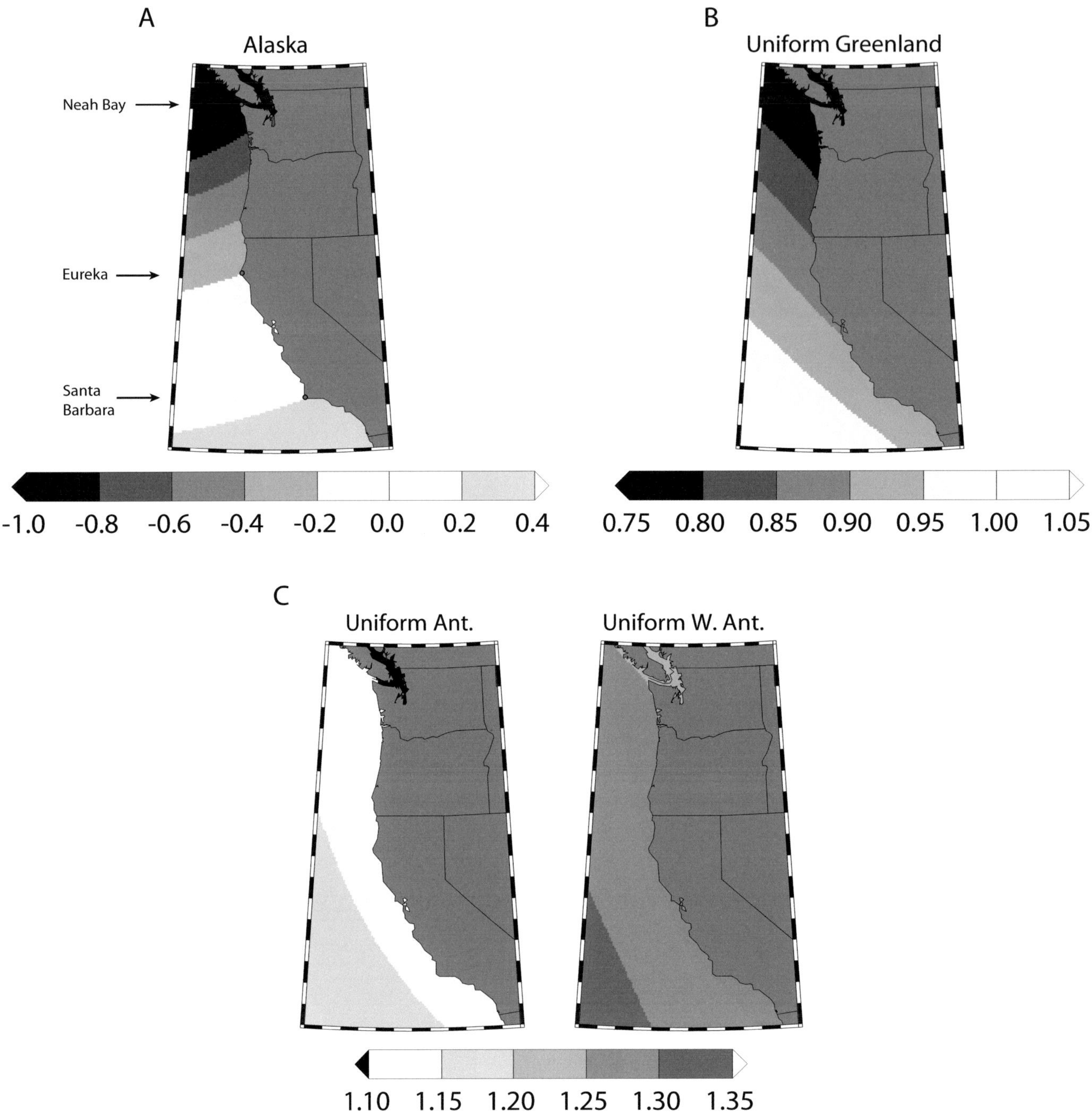

FIGURE 4.9 Sea-level responses in the northeast Pacific to ice loss from three major ice masses. The responses are shown as scale factors, which are the local sea-level equivalent divided by the global mean sea-level equivalent. (*A*) Response to melt from the Alaskan glacier system, as modeled in Tamisiea et al. (2003). (*B*) Response to uniform melting over the entire grounded portion of the Greenland Ice Sheet. (*C*) Response to melting across the entire Antarctic Ice Sheet (left) or the West Antarctic Ice Sheet (right). All of the calculations underlying this figure treat the Earth as elastic; that is, the timescale of response is assumed to be sufficiently rapid that viscous effects can be neglected. SOURCE: Courtesy of Jerry Mitrovica and Natayla Gomex, Harvard, based on calculations described in Mitrovica et al. (2011).

the committee used ice loss rates averaged from data reported in Appendix C. To simplify the analysis, scale factors were picked from Figure 4.9 for three representative locations along the U.S. west coast: the north coast (approximately Neah Bay, Washington), the central coast (approximately Eureka, California), and the south coast (approximately Santa Barbara, California).

In the absence of a sea-level fingerprint effect, the expected sea-level rise along the U.S. west coast from ice loss in Alaska, Greenland, and Antarctica would be 0.79 mm yr^{-1}, the sum of the ice mass loss rates in Table 4.1. The overall effect of the fingerprint is to lower sea-level rise along the entire west coast. Although melting of Alaska glaciers contributes less water to the oceans than melting of the Greenland Ice Sheet, the Alaska glaciers are closer to the U.S. west coast and have a greater effect on relative sea level in the region. The adjusted rates of relative sea-level rise for the three sources (found by multiplying the loss rate by the fingerprint scale factors) are 0.46 mm yr^{-1} for the north coast, 0.60 mm yr^{-1} for the central coast, and 0.68 mm yr^{-1} for the south coast. This adjusted rate of sea-level rise is 42 percent lower than the rate for melting of the three ice sources (0.79 mm yr^{-1}) for the north coast, 24 percent lower for the central coast, and 14 percent lower for the south coast.

This simple calculation provides only an approximate estimate of the magnitude and sign of relative sea-level change due to gravitational and deformational effects of modern land ice melting. Uncertainties in the rate of ice loss and, to a lesser extent, the neglect of fingerprints of other sources of land ice can lead to significant uncertainties in the adjusted rates of relative sea-level rise. In particular, the steep gradient caused by Alaska's proximity to the study region, combined with the high uncertainty in the rate of ice loss from Alaska compared to the ice sheets, yield a wide range of possible adjustments to relative sea-level rise (see Appendix C). When the uncertainty in loss rates from the three sources is considered, the adjusted rate of relative sea-level rise due to melting of these ice masses ranges from 0.1–0.9 mm yr^{-1} for the north coast, 0.1–1.1 mm yr^{-1} for the central coast, and 0.1–1.3 mm yr^{-1} for the south coast.

Summary

The large mass of glaciers and ice sheets creates a gravitational pull that draws ocean water closer. As the ice melts, the gravitational pull decreases, ice melt enters the ocean, and the land and ocean basins deform as a result of this loss of land ice mass. These gravitational and deformational effects produce a spatial pattern of regional sea-level change commonly referred to as a sea-level fingerprint. The land ice masses that most affect sea level along the California, Oregon, and Washington coasts are in Alaska, which is nearby, and Greenland and Antarctica, which are large. Melting in Alaska and, to a lesser extent, Greenland, causes relative sea level to fall at decreasing rates from northern Washington to southern California. Melting in Antarctica causes a uniform sea-level rise along the entire west coast of the United States. The net result is a reduction in the contribution of Alaska, Greenland, and Antarctica ice melt to relative sea-level rise off Washington, Oregon, and California. The magnitude of this reduction decreases from about 42 percent along the north coast (Neah Bay) to 24 percent along the central coast (Eureka) to 14 percent along the south coast (Santa Barbara) for 1992–2008.

TABLE 4.1 Ice Loss Rates, Fingerprint Scale Factors, and Adjusted Rates of Relative Sea-Level Rise for Three West Coast Locations

		North Coast		Central Coast		South Coast	
Ice Source	Ice Mass Loss Rate (mm yr^{-1} SLE)[a]	Scale Factor	Adjusted Sea-Level Rise (mm yr^{-1})	Scale Factor	Adjusted Sea-Level Rise (mm yr^{-1})	Scale Factor	Adjusted Sea-Level Rise (mm yr^{-1})
Alaska	0.16	-0.8	-0.13	-0.2	-0.03	0.2	0.03
Greenland	0.35	0.75	0.26	0.87	0.30	0.92	0.32
Antarctica[b]	0.28	1.17	0.33	1.17	0.33	1.17	0.33
Sum	0.79		0.46		0.60		0.68

[a] Based on the average of published rates for 1992–2009 for Greenland and Antarctica and 1992–2008 for Alaska, as described in Appendix C.
[b] Average of east and west Antarctic values.

VERTICAL LAND MOTION ALONG THE U.S. WEST COAST

Vertical land movements that affect relative sea level may be caused by geologic processes (e.g., glacial isostatic adjustment, tectonics, compaction) or anthropogenic activities (e.g., groundwater or oil extraction). Each of these processes can in principle be modeled or observed (Box 4.2), although data coverage is sparse and uncertainties are large. The estimated rates of vertical land motions resulting from these processes are on the order of several mm yr^{-1}—about the same as the rate of global sea-level rise—with magnitudes that vary over spatial scales ranging from one to thousands of km.

Glacial Isostatic Adjustment

The solid earth and oceans continue to respond to the decay of ice sheets since the last deglaciation through glacial isostatic adjustment (GIA). The loss of ice mass produces uplift in regions under the former ice masses, including northern Washington, and subsidence in areas at the ice margin and beyond, including the rest of Washington, Oregon, and California (Box 1.2). In addition, the transfer of melt water to the oceans and the consequent subsidence of the ocean basins in response to the increased water load produce a change in the absolute sea level (or geoid, an equipotential surface of the Earth's gravity field that coincides with the mean sea surface). Both processes are manifested in geological records of relative sea-level rise, geodetic observations, and GIA models.

GIA models commonly focus on predictions of sea level because many of the time series used as constraints are from paleo sea-level data (e.g., Engelhart et al., 2011). The sea-level predictions in GIA literature (e.g., Peltier, 2004) are typically a measure of relative sea-level change, according to the following equation:

Relative sea-level change =
Change in absolute sea level -
Change in height of the solid earth surface,

where the changes in absolute sea level and the height of the solid earth surface are measured relative to a common datum (e.g., the Earth's center of mass). In GIA models, the solution is obtained using the sea-level equation (Farrell and Clark, 1976; Peltier, 1976;

BOX 4.2
Geodetic Observations of Vertical Land Motion

Ground-based and space-based geodetic techniques are used to observe vertical land motion at sub-cm vertical precision. Leveling measures the vertical component of land motions from ground stations spaced hundreds to thousands of meters apart. Height differences between points are measured by setting a level on a tripod and orienting it so that the line of sight is horizontal. For short distances between benchmarks (e.g., 1 km, similar to the spacing used for tide gage leveling), a vertical accuracy of about 1 mm can be achieved (see Appendix D). For longer lines (e.g., 10 km), such as are used for tectonic studies, expected accuracies are about 2 mm.

Satellite-based systems, including the Global Navigation Satellite System (GNSS) and Interferometric Synthetic Aperture Radar (InSAR), have been available for selected regions since the 1990s. The GNSS comprises constellations of navigation satellite systems, including 24 Global Positioning System (GPS) satellites, circling the Earth with accurately determined orbits and broadcasting their precise locations and velocities. The global network of GNSS stations, along with other space geodetic techniques (e.g., satellite laser ranging, very long baseline interferometry, Doppler Orbitography and Radiopositioning Integrated by Satellite [DORIS]), provide the fundamental reference frame that makes accurate positioning and time transfer possible. The radio signals sent by the satellites are received at fixed ground stations or low-orbiting satellites. Because errors in the vertical component are typically twice as large as errors in the horizontal components, only continuous GPS (CGPS) stations are used routinely to measure vertical land motion. The National Science Foundation's Plate Boundary Observatory significantly increased the number of CGPS stations in the western United States. Stations along the west coast are spaced ~25–50 km apart. If time series are long (> 5 years) and the location of the station is accurately known, vertical resolution can reach ~1–2 mm yr^{-1}.

InSAR uses phase differences between radar images from repeat satellite passes to infer changes in the round-trip travel time of the radar signals between the earth surface and the satellite. These changes can be used to generate interferograms to infer line-of-sight surface deformation. The high-resolution image swath size is 60–100 km, and the spatial resolution of the measurement tends to be on the order of 40–100 m (pixel size). The vertical resolution of the measured surface change is less than 1 cm.

Clark et al., 1978), which assumes that the volume of water in the earth system must be conserved. Note that the GIA literature that uses the sea-level equation frequently interchanges the terms "absolute sea level," "sea surface height," and "geoid," which creates a problem when comparing and discussing predicted GIA contributions to altimetry and GRACE observations (see the discussion in Tamisiea, 2011). "Height change," "radial displacement," and "vertical motion" also are used interchangeably.

The committee compared the GIA predictions of relative sea-level change at 21 tide gage locations in California, Oregon, and Washington from an ensemble of 16 models (Figure 4.10). The time period of these models is ± 250 years relative to the present day. The models differ significantly from one another, depending on the earth rheology parameters and deglaciation model used (Table 4.2). Most GIA models employ a self-gravitating, spherically symmetric Earth model, with Maxwell rheology. Some use laterally varying viscosity and mantle thicknesses (e.g., Wu, 2006).

The new ICE-5G reconstruction of the surface topography and land-ice distribution at the last glacial maximum differs significantly from its ICE-4G precursor at all Northern Hemisphere locations that were glaciated (see Peltier, 2002a; Tarasov and Peltier, 2002). ICE-5G and ICE-6G (Peltier, 2010) contain a similar mass for the Laurentide Ice Sheet and cover the same surface area of the North American continent. They differ in the relative thickness of the ice sheet, which in the case of ICE-6G has been adjusted to eliminate the misfits between the vertical motion predictions of the model and the GPS observation analyses by Argus and Peltier (2010). Changes include a thickening of the ice cover over Labrador, Yellowknife, and the northwestern border between British Columbia and Alberta, as well as a thinning of the ice cover to the south of Hudson Bay.

All the GIA models shown in Figure 4.10 predict a similar pattern of variability in relative sea-level change along the Pacific coast, rising from 32° latitude to a maximum around 47° latitude, and then declining

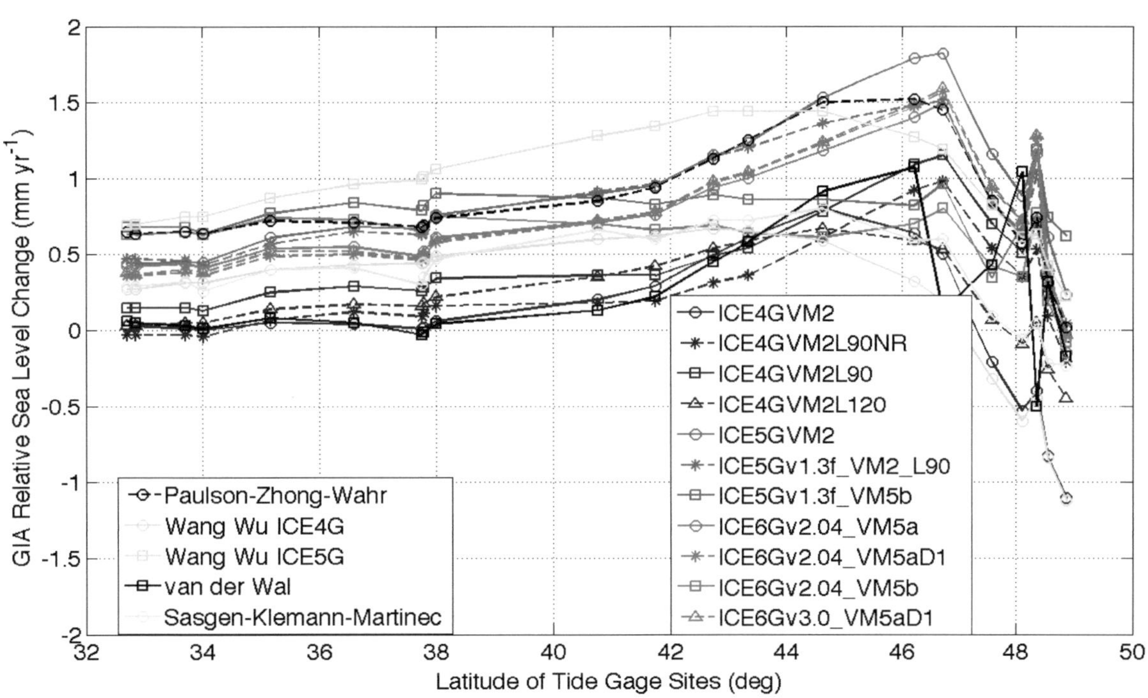

FIGURE 4.10 Ensemble of 16 GIA models showing the predictions of relative sea-level rise (expressed as change in absolute sea level minus change in height of the solid earth surface) at the latitudes of 21 tide gages off the California, Oregon, and Washington coasts. The time period of these models is ± 250 years relative to the present day. SOURCES: ICE4GVM2 (Peltier, 1998) and ICE-5GVM2 (Peltier, 2004) models and their variations are from <http://www.sbl.statkart.no/projects/pgs/authors>. Other GIA models (Wang and Wu, 2006; Paulson et al., 2007; van der Wal et al., 2009; Sasgen et al., 2012; H. Wang, personal communication) were provided by the respective authors. Predicted values from ICE5G and ICE6G models and their variations were computed for this study by Richard Peltier, University of Toronto.

TABLE 4.2 Earth Rheology Parameters Used in Selected GIA Models

GIA Model	Ice History Model	Lithosphere Thickness (km)	Viscosity ($\times 10^{-21}$ Pa s) Upper Mantle	Lower Mantle
ICE4GVM2 (Peltier, 2002a)	ICE-4G	90	0.4~1.5	1.3~3.9
ICE5GVM2[a] (Peltier, 2004)	ICE-5G	90	0.4~1.5	1.3~3.9
ICE5GVM4[a] (Peltier, 2004)	ICE-5G	90	0.4~0.9	0.9~3.9
Paulson-Zhong-Wahr[a] (Paulson et al., 2007)	ICE-5G	98	0.9	3.6
Sasgen-Klemann-Martinec[a] (Sasgen et al., 2012)	HUY, NAWI[b]	100	0.52	5.9
van der Wal[a] (van der Wal et al., 2009)	ICE-5G	98	0.9	3.6
Wang Wu ICE4G (Wang and Wu, 2006)	ICE-4G	115	0.6	LM1 = 3, LM2 = 6, β = 0.4[c]
Wang Wu ICE5G (H. Wang, personal communication)	ICE-5G	115	0.6	LM1 = 3, LM2 = 6, β = 0.4[c]

[a] Models that considered rotational feedback.
[b] HUY is the Antarctica ice model (Huybrechts, 2002), scaled to 12 m of sea-level rise since the last glacial maximum. NAWI is the Northern Hemisphere ice model (Huybrechts, 2002).
[c] Laterally varying mantle viscosity. LM1 is a shallow lower mantle, and LM2 is a deep lower mantle. Lateral variation is inferred from lateral shear wave velocity anomalies given in the seismic tomographic model S20A with a scaling factor β.

sharply. The strong latitudinal gradient in Washington illustrates the importance of glacial isostatic adjustment in regions under or at the margins of the extinct Laurentide Ice Sheet. In Cascadia, uplift is expected at the far north locations, which had been covered by the ice sheet, and subsidence is expected at the other locations, which are along the former margins of the ice sheet. In Oregon and California, the variance among models is almost as large as any apparent trend. The mean relative sea-level rise from the GIA model ensemble at each tide gage location is given in Table 4.3.

It should be noted that some studies suggest that the global earth rheology parameters (e.g., mantle viscosity) used to study the GIA process may not be suitable for subduction zones such as Cascadia. For example, James et al. (2000) used a regional, rather than global, deglaciation history to analyze GIA in southern Vancouver Island. Local paleo sea-level data show rapid uplift 12,000 years before present, which best fits a mantle with much lower mantle viscosity (~10^{19} Pa s) than is used in the GIA models shown in Table 4.3. James et al. (2000) extrapolated these results, concluding that vertical land motion from glacial isostatic adjustment along the Cascadia Subduction Zone is negligible compared to the influence of tectonics.

Tectonics

The U.S. west coast is characterized by two tectonically distinct regions: (1) the Cascadia Subduction Zone, where lithospheric plates are colliding north of Cape Mendocino, California, and (2) the San Andreas Fault Zone, where the plates are sliding past one another south of Cape Mendocino (Figure 1.8). Vertical land motions in both regions are caused by strain buildup along faults and release during an earthquake. Vertical land motions associated with the Cascadia Subduction Zone (e.g., Hyndman and Wang, 1993) are generally larger than those associated with the San Andreas Fault Zone (Argus and Gordon, 2001). Surficial crustal motions along the San Andreas Fault Zone are primarily horizontal, although convergence in some areas can produce locally significant rates of vertical deformation (e.g., Argus et al., 1999; Argus and Gordon, 2001). South of the San Francisco Bay area, the principal fault trace extends inland as much as 50–100 km, further reducing its effect on coastal vertical land motion.

The history of crustal strain accumulation and release above subduction zone faults over hundreds of years is described by the earthquake deformation cycle (Nelson et al., 1996; Satake and Atwater, 2007). During an earthquake (known as the coseismic period), vertical land motion can change almost instantly by more than a meter (see "Rare Extreme Events" in Chapter 5). Between earthquakes (known as the interseismic period), rates of vertical land motion can be on the order of mm yr^{-1} and thus can have a significant impact on the relative sea level. Vertical land motions for the Cascadia Subduction Zone and San Andreas Fault Zone are described below.

TABLE 4.3 GIA Predicted Relative Sea-Level Rise for ± 250 Years Relative to the Present Day Using an Ensemble of 16 GIA Models at 21 West Coast Tide Gage Locations

Location	Latitude	Longitude	GIA Predicted Relative Sea-Level Rise (mm yr^{-1}) Mean	Standard Deviation
Cherry Point, WA	48.87	-122.75	-0.16	0.44
Friday Harbor, WA	48.55	-123.00	0.14	0.46
Neah Bay, WA	48.37	-124.62	0.58	0.64
Port Townsend, WA	48.12	-122.75	0.40	0.48
Seattle, WA	47.60	-122.33	0.53	0.44
Toke Point, WA	46.72	-123.97	1.03	0.53
Astoria, OR	46.22	-123.77	1.07	0.43
South Beach, OR	44.63	-124.05	1.00	0.34
Charleston II, OR	43.35	-124.32	0.86	0.32
Port Orford, OR	42.73	-124.50	0.81	0.32
Crescent City, CA	41.75	-124.20	0.67	0.31
N. Spit, Humboldt Bay, CA	40.77	-124.22	0.63	0.32
Point Reyes, CA	38.00	-122.98	0.53	0.30
San Francisco, CA	37.80	-122.47	0.47	0.29
Alameda, CA	37.77	-122.30	0.44	0.29
Monterey, CA	36.60	-121.88	0.48	0.28
Port San Luis, CA	35.17	-120.75	0.45	0.27
Santa Monica, CA	34.02	-118.50	0.34	0.25
Los Angeles, CA	33.72	-118.27	0.36	0.25
La Jolla, CA	32.87	-117.25	0.34	0.25
San Diego, CA	32.72	-117.17	0.35	0.25

NOTE: Relative sea-level change is the change in absolute sea level minus the change in height of the solid earth surface. Relative sea-level rise has a negative sign compared to uplift of the earth surface due to GIA.

Cascadia Subduction Zone

Along much of the Oregon and Washington coasts, the earthquake cycle yields a characteristic pattern of vertical land movements (Figure 4.11). In the first stage of the cycle, slow interseismic strain accumulation over hundreds of years causes the upper plate to bend upward, leading to gradual uplift along the coasts above this part of the subduction zone. In areas closer to the plate boundary (usually the continental shelf) and further inland, the slow bending of the upper plate causes gradual subsidence. In the second stage of the cycle, the plate-boundary megathrust fault slips in a great earthquake, releasing hundreds of years of accumulated strain along many hundreds of kilometers of the plate boundary. During the earthquake, the former slow vertical deformation of the upper plate is reversed: coastal areas suddenly subside as much as 2 m and formerly subsiding areas landward and seaward are suddenly uplifted.

Current rates of interseismic vertical deformation can be estimated using dislocation models constrained by geodetic, thermal, and seismic data (e.g., Okada, 1985; Hyndman and Wang, 1993, 1995; Flück et al., 1997; Wang et al., 2003). To estimate interseismic deformation along the Washington and Oregon coasts, the committee used results from the CAS3D-2 model (He et al., 2003; Wang et al., 2003; Wang, 2007), a three-dimensional, viscoelastic, spherical earth, finite element model that assumes negligible present-day influence of GIA (following the work of James et al., 2000). The model has been further constrained by comparisons between geological estimates of coseismic subsidence of the 1700 earthquake and predictions from elastic dislocation models of slip on the Cascadia subduction zone (Leonard et al., 2004, 2010; Hawkes et al., 2011).

Table 4.4 shows the predicted rates of vertical land motion for the Cascadia Subduction Zone for 2010–2030 from the CAS3D-2 model assuming a continuation of the interseismic phase of the earthquake deformation cycle. The projections suggest that coastal sites, which are closest to the offshore subduction boundary, should be experiencing uplift, whereas more inland locations (Anacortes and Seattle) should

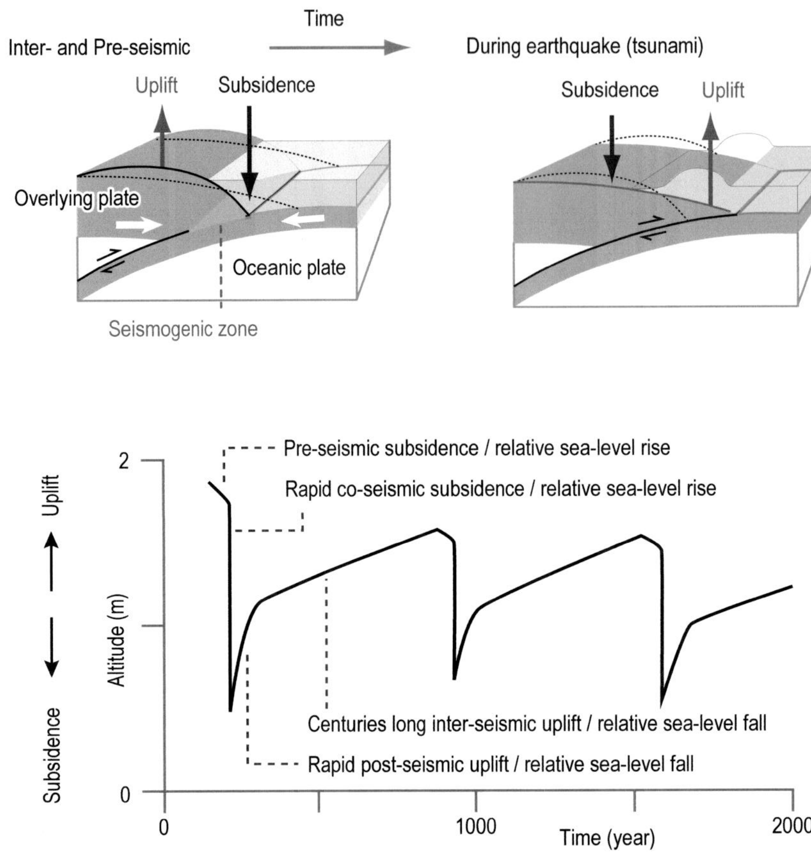

FIGURE 4.11 (*Top*) Deformation associated with a subduction-zone thrust fault on a coastline during an earthquake cycle. (*Bottom*) Idealized seismic cycle for a subduction zone, showing a long period of uplift, followed by small-scale subsidence and then a sudden drop in land elevation during a great earthquake. SOURCE: Modified from Horton and Sawai (2010).

TABLE 4.4 Vertical Land Motion Rates Predicted by the CAS3D-2 Model for 2010–2030

Location	Latitude	Longitude	Rate of Vertical Land Motion (mm yr^{-1})
Anacortes, WA	48.56	-122.64	-0.87
Seattle, WA	47.85	-122.73	-0.59
Long Beach, WA	46.58	-123.83	1.87
Pacific City, OR	45.38	-123.94	1.69
Waldport, OR	44.42	-124.02	1.66
Coos Bay, OR	43.36	-124.30	2.33
Eureka, CA	40.87	-124.15	2.98

SOURCE: Rates provided by Kelin Wang, Geological Survey of Canada, using the CAS3D-2 model (He et al., 2003; Wang, 2007). The model deformation history includes a coseismic rupture of the entire Cascadia subduction fault, representing the 1700 M 9 great earthquake, followed by locking of the fault, modeled using the conventional backslip approach (Savage, 1983). A mantle wedge viscosity of 10 Pa s was used, consistent with the results of postglacial rebound analyses at northern Cascadia and values adopted at other subduction zones.

be experiencing subsidence. Comparisons of the model projections with GPS data are discussed below (see "Current Rates of Vertical Land Motion Along the U.S. West Coast"). Model projections further forward in time are given in Chapter 5.

San Andreas Fault Zone

Unlike the Cascadia Subduction Zone, vertical land motions along the San Andreas Fault Zone cannot be characterized by a single tectonic model. The San Andreas Fault Zone comprises multiple sub-parallel faults, each with limited extent and unique seismotectonic character. Although crustal displacement is primarily horizontal (Figure 4.12), local vertical motions result from rock uplift associated with restraining bends (e.g., Anderson, 1990) and active contractional processes associated with the Transverse ranges and the Ventura and Los Angeles basins (Namson and Davis, 1991; Donnellan et al., 1993; Yeats, 1993; Shaw and Suppe, 1994, 1996; Yeats and Huftile, 1995; Dong et al., 1998; Orme, 1998; Argus et al., 1999, 2005; Hager et al., 1999; Shaw and Shearer, 1999; Argus and Gordon, 2001; Bawden et al., 2001). A comprehensive analysis of tectonically induced vertical land motions for the San Andreas Fault Zone has not been done.

Sediment Compaction

Compaction may rearrange the mineral matrix of sediment, reducing its volume (Kaye and Barghoorn, 1964; Allen, 2000; Brain et al., 2011). The amount of compaction depends on a number of factors, including the mechanical and chemical properties of the sediment (e.g., composition, porosity), the water content, and the loading history (Brain et al., 2011). For example, deposits with a high sand fraction undergo little compaction, whereas peat may compact as much as 90 percent by volume (Jelgersma, 1961).

Early studies of wetlands in North America (Kaye and Barghoorn, 1964) and Europe (Jelgersma, 1961) illustrated the importance of sediment compaction to relative sea-level rise. However, only a few studies have quantified compaction rates of coastal sediments. Törnqvist et al. (2008) analyzed wetland sediments from the Mississippi Delta and found compaction rates of 5 mm yr^{-1} on millennial timescales and more than 10 mm yr^{-1} in some areas on decadal to century timescales. These high rates of compaction were thought to contribute significantly to the high rates of relative sea-level rise (10 mm yr^{-1} over the past century) in the Mississippi Delta. Horton and Shennan (2009) found compaction rates of 0.4 ± 0.3 mm yr^{-1} during the past 4,000 years in eastern England, with higher values in large estuaries and considerable local variability depending on sediment types and drainage histories. Galloway

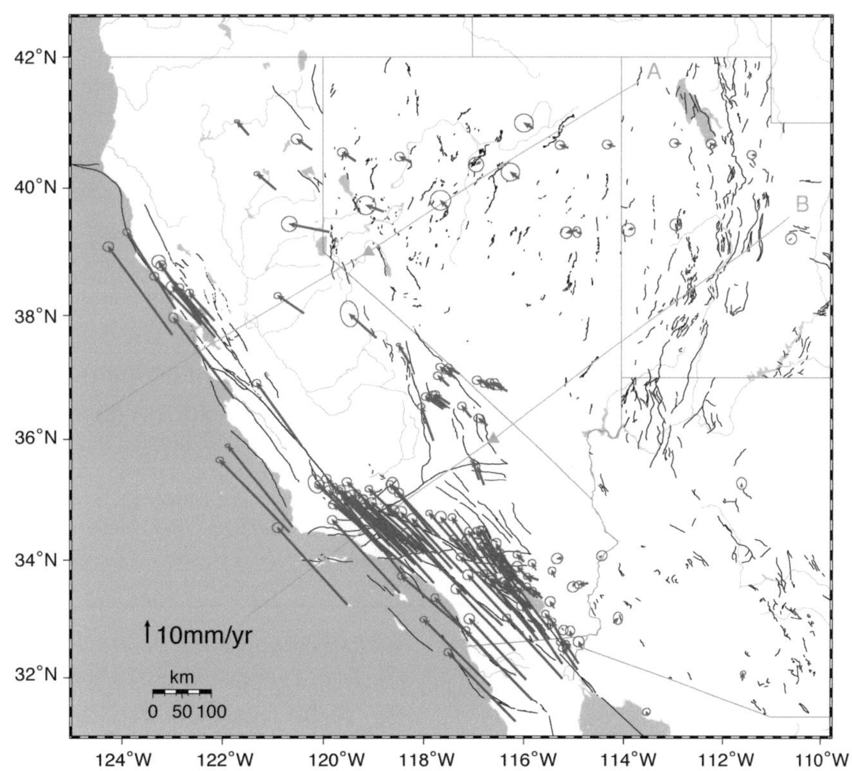

FIGURE 4.12 Faults (black lines) and GPS-defined horizontal velocities (red arrows) for sites in the western United States relative to stable North America. Circles are error ellipses at the 95 percent confidence level. SOURCE: Bennett et al. (1999).

et al. (2001) found that compaction of organic soils in the Sacramento Bay Delta (2–7 cm yr^{-1}), combined with reclamation and agriculture, has resulted in islands sinking below sea level (see also "California Bay Delta Case Study" in Chapter 6).

Comprehensive studies of compaction rates for the types of geomorphic environments that dominate the U.S. west coast (see "Geographic Variation Along the U.S. West Coast" in Chapter 1) are not available. Most of these environments, particularly the peat- and mud-rich estuaries and tidal marshes, will subside as a result of compaction.

Groundwater and Petroleum-Related Drawdown and Recharge

Withdrawal of groundwater and petroleum can lower large areas of the land surface. Subsurface fluid extraction depressurizes underground reservoirs, altering the arrangement of in situ stresses within the reservoir and surrounding rock or sediment (Donaldson et al., 1995). The elastic compaction can be recovered if the fluid level rises again (e.g., Schmidt and Bürgmann, 2003), but the inelastic compaction becomes permanent, resulting in subsidence (Sun et al., 1999). Some of the best documented examples of subsidence due to groundwater withdrawal along the U.S. west coast are in California (Figure 4.13). Intense cultivation in the Santa Clara Valley during the first half of the 1900s caused the land surface to subside up to 4 m in San Jose and 0.6–2.4 m near the southern end of San Francisco Bay, putting 44 km^2 below the high-tide level (Galloway et al., 2001). In the San Joaquin Valley, one of the world's most productive agricultural regions, the land surface dropped 0.3–9 m over 75 years, mainly due to groundwater pumping and compaction. Since 1969, groundwater recharge and the supplemental use of surface water for irrigation has slowed land subsidence in both valleys.

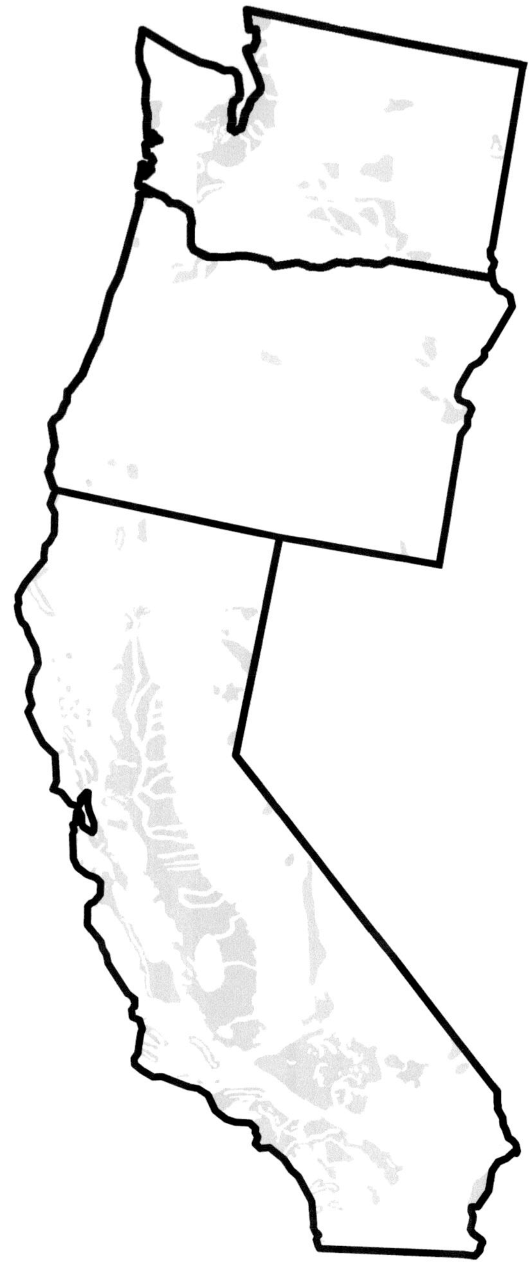

FIGURE 4.13 Areas in Washington, Oregon, and California where significant subsidence has been attributed to groundwater withdrawal (blue). The impact of groundwater withdrawal has been greater in California than in Oregon or Washington. SOURCE: Modified from Galloway et al. (2001).

In some cases, subsidence is partly offset by groundwater recharge. For example, long-term subsidence in the Santa Ana Basin (Los Angeles area) is ~12 mm yr^{-1}, but groundwater recharge produces seasonal vertical oscillations of up to 55 mm (Bawden et al., 2001).

Petroleum production requires the withdrawal of subsurface liquid hydrocarbons and also significant quantities of groundwater (Yuill et al., 2009). Ground surface subsidence related to petroleum withdrawal has been documented in a number of areas, including the California San Joaquin Valley, Las Vegas, New Orleans, and Houston. The best documented example is the Wilmington oil field in Long Beach, California, which subsided up to 9 m over 27 years (Mayuga and

Allen, 1969; Nagel, 2001). However, use of secondary recovery techniques, such as pumping seawater into the reservoirs to increase oil production, can stabilize compaction and halt subsidence. Large active oil fields along the coastal west U.S. coast are located mainly in the area between Santa Barbara and the Los Angeles Basin.

Current Rates of Vertical Land Motion Along the U.S. West Coast

Observations of vertical land motion in coastal California, Oregon, and Washington are given in Table 4.5. The values in the table represent the total vertical land motion, which is often caused by a combination of processes. For example, in the Los Angeles Basin, subsidence due to hydrocarbon and groundwater withdrawal, together with faulting, raised or lowered the surface elevation by upwards of 10 mm yr^{-1} from 1992 to 2000, with seasonal oscillations as high as 55 mm yr^{-1} (Box 4.3).

The spatial distribution of published data on vertical land motions is not optimal for assessing sea-level rise along the west coast. Consequently, the committee characterized the spatial variability of vertical land motion using the Scripps Orbit and Permanent Array Center velocity model and continuous GPS (CGPS) velocity data taken within ~15 km of the coast. The CGPS data provide an accurate, self-consistent vertical land motion estimate with well-defined and conservative error estimates (see Appendix A). The vertical land motion rates are shown in Figure 4.14. Most of the coastal CGPS vertical land motion rates fall within ± 3 mm yr^{-1} (Figure 4.14b). The average rates with obvious outliers removed (Figure 4.14c, d) are similar to longer-term estimates from leveling data for Cascadia (Burgette et al., 2009) and the San Andreas region (Appendix D). Annual rates of vertical land motion are generally positive in Washington and Oregon and generally negative in California (Figure 4.14a). This spatial pattern suggests that the tectonic boundary at the Mendocino Triple Junction is a fundamental and, most likely temporally stationary, boundary for vertical land motion. Uplift in Washington and Oregon is consistent with the buildup of interseismic strain in the Cascadia Subduction Zone as described by the CAS3D-2 model (He et al., 2003; Wang, 2007), rather than the subsidence predicted by GIA models. Subsidence in California is consistent with glacial isostatic adjustment; most GIA models predict subsidence south of the Mendocino Triple Junction (gray band in Figure 4.14b; see also Sella et al., 2007; Mazzotti et al., 2008; Argus and Peltier, 2010). As noted above (Box 4.3, Table 4.5), however, large vertical land motion signals associated with local tectonics and/or subsurface fluid movements can locally overwhelm the regional tectonic signal. This effect appears to be most prevalent toward southern California, although the paucity of

TABLE 4.5 Current Rates and Causes of Vertical Land Motion Along the U.S. West Coast

Source	Location	Method	Period (yr)	Rate of Vertical Land Motion (mm yr^{-1})
Cascadia Subduction Zone				
Mazzotti et al. (2008)	Cascadia Subduction Zone	GPS	1993–2003	1.1–3.5
Burgette et al. (2009)	Cascadia Subduction Zone	Leveling	1925–2006	-0.28–3.29
San Andreas Fault Zone				
Cooke and Marshall (2006) and	Palos Verdes Fault	Geodesy and modeling	Holocene–Quaternary	-0.5–0.4
Wills et al. (2006)	Santa Monica Fault			0.5–1.0
	Los Angeles Basin interior faults			0.26–5.0
Bürgmann et al. (2006)	San Andreas System	InSAR	1992–2000	-2.0–1.5
California Aquifers and Oil Fields				
Bawden et al. (2001)	Santa Ana Aquifer, long term	InSAR	1997–2000	-12
	Santa Ana Aquifer, seasonal			± 55
Argus et al. (2005)	Santa Ana Aquifer, seasonal	InSAR and GPS	1992–1999	-62–35
	Long Beach Oil Field			5
	Huntington Beach Oil Field			-8
	Wilmington Oil Field			-6–9

BOX 4.3
Spatial Variability of Vertical Land Motion and Relative Sea-Level Change in Los Angeles

Vertical land motions in the Los Angeles Basin vary on small spatial scales because of subsidence from groundwater and hydrocarbon withdrawal and active thrust faulting (Bawden et al., 2001; Lanari et al, 2004; Argus et al., 2005). Brooks et al. (2007) used InSAR to create a vertical land motion map of the Los Angeles Basin. The figure shows the rapid spatial change in land elevation at sub-15 km scales in this area.

Brooks et al. (2007) also used land motion rates to adjust local tide gage records to produce a profile of relative sea-level change along the coast. Vertical land motion differs on the west and east side of the Palos Verdes Peninsula. To the west, relative sea level was nearly constant from 1992 to 2000, with most values less than zero. To the east, approaching the Long Beach/Wilmington oil field, relative sea-level rates varied from -1.7 to 1.3 mm yr^{-1} and by as much as ~3 mm yr^{-1} over distances as short as ~5 km. The Brooks et al. (2007) results show the danger of assuming that a tide gage is representative of relative sea level for a region undergoing uplift or subsidence. Interpretation of the Los Angeles Harbor tide gage alone would miss the spatial variability in sea level to the east and assume the wrong sign of relative sea-level change to the west.

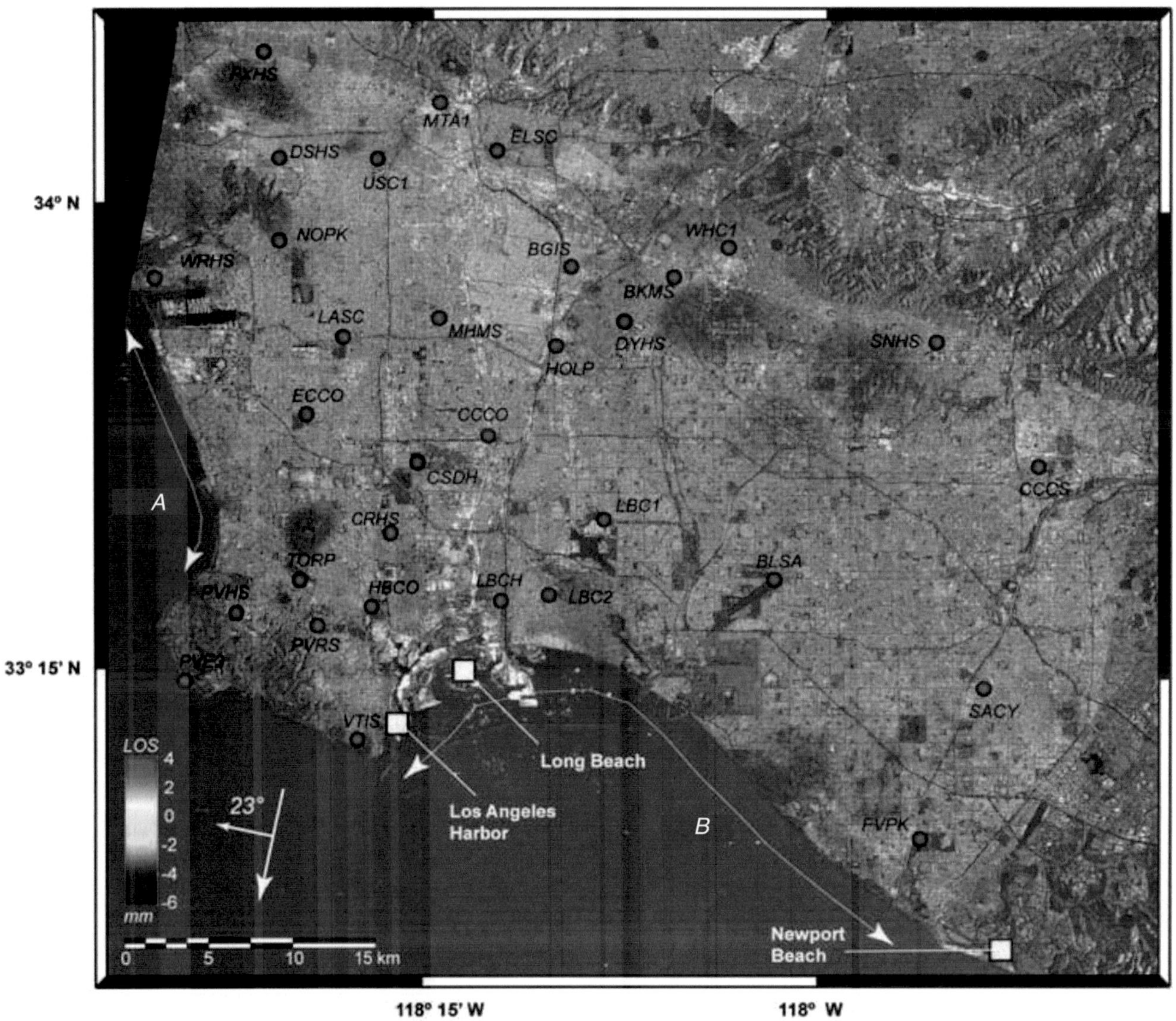

FIGURE Land motion (line-of-sight, 23 degrees inclined from vertical) from 1992 to 2000 in the Los Angeles Basin determined from InSAR (colors coded in mm yr^{-1}) and GPS (red circles), showing variability due to tectonics and hydrocarbon and groundwater fluctuations. Tide gages are shown as yellow squares. SOURCE: Brooks et al. (2007).

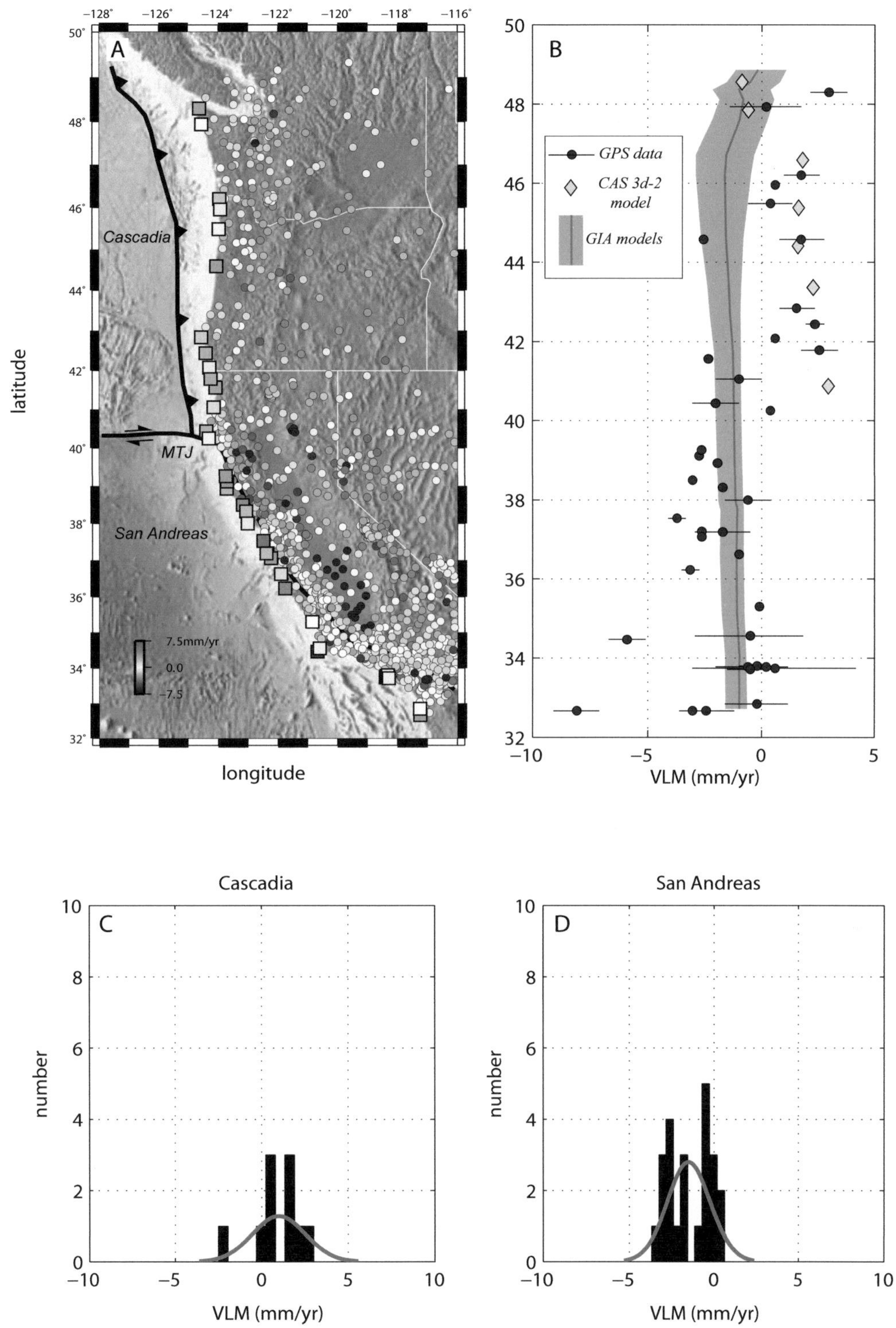

FIGURE 4.14 Continuous GPS vertical land motion (VLM) rates. (*A*) Map of the west coast of the United States showing major tectonic boundaries and locations of GPS stations color-coded for vertical land motion rates. Squares are stations within ~15 km of the coast. Circles are other stations, which are shown to demonstrate the overall spatial variability of vertical land motion in the western United States. MTJ = Mendocino Triple Junction. (*B*) Vertical land motion versus latitude for the coastal GPS stations (squares in panel A), compared with predictions of current uplift from the CAS3D-2 model (green diamonds; from Table 4.5) and the ensemble of GIA models (gray shading; from Table 4.3). GPS errors are 1 standard deviation. (*C*) Histogram and normal density function for the Cascadia coastal stations in panels A and B. (*D*) Histogram and normal density function for the San Andreas coastal stations in panels A and B. In both areas, obvious outliers have been removed. SOURCE: GPS data from the Scripps Orbit and Permanent Array Center, <http://sopac.ucsd.edu/processing/refinedModelDoc.html>.

data adequate to sense the km-scale variations in vertical land motion precludes complete characterization of these strong local signals along the entire west coast.

Summary

The west coast of the United States is undergoing active vertical deformation due to a combination of tectonics, sediment compaction, fluid withdrawal and recharge, and glacial isostatic adjustment. Assessing their relative contribution to the observed vertical land motion is complicated by a shortage of data and by the wide spatial and temporal variability of the various processes. Continuous GPS measurements over the past two decades, in concert with 20th century leveling studies, show that the coast north of Cape Mendocino is rising on the order of ~1.5–3.0 mm yr^{-1}, likely as a result of building interseismic strain along the Cascadia Subduction Zone. In contrast, the California coast south of Cape Mendocino is subsiding at a mean rate of ~1 mm yr^{-1} or more, although GPS-measured vertical land motions vary widely (-3.7–0.6 mm yr^{-1}). The boundary between uplift and subsidence takes place at the Mendocino Triple Junction, highlighting the importance of regional tectonics in relative sea-level rise. Subsidence south of Cape Mendocino is consistent with models of glacial isostatic adjustment. However, more detailed analysis of potential reference frame bias and sensitivity tests of GIA models have to be carried out to determine whether GIA is responsible for the regional subsidence. Local tectonics, sediment compaction, and fluid withdrawal and recharge can cause much higher rates of subsidence or uplift than the regional mean, especially in California, but at spatial scales too small (as little as 1 km) to have a significant impact on sea-level change in the region.

WEST COAST TIDE GAGE RECORDS

The sea level along the west coast of the United States reflects contributions from both the global sea level and the local and regional processes discussed above. Tide gage data can be used to estimate rates of relative sea-level change, but only a few such estimates have been made for the west coast of the United States. Douglas (1991) compared tide gage records for 1930–1980 and found large differences in rates of sea-level rise between coastal California and northernmost California, Oregon, and Washington, consistent with a major tectonic influence (Table 4.6). Only a few other tide-gage-based estimates of sea-level change along the U.S. west coast have been published (e.g., Tebaldi et al., 2012), and most are based on the Douglas (1991) data (e.g., Peltier, 2001) or consider records from only a few gages (e.g., Nakada and Inoue, 2005; Bromirski et al., 2011; Table 4.6).

The committee obtained records from 28 tide gages along the California, Oregon, and Washington coasts archived at the Permanent Service for Mean Sea Level. Of these, 12 are currently operating, contain no long gaps, and have been recording sea level for at least 60 years, and thus were considered suitable for determining long-term trends in sea-level rise. For each gage, the rate of relative sea-level rise was determined by fitting a straight line through the monthly mean data plotted as a function of time (see Appendix A for details). The committee's estimated rates of relative sea-level change at the 12 tide gages are given in Table 4.6 and shown geographically in Figure 4.15. Most of the gages north of Cape Mendocino (Crescent City to Neah Bay) indicate that relative sea level is falling, which is consistent with uplift associated with the buildup of interseismic strain along the Cascadia Subduction Zone, whereas most of the gages south of Cape Mendocino show that relative sea level is rising, which is consistent with land subsidence. Some gages (e.g., Friday Harbor, Seattle) deviate from these regional sea-level trends, likely as a result of local tectonic, compaction, or fluid withdrawal or recharge effects. The average rate of relative sea-level rise is 0.03 ± 1.49 mm yr^{-1} north of Cape Mendocino and 1.38 ± 0.64 mm yr^{-1} south of Cape Mendocino for the past 6–10 decades.

The change in relative sea level is what coastal residents experience and state and local managers factor into planning. To compare west coast sea-level trends with the global sea-level trend, it is necessary to adjust the relative rates of sea-level rise for changes in atmospheric pressure and vertical land motions, both of which affect the local water level (see Appendix A). Figure 4.16 illustrates the effect of these corrections on the sea-level trend for 108 years of monthly tidal data for Seattle, Washington. The slope of the blue straight line gives the rate of relative sea-level rise, in

TABLE 4.6 Rates of Relative Sea-Level Rise Estimated from U.S. West Coast Tide Gages

Source	Tide Gage	Period	Rate of Sea-Level Rise (mm yr^{-1})
Convergent margin			
Douglas (1991)	Friday Harbor, WA	1930–1980	0.6
This report		1934–2008	1.04
Douglas (1991)	Neah Bay, WA	1930–1980	-1.6
This report		1934–2008	-1.77
Douglas (1991)	Seattle, WA	1930–1980	2.5
Bromirski et al. (2011)		1930–1980	2.47
This report		1900–2008	2.01
Douglas (1991)	Astoria, OR	1930–1980	-0.4
This report		1925–2008	-0.38
Douglas (1991)	Crescent City, CA	1930–1980	-0.9
This report		1933–2008	-0.73
Transform margin			
Douglas (1991)	San Francisco, CA	1930–1980	1.8
Bromirski et al. (2011)		1930–1980	1.91
This report		1900–2008	1.92
This report	Alameda, CA	1939–2008	0.70
This report	Port San Luis, CA	1945–2008	0.68
This report	Santa Monica, CA	1933–2008	1.41
Douglas (1991)	Los Angeles, CA	1930–1980	0.2
This report		1923–2008	0.84
Douglas (1991)	La Jolla, CA	1930–1980	1.8
This report		1924–2008	2.08
Douglas (1991)	San Diego, CA	1930–1980	1.7
Bromirski et al. (2011)		1930–1980	1.80
This report		1906–2008	2.04

this case, 2.01 mm yr^{-1}. The green line in Figure 4.16 shows that the atmospheric correction is small. The atmospheric adjusted sea-level rise (using data from the National Ocean and Atmospheric Administration's Earth System Research Laboratory) is 2.10 mm yr^{-1}, about 4 percent higher than the relative sea-level rise.

Ideally, vertical land motions would be corrected using GPS data collected at the tide gage. However, none of the tide gage stations analyzed in this report include GPS instruments. Consequently, the committee followed the practice of using data from the closest CGPS station, as long as it is within 15 km of the gage (e.g., Mazzotti et al., 2007; Wöppelmann et al., 2007). Data from all CGPS stations within a 15 km radius of the gage also were analyzed to assess the spatial variability of vertical land motions near the tide gages.

The CGPS data were obtained from the Scripps Orbit and Permanent Array Center. Although GPS records extend back only a few decades, the committee assumed that current motions are representative of motions over the entire history of the tide gage and that land motion does not vary between the gage and the GPS station. It is likely that rates of vertical land motion near at least some of the tide gages have varied over the past century because of earthquakes, groundwater extraction and recharge, or other processes, but the absence of detailed geologic histories for each gage precluded a more sophisticated approach. The vertical land motion correction to the sea-level record was often relatively large, changing rates in one case by almost 150 percent (see Table A.2 in Appendix A). For five of the gages analyzed, correcting for vertical land motion changed the sign of sea-level change.

The rate of sea-level rise at the tide gage, adjusted for vertical land motion and atmospheric pressure, is the slope of the red line in Figure 4.16, which is 2.3 mm yr^{-1} for Seattle, about 15 percent higher than the rate of relative sea-level rise. Along the coast, the mean adjusted rates of sea-level rise are 1.59 ± 0.80 mm yr^{-1} north of Cape Mendocino and 1.02 ± 1.73 mm yr^{-1} south of Cape Mendocino, both of which are lower than global mean sea-level rise.

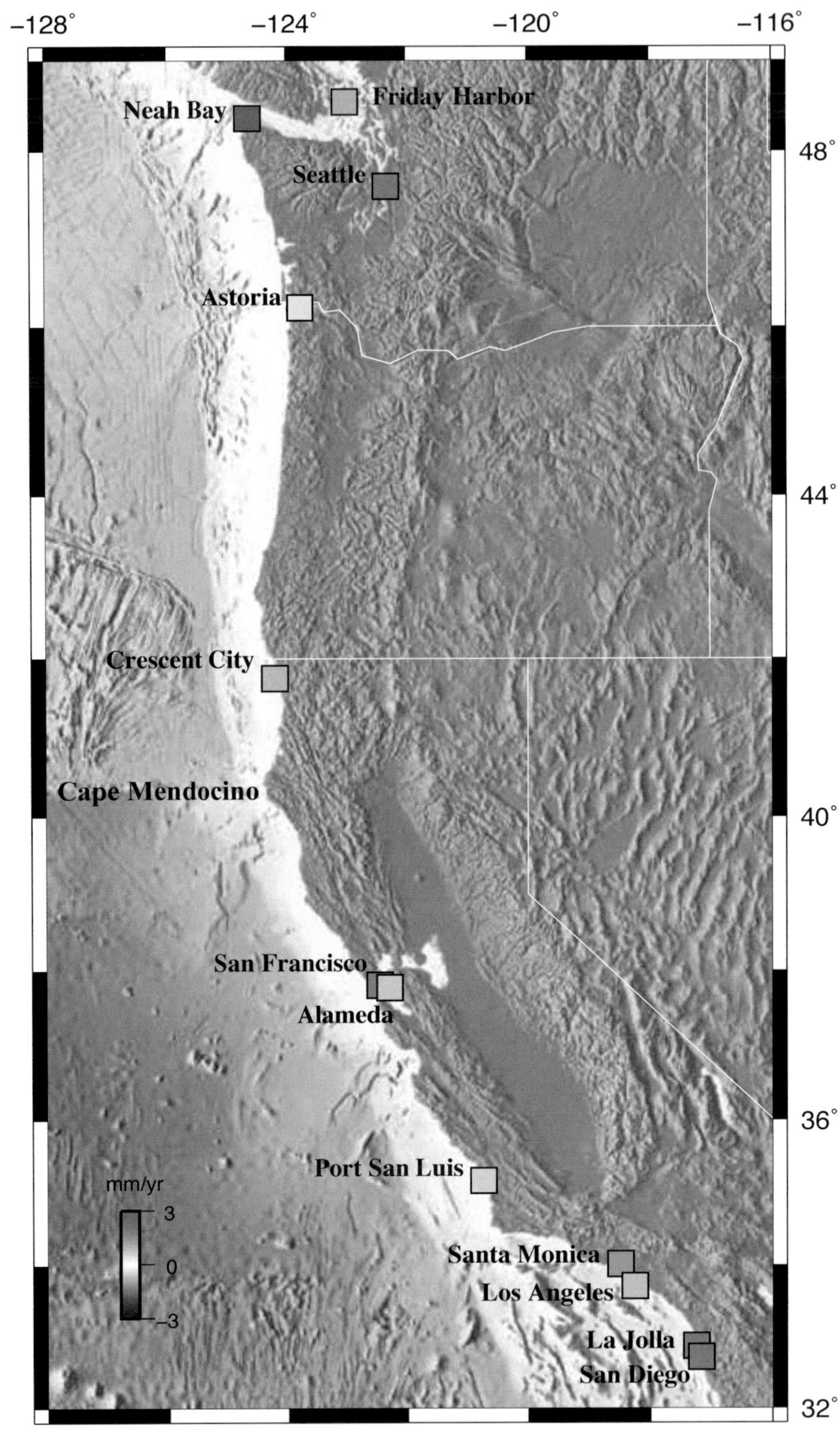

FIGURE 4.15 Rates of relative sea-level change estimated from long tide gage records (63–108 years) analyzed in this report.

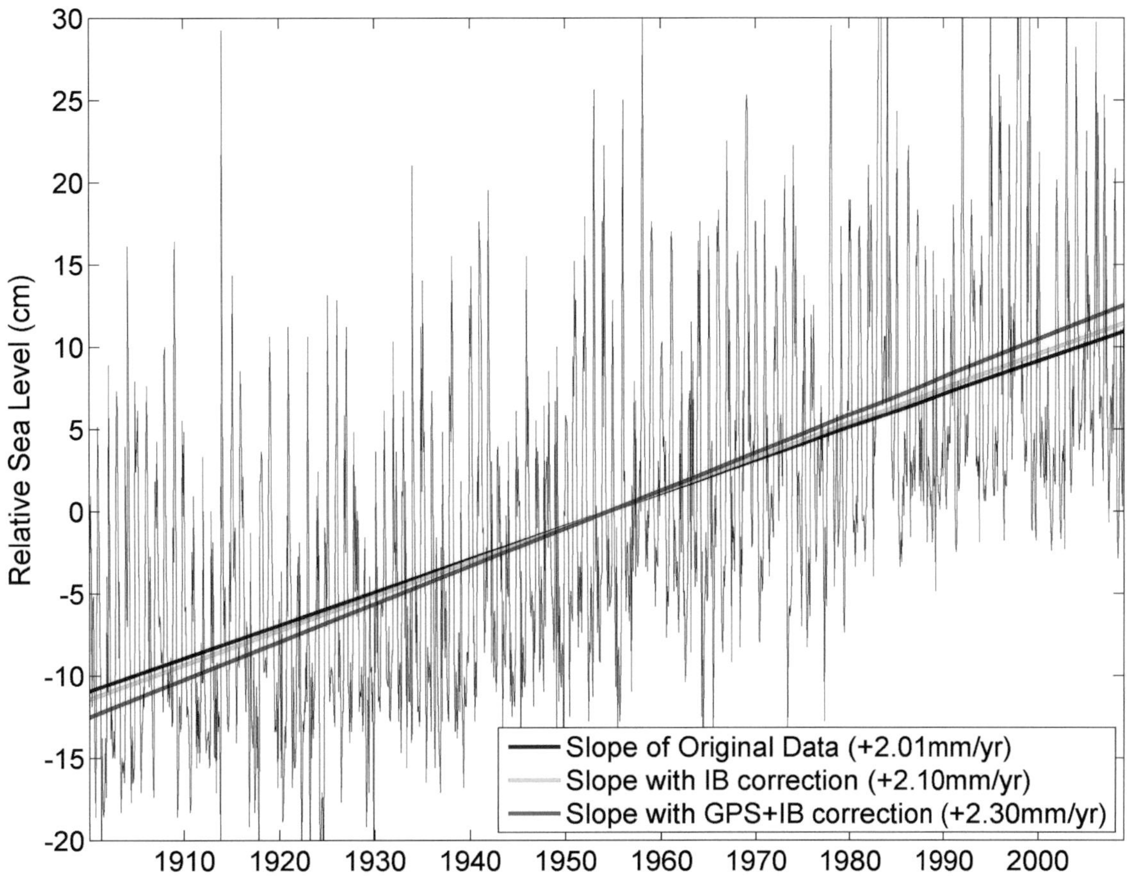

FIGURE 4.16 Monthly sea level for Seattle, Washington, from 1900 to 2008. Straight-line fits to the data show the relative sea-level rise (blue line), the sea-level rise adjusted for atmospheric pressure (green line), and the sea-level rise adjusted for vertical land motion and atmospheric pressure (red line).

CONCLUSIONS

Sea level at any given place and time depends on the global sea level and the net contribution of atmospheric, oceanographic, geologic, and anthropogenic processes operating in the area. Processes that raise relative sea level in the northeastern Pacific Ocean include warm phases of climate oscillations (El Niños, positive phase of the PDO) and land subsidence due to glacial isostatic adjustment, sediment compaction, and the withdrawal of groundwater or hydrocarbons. Processes that lower relative sea level include cool phases of climate oscillations (La Niñas, negative phase of the PDO), gravitational and deformational effects of modern melting of glaciated land masses, and land uplift due to tectonics or fluid recharge.

The highest sea levels recorded along the west coast are usually associated with El Niño events, which can elevate coastal sea level by 10–30 cm for several winter months. Cool climate phases have less influence on local sea level than warm climate phases. Changes between warm and cool climate phases, which occur on seasonal to multidecadal timescales, cause large-amplitude variations in the relative sea-level trend.

Modern melting of glaciers and ice sheets adds new water to the ocean basins and produces gravitational and deformational effects that create regional patterns of relative sea-level change. The glaciated land masses that most effect sea level along the west coast of the United States are Alaska, which is close, and Greenland and Antarctica, which are large. The gravitational and deformational effects reduce the contribution of melting of these three ice sources to relative sea-level rise for 1992–2008 by about 42 percent along the north coast (Neah Bay, Washington), 24 percent along the central coast (Eureka,

California), and 14 percent along the south coast (Santa Barbara, California).

Vertical land motions along the west coast of the United States are caused by a complex combination of tectonics, glacial isostatic adjustment, sediment compaction, and fluid withdrawal and recharge. The area straddles two tectonic regimes: (1) the Cascadia Subduction Zone, where the buildup of interseismic strain is causing coastal uplift north of Cape Mendocino, California; and (2) the San Andreas Fault Zone, where the lateral motion of the lithospheric plates produces relatively little vertical land motion south of Cape Mendocino. Glacial isostatic adjustment is producing uplift in northernmost Washington, which had been covered by the former Laurentide Ice Sheet, and subsidence in areas peripheral to the center of the former ice mass, including the rest of Washington, Oregon, and California. Land levels in some areas also are rising or sinking because of local tectonics, compaction of wetland sediments, and/or fluid withdrawal or recharge. Continuous GPS measurements over the past two decades and leveling studies over the past eight or nine decades shows that the coast north of Cape Mendocino is rising at rates of 1.5–3.0 mm yr^{-1} and the coast south of Cape Mendocino is subsiding at a mean rate of about 1 mm yr^{-1}, although with considerable spatial variability (-3.7–0.6 mm yr^{-1}).

Tide gage records along the west coast of the United States indicate that relative sea-level change is variable along the coast. Most gages north of Cape Mendocino show relative sea-level fall for the past 6–10 decades, consistent with coastal uplift along the Cascadia Subduction Zone. Most gages south of Cape Mendocino show relative sea-level rise, consistent with land subsidence. When adjusted for vertical land motions and for atmospheric pressure effects, the rates of relative sea-level rise along the U.S. west coast are lower than the rate of global mean sea-level rise.

Although rates of sea-level rise are relatively low along the west coast of the United States, the combination of sea-level rise and winter storms increases the potential for significant coastal damage. Historically, most coastal damage has occurred when storm surges and large waves coincided with high astronomical tides and El Niños—a combination that can raise short-term sea level above sea levels projected for 2100. All climate models project ample winter storm activity, but a clear consensus has not yet emerged on whether storm frequency or intensity will change in the northeast Pacific. Several climate models predict a northward shift in the North Pacific storm track over the 21st century, and some observational studies report that a northward shift has been detected. However, most observational records are not long enough to determine whether a shift has begun.

Several observational studies have reported that high waves have been getting higher and that winds have been getting stronger in the northeastern Pacific over the past few decades. The magnitude and cause of these changes are under investigation; at least part of the observed increase likely reflects natural climate variability. But even if storminess does not increase in the future, sea-level rise will magnify the adverse impact of storm surges and high waves on the coast of California, Oregon, and Washington.

5

Projections of Sea-Level Change

Observations provide unequivocal evidence that global mean sea level has been rising over the past century, but that the rate of sea-level rise has significant regional variability. The key question for planners is how much sea level will rise in their region in an increasingly warm future world. Most projections are based on knowledge of the current contributions to sea-level change and assumptions about future warming and the behavior of key geophysical processes. This chapter describes methods for projecting global and regional sea-level rise, summarizes recent results, and presents the committee's own projections for the years 2030, 2050, and 2100, relative to year 2000. The chapter begins with a discussion of the global projections, then describes how these are adjusted using local and regional information from the U.S. west coast to develop projections for California, Oregon, and Washington. The chapter concludes with a discussion of rare extreme events that could induce a large, rapid change in sea level along the west coast of the United States.

RECENT GLOBAL SEA-LEVEL PROJECTIONS

Projections of future global sea-level change are commonly made using models of the primary processes that contribute to global sea-level change—the transfer of fresh water from the melting cryosphere to the oceans, and changes in water density (steric changes) arising mainly from the thermal expansion of ocean water as it warms. Although the steric contribution can be computed from ocean models, the cryospheric contribution cannot yet be modeled satisfactorily. Given this shortcoming, some investigators use current observations to extrapolate the future behavior of the cryosphere. Another approach to projections, called the "semi-empirical" approach, is based on the observed relationship between sea-level change and global temperature change, and takes no account of the individual contributions to sea-level rise or their physical constraints. Recent projections of global sea-level rise from these different approaches are summarized below.

Models of Physical Processes

The Intergovernmental Panel on Climate Change (IPCC) Fourth Assessment Report projected global sea-level rise to 2100, relative to the year 2000, using numerical models forced by different emission scenarios, as well as simplified climate models. The scenarios represent a range of driving forces and emissions developed using different assumptions about demographic, social, economic, technological, and environmental developments (Box 5.1). The IPCC (2007) projected the individual contributions of steric changes and melting of glaciers and ice caps, the Greenland Ice Sheet, and the Antarctic Ice Sheet to future sea-level change for each emission scenario, then summed the contributions. Changes in water stored in other land reservoirs or extracted from the ground or aquifers were considered too uncertain to project. The IPCC (2007) projections are given in Table 5.1 and are discussed below.

BOX 5.1
IPCC (2000) Emission Scenarios

The IPCC Fourth Assessment Report projected global sea levels over the next 100 years based on 6 families of emission scenarios described in IPCC (2000). The A1 scenario family assumes high economic growth, low population growth that peaks mid century, and the rapid introduction of more efficient technologies. Within this family are scenarios designated as A1FI (fossil fuel intensive), A1B (balanced fuel), and A1T (predominantly nonfossil fuels). The A2 scenario family assumes slower economic growth and technological change, but high population growth. The B1 scenario family assumes the same low population growth as the A1 scenarios, but a shift toward a lower-emission service and information economy and cleaner technologies. Finally, the B2 scenario family assumes moderate population growth, intermediate economic growth, and slower and more diverse technological change than in the B1 and A1 scenarios. The A1FI scenario yields the highest carbon dioxide (CO_2) emissions by 2100, and the B1 scenario yields the lowest CO_2 emissions.

TABLE 5.1 IPCC (2007) Projected Contributions to Global Sea-Level Change, Relative to 2000

	Projections for 2100 (cm)					
Term	B1	B2	A1B	A1T	A2	A1FI
Thermal expansion	10–24	12–28	13–32	12–30	14–35	17–41
Glaciers and ice caps	7–14	7–15	8–15	8–15	8–16	8–17
Greenland Ice Sheet SMB	1–5	1–6	1–8	1–7	1–8	2–12
Antarctica Ice Sheet SMB	-10– -2	-11– -2	-12– -2	-12– -2	-12– -3	-14– -3
Sea-level rise	18–38	20–43	21–48	20–45	23–51	26–59
Scaled-up ice sheet discharge	0–9	0–11	-1–13	-1–13	-1–13	-1–17

SOURCE: Adapted from Table 10.7 in Meehl et al. (2007).
NOTE: SMB = surface mass balance.

Steric Contributions

The IPCC Fourth Assessment Report used general circulation models (GCMs) of the ocean and atmosphere to estimate global steric response. Because the GCM simulations were only available for three emission scenarios, simplified climate models were used for the other three scenarios. Global ocean models compute both temperature and salinity, so their outputs can be used directly to calculate changes in sea level due to thermal expansion (thermosteric changes) and changes in salinity (halosteric changes). Thermosteric contributions from the ocean general circulation models used in the IPCC Fourth Assessment Report are shown in Figure 5.1 (Meehl et al., 2007). Note that the model results vary with time and emission scenario. The IPCC (2007) projected that thermal expansion would account for 55–69 percent of sea-level rise in 2100 (Table 5.1).

Cryospheric Contributions

The IPCC (2007) estimated the cryosphere response using models of ice sheet surface mass balance in Greenland and Antarctica and empirical models of the mass balance response of glaciers and ice caps to temperature and precipitation forcing. They projected that glaciers and ice caps would be the largest source of new water to the oceans throughout the 21st century (Table 5.1). The ice sheets were projected to contribute less new water than glaciers and ice caps, mainly because the Antarctic contribution was expected to be negative (i.e., mass gained because of increased snowfall would withdraw water from the ocean). However, recent observations of Antarctica show the opposite—a growing Antarctic contribution to sea-level rise due to the rapid transfer of ice from land to the ocean by glacier flow and iceberg calving, referred to here as "rapid dynamic response" (see "Glaciers, Ice Caps, and Ice Sheets" in Chapter 3).

At the time data were synthesized for the IPCC Fourth Assessment Report (until mid-2006), rapid transfers of ice at a global level were only beginning to be observed. In addition, the relatively simple treatment of land ice dynamics in the climate models precluded simulation of rapid dynamics. Although stand-alone

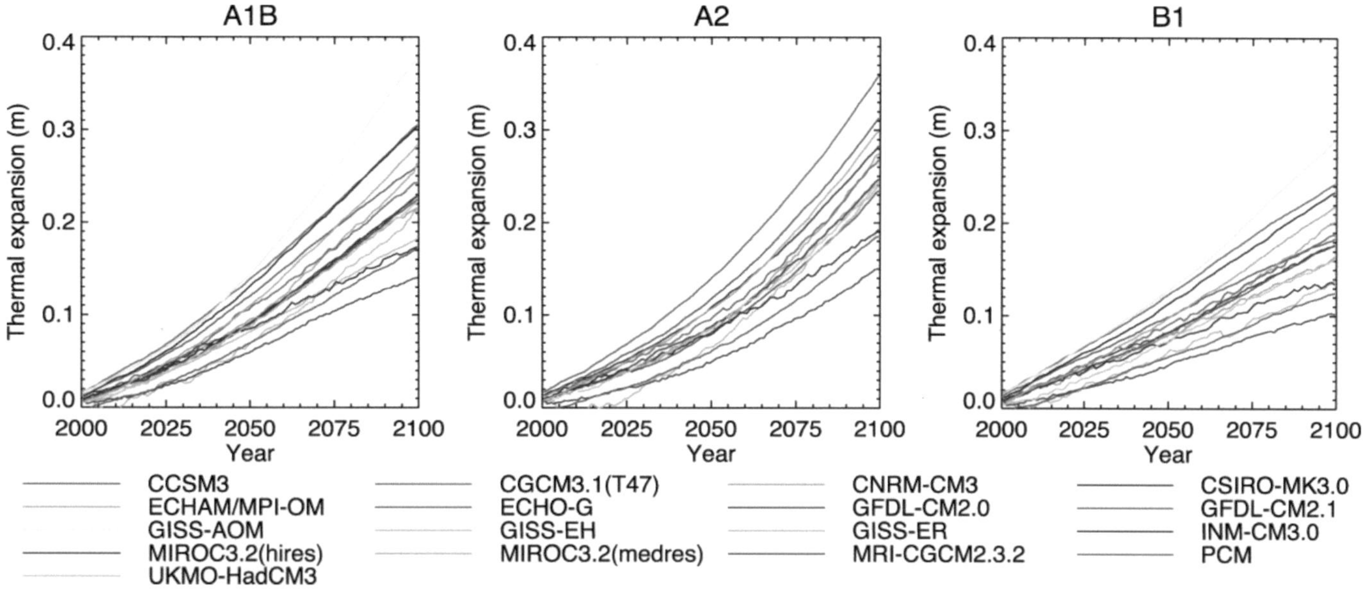

FIGURE 5.1 Thermal expansion contribution to global sea-level rise calculated by a range of models for three emission scenarios: A1B, A2, and B1. SOURCE: Figure 10.31 from Meehl et al. (2007).

ice sheet models with far more sophisticated dynamic capabilities have long been in use, they are difficult to drive in a realistic fashion with only climatic forcing variables, and are still not a feature of integrated atmosphere-ice-ocean models. Consequently, the IPCC (2007) treated ice sheets as fixed geographic features that could gain and lose mass through accumulation and ablation, but would not otherwise change size or undergo variations in flow. The IPCC (2007) attempted to account for rapid transfers of ice from land to ocean by scaling up certain components of the modeled results, shown in Table 5.1 under "Scaled-up ice sheet discharge." However, the estimates were not based on physical models of ice sheet processes, and they were not included in the projections of global sea-level rise. The IPCC (2007) projections of the cryospheric contribution to sea-level rise are widely regarded as too low (e.g., Kerr, 2007; Pfeffer et al., 2008; van der Veen and IMASS, 2010; AMAP, 2011; Price et al., 2011).

Extrapolation of Land Ice Contributions

As noted above, some aspects of the cryospheric system are not yet understood well enough to be confidently represented in physical models, and many of the observations needed to characterize boundary and initial conditions or other model parameters are not available. Consequently, some investigators use extrapolation methods to project the cryospheric contribution to sea-level rise. Extrapolations carry past and present-day observed rates of change forward in time at rates that remain constant or vary according to assumed rules.

A number of recent studies have projected the future contributions of land ice to sea-level change by extrapolating observed trends in ice loss rates. Meier et al. (2007) extrapolated loss rates for the Greenland and Antarctic ice sheets and for aggregate glaciers and ice caps, and estimated that land ice would contribute ca. 8–16 cm to sea-level rise by 2050 and 17–56 cm by 2100 under plausible future conditions. The lower estimate assumed that present-day loss rates continued unchanged in the future. The higher estimate assumed that the present-day loss rate continued to increase in the future. Future sea-level rise could be less than the lower estimate only if global loss rates actually decreased in the future, an unlikely outcome of most climate and mass balance and ice dynamics modeling. Whether the higher estimate, which was not proposed as a firm upper limit, bounds the true upper range of outcomes is uncertain.

Pfeffer et al. (2008) made extrapolations that were intended to constrain the upper limits of glacier and ice sheet contributions to sea level. Rather than project present-day observed rates forward in time, the authors calculated what loss rates would be required to achieve certain hypothesized future sea-level values. For example, a hypothesized 2 m rise in global sea level by 2100 from the Greenland Ice Sheet alone would require the average velocity of Greenland's outlet glaciers to immediately rise to 49 km yr^{-1}, which is highly unlikely and thus not a plausible future scenario. The authors also hypothesized a range of accelerated but reasonable glacier dynamic behavior for the Greenland and Antarctic ice sheets and for glaciers and ice caps, and they projected land ice contributions ranging from 0.8 m to 1.7 m by 2100, with roughly equal contributions from Greenland, Antarctica, and glaciers and ice caps.

Using comprehensive data extending back to 1992, Rignot et al. (2011a) constructed detailed mass loss time series for both the Greenland and Antarctic ice sheets, and extrapolated linear trends fit to that data to estimate future sea-level contributions from the ice sheets. The observed mass loss trends for 1992–2009 were 21.9 ± 1 GT yr^{-2} for the Greenland Ice Sheet and 14.5 ± 2 GT yr^{-2} for the Antarctica Ice Sheet. Extrapolating these loss trends forward to 2100, Rignot et al. (2011a) estimated a sea-level contribution from the ice sheets of 15 ± 2 cm by 2050 and 56 cm (with no stated uncertainty) by 2100. To arrive at a total land ice projection, Rignot et al. (2011a) used the glacier and ice cap values calculated by Meier et al. (2007). The uncertainty attached to the projection reflected the quality of fit of the linear regression of the trend to the loss rate data, rather than uncertainty of the data. The authors suggested that their calculations provide an indication of the potential contributions of ice sheets to sea level in the next century, but should not be regarded as projections, given the uncertainty in future acceleration of ice mass loss and the simplicity of their model.

The extrapolation methods used by Meier et al. (2007) and Rignot et al. (2011a) assume geostatistical stationarity—that the statistical characteristics during the period of observation remain valid over the period of extrapolation. For unvarying processes or for short extrapolation periods relative to the observation period, this assumption is justifiable. For time-varying processes or for long extrapolation periods, this assumption is more questionable. Glaciers, ice caps, and ice sheets may undergo changes in the next century that are quite unlike the changes recorded over the past few decades, such as an increase or decrease in the speed of marine-ending outlet glaciers. Analyses to evaluate the effects of non-stationarity (time-varying processes) and to qualitatively estimate the timescale for which extrapolations are valid are described in the committee's projections of global sea-level rise (see "Cryosphere Contributions" below).

Semi-Empirical Models

Projections of 21st century sea-level rise are subject to uncertainties arising from the nonlinear responses of the Greenland and Western Antarctic ice sheets (Pfeffer et al., 2008; Rahmstorf, 2010), steric changes (Domingues et al., 2008; Leuliette and Miller, 2009), and contributions from mountain glaciers (Meier et al., 2007). One way to avoid the difficulties of accurately estimating these individual contributions is to postulate a simple link between observed sea-level rise and observed global temperature changes in the past (Rahmstorf, 2010). Such semi-empirical models are based on the simple physical concept that sea level rises faster as the Earth gets warmer. This concept is supported by observations on long timescales.

Early semi-empirical models assumed a linear relationship between global temperature and sea-level rise (e.g., Gornitz and Lebedeff, 1987), but subsequent refinements have included corrections for the time-response characteristics of sea level to temperature forcing. A frequently cited semi-empirical model to project future sea-level rise was developed by Rahmstorf (2007), who related rising sea level to global near-surface air temperature as follows:

$$dH/dt = a\,(T(t) - T_0),$$

where H is the sea level, T is the mean global temperature, T_0 is the baseline temperature at which sea level is stable, and a is the sea-level sensitivity, which measures how much the rate of sea-level rise accelerates per unit change in global temperature. The model postulates that if the temperature rises above T_0, sea level will rise indefinitely at a rate determined by the magnitude of the temperature rise, so a linear rise in temperature with time leads to a quadratic change in sea level. The unknown parameters a and T_0 are determined from global sea-level reconstructions (e.g., Church and White, 2006) and global temperature data archived by the National Aeronautics and Space Administration Goddard Institute for Space Studies. Rahmstorf (2007) found that the parameter a is 3.4 mm $yr^{-1}/°C$. Projecting the equation forward using the IPCC (2000) scenarios for temperature change yielded a rise in sea level between 0.38 m and 1.2 m by 2100.

A subsequent revision to the model (Vermeer and Rahmstorf, 2009) included an extra term b to allow sea level to respond directly to temperature change:

$$dS/dt = a\,(T - T_0) + b\,dT/dt.$$

To gain confidence in the model, the authors calibrated the a and b coefficients with temperature data from 1880 to 2000, then verified the model over a 1,000-year time frame using sea-level proxy data for the past millennium. With this model and the IPCC (2000) emission scenarios, Vermeer and Rahmstorf (2009) projected that sea level would rise between 0.81 m and 1.79 m by 2100. Their projections for three of the IPCC (2000) emission scenarios are shown in Figure 5.2.

Grinsted et al. (2009) used a much longer temperature record and a different semi-empirical model to project sea-level rise:

$$dS/dt = (S_{eq} - S)/\tau,$$

where τ is the response time on the order of centuries, S_{eq} is the equilibrium sea level at a fixed global temperature, and S is the global mean sea level relative to the mean over a well-documented time interval. S_{eq} is assumed to change linearly with temperature. As the atmospheric temperature rises, the sea level rises at a rate that depends both on the magnitude of the total warming (which determines $S_{eq} - S$) and the response time τ. Grinsted et al. (2009) calibrated their equation using several historical global temperature data sets, then used the IPCC (2000) scenarios to project into the future. For all IPCC (2000) emission scenarios, they projected that sea level would rise between 0.21 m and 2.15 m by 2100.

All of the semi-empirical model projections are higher than the IPCC (2007) projections by a factor of two or even three (Rahmstorf, 2010; Figure 5.2). The two projection approaches rest on different foundations—GCMs on the physical processes that

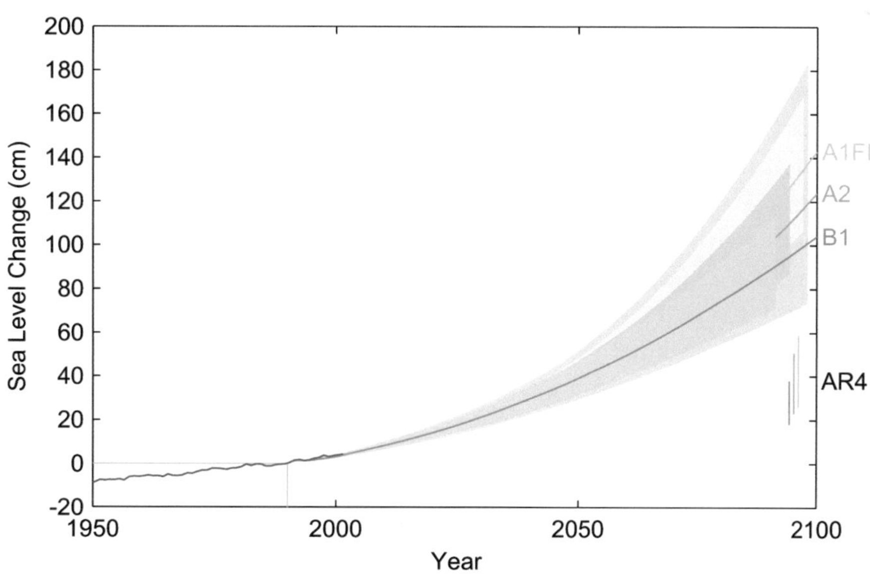

FIGURE 5.2 Projections of sea-level rise from 1990 to 2100, based on the Vermeer and Rahmstorf (2009) semi-empirical model and three IPCC (2000) emission scenarios (A1FI, A2, and B1). Uncertainty ranges are 1 standard deviation from the model means, and the gray shading is an added ± 7 percent, representing uncertainty in the fit of the data. The corresponding sea-level projections by IPCC (2007; labeled AR4) are shown for comparison in the bars on the bottom right. Also shown are the observations of annual global sea level (red line). SOURCE: Vermeer and Rahmstorf (2009).

cause sea level to rise and semi-empirical models on the observed relationship between temperature and sea level—so it is not surprising that they do not agree. Moreover, the IPCC (2007) projections are likely underestimates because they do not account fully for cryospheric processes. The highest projections made by semi-empirical models (more than 2 m of sea-level rise) are likely overestimates because they would require unrealistically rapid acceleration of glaciological processes (Pfeffer et al., 2008).

An advantage of semi-empirical models is that, by parameter fitting, they reproduce the observed past sea-level rise. However, the simple empirical connection found for the past may not hold in the future. In particular, the ice sheets appear to have been negligible sea-level contributors during the observational periods used by Gornitz and Lebedeff (1987), Rahmstorf (2007), and Vermeer and Rahmstorf (2009), but ice sheet dynamic response is widely regarded as the most uncertain aspect of sea-level change. Indeed, some events, such as ice shelf melting triggering an instability of the West Antarctic Ice Sheet, would not be factored into semi-empirical models.

COMMITTEE PROJECTIONS OF GLOBAL SEA-LEVEL RISE

The committee was charged with projecting both the individual contributions to global sea-level rise (e.g., thermal expansion, melting of land ice) and the total global sea-level rise for the years 2030, 2050, and 2100 (Task 1, see Box 1.1). Given the state of knowledge and the limited time and computational capability available for a National Research Council study, the committee chose a combination of approaches for its projections. The output of GCMs was used to project the steric contribution (primarily thermal expansion) to global sea-level rise over the three time frames. For the land ice projections, the committee extrapolated mass balance estimates. Like the IPCC (2007), the committee did not project land hydrology contributions because uncertainties are too large, and a recent comprehensive assessment (Milly et al., 2010) found that the primary sources (groundwater depletion) and sinks (reservoir storage) appear to effectively cancel out. The individual components were then summed and compared with results from semi-empirical methods. The projections are for individual years (2030, 2050, and 2100, relative to 2000), and were derived using single-year values from low-order curves, except for the steric values, as explained below. The projections are given in Table 5.2 and discussed below.

Steric Contribution

The most recent GCM results for the steric contribution that were available to the committee were from the Coupled Model Intercomparison Project Phase 3 (CMIP3), which were used in the IPCC Fourth Assessment Report. Although outputs from a new generation of GCMs are beginning to be available, performing computations of derived quantities like global sea-level changes from these new outputs is beyond the charge and capability of the committee. Consequently, the committee drew on the work of Pardaens et al. (2010), who analyzed an ensemble of IPCC (2007) model projections using the A1B emission scenario (Figure 5.3). Drs. Pardaens and Gregory[1] provided the gridded annual mean sea-level data used in their paper, and the committee analyzed the combine steric and ocean dynamic height data for the globe.

The models in Pardaens et al. (2010) yielded time series of annual mean sea level spanning roughly the 21st century: the first year in the various model simulations ranged from 2000 to 2004, and the final year was 2099. The committee performed a quadratic fit on each model's time series at each grid point and, using the values on the quadratic curves, obtained steric sea-level changes for 2030, 2050, and 2100 relative to year 2000 for each model. The results are presented in the first row of Table 5.2.

Uncertainties

The committee endeavored to incorporate and describe as accurately as possible the known sources of uncertainty in the steric projections. These uncertainties are related to future greenhouse gas and aerosol emissions and concentrations (human forcing), the response of global temperatures to human forcing, and the response of the ocean to those global temperature distributions. The IPCC (2007) treated uncertainty in

[1] See <http://www.met.rdg.ac.uk/~jonathan/data/ar4_sealevel/>.

TABLE 5.2 Committee's Global Sea-Level Rise Projections (in cm) Relative to Year 2000

	2030		2050		2100	
Term	Projection	Range	Projection	Range	Projection	Range
Steric[a]	5.4 ± 1.6	1.7–11.0 (B1–A1FI)	9.9 ± 2.4	4.0–18.9 (B1–A1FI)	24.2 ± 5.9	9.6–46.2 (B1–A1FI)
Glaciers and ice caps[b]	2.9 ± 0.1	2.7–3.6	5.5 ± 0.2	5.1–7.3	14.3 ± 0.7	12.9–19.4
Greenland[b]	2.3 ± 0.2	1.8–4.0	5.6 ± 0.7	4.3–10.2	20.1 ± 2.7	14.8–33.8
Antarctica[b]	2.9 ± 0.7	1.5–5.1	7.0 ± 2.1	3.0–13.3	24.0 ± 8.3	7.7–46.2
Total Cryosphere[b]	8.1 ± 0.8	6.6–12.2	18.0 ± 2.2	13.7–29.4	58.4 ± 8.8	40.9–94.1
Sum[c]	13.5 ± 1.8	8.3–23.2	28.0 ± 3.2	17.6–48.2	82.7 ± 10.6	50.4–140.2
Semi-empirical[d]	18	14–22 (B1–A1FI)	37	28–47 (B1–A1FI)	121	78–175 (B1–A1FI)

[a] For the steric contribution, the projection is for scenario A1B from Pardaens et al. (2010), ±1 standard deviation computed for 20-year windows across models, and the range was determined by scaling the A1B projections for 2100 to the low value of B1 and the high value of A1FI for A1B, from Table 5.1.
[b] The cryospheric projection is an extrapolation from observed changes, ±1 standard deviation. The range column includes an additional dynamic contribution, described in Appendix E, which is used only for the high-end estimates.
[c] The low value of the range for each year (2030, 2050, 2100) was computed by subtracting twice the standard deviation from the mean in the projection column, and adjusting to the difference between A1B and B1. The high value of the range was computed by adding twice the standard deviation to the mean, adjusting to the difference between A1FI and A1B, and adding the dynamical imbalance contribution.
[d] Data from Vermeer and Rahmstorf (2009).

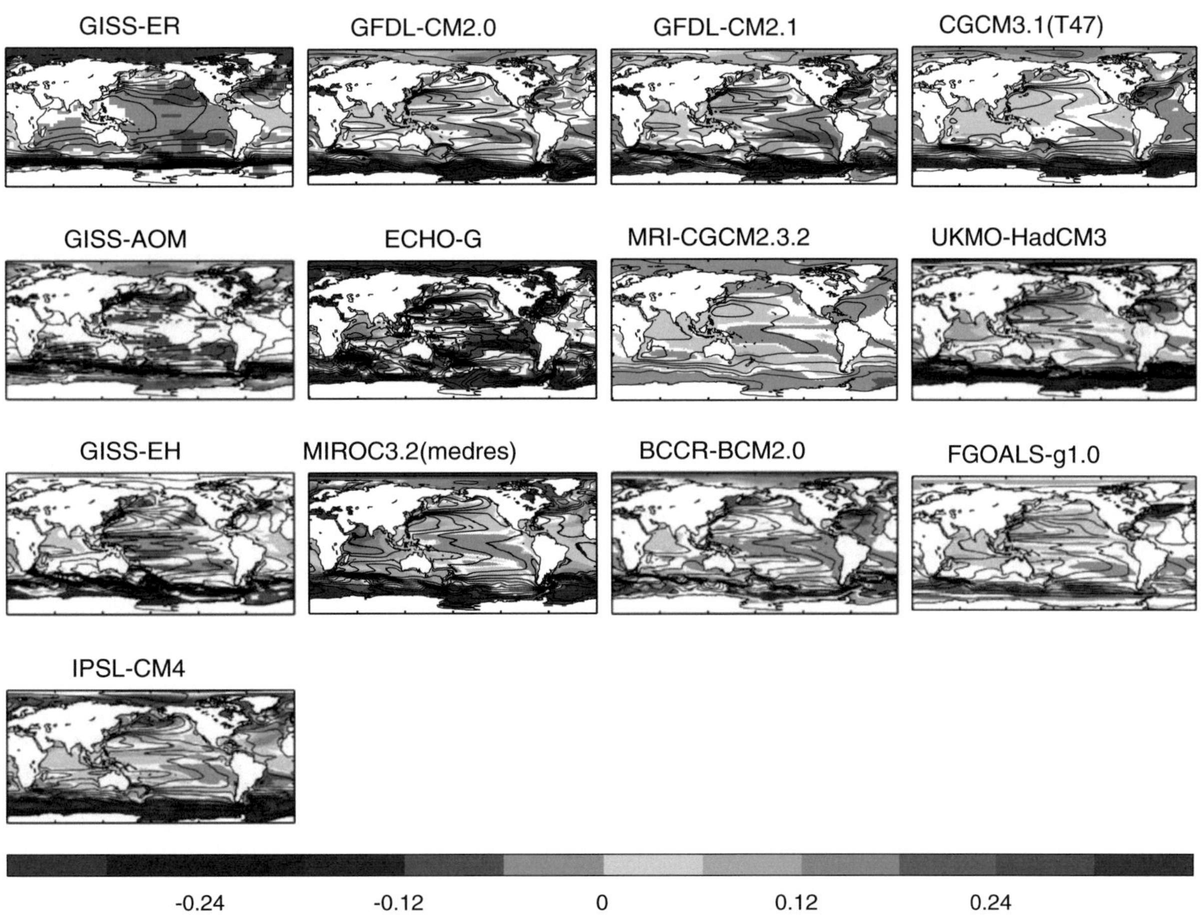

FIGURE 5.3 Combined steric and wind-driven sea-level changes (1980–1999 to 2080–2099, units in m) for the indicated models, relative to each model's global mean. The overlying contour lines are of the sea-level distribution in the baseline control simulations, averaged over a 120-year period (contours are every 0.2 m). SOURCE: Pardaens et al. (2010).

human forcing by calculating results for six emission scenarios (Box 5.1, Table 5.1). The most common approach for treating uncertainty in the global temperature response is to use the results of many climate models, which also provides a range of projected values for the global ocean response. The IPCC (2007) global model simulations for the thermal expansion and dynamical components are available for only the B1, A1B, and A2 emission scenarios, and Pardaens et al. (2010) performed their calculations for only the A1B scenario. To provide a range of projections for all six scenarios, the committee used the ratios of thermal expansion projections from Table 5.1. For example, the global projection for the low value of B1 in 2030 was computed from the digital values in Pardaens et al. (2010) using the quadratic fit for 2030 as described above. To determine the range, this value was multiplied by the ratio of the low value of the thermal expansion term in 2100 for B1 to the low value for A1B (0.10/0.13). This approach slightly underestimates the B1 values for 2030 and 2050 because sea-level change under the B1 scenario is fairly linear (see Figure 5.1), but the difference is estimated to be within rounding error (a few mm).

Cryosphere Contribution

The committee projected the cryosphere contribution to global sea-level rise using adaptations of the Meier et al. (2007) extrapolation techniques and the Pfeffer et al. (2008) methods for evaluating uncertainty and establishing projection boundaries. The committee's extrapolations were based on selected observational data for glaciers, ice caps, and the Greenland and Antarctic ice sheets. The most comprehensive time series of mass loss of the Greenland and Antarctic ice sheets is the Rignot et al. (2011a) compilation, which combines modeled surface mass balance and measured and modeled ice discharge to produce net balances for both ice sheets for 1992–2010, the earliest date from which continuous observations are available. For glaciers and ice caps, the committee used data from Dyurgerov and Meier (2005), Cogley (2009), and Dyurgerov (2010). At the time this report was being written, Dyurgerov and Meier's (2005) mass balance data from glaciers and ice caps for the 1960–2005 period were the most recent global compilation of continuous records. Dyurgerov and Meier (2005) used known area and area-altitude distributions by region to scale up limited point mass balance observations. Their analysis considered surface mass balance changes directly modulated by climate and excluded losses by calving. Dyurgerov (2010) reevaluated the data used in Dyurgerov and Meier (2005) and made significant corrections to changing glacier areas during the period of observations. Cogley (2009) presented an independent data set, evaluated in 5-year increments from ca. 1850 to 2009, that includes both climatically forced and calving losses. The data for glaciers and ice caps were averaged using techniques that weight the data according to its quality as measured by the magnitude of the stated uncertainty (see Appendix E for details).

The base-rate extrapolation assumes that present-day observed trends in loss rates continue in the future. To investigate the effect of varying rapid dynamic discharge on these projections, the committee performed model experiments to calculate the effects of both acceleration and deceleration in ice discharge relative to observed present-day rates. Both possibilities have been examined in the literature. For example, Pfeffer et al. (2008) discussed the consequences of large-scale losses from both Greenland and Antarctica in hypothetical terms, and Rignot et al. (2011a) projected a large dynamic contribution to sea-level rise from the ice sheets on the basis of past observations. On the other hand, observations in Greenland (e.g., Moon et al., 2012) show that recently active outlet glaciers are slowing down, suggesting that rapid dynamics may have an episodic or periodic nature and that future increases in sea level from rapid dynamics may not be as dramatic as have been postulated elsewhere. Price et al. (2011) used a high-order numerical model to explore the effect of outlet glacier dynamics, their influence on upstream ice dynamics, and time variations in outlet glacier dynamics on future losses from the Greenland Ice Sheet.

Increased ice discharge beyond presently observed rates was simulated by extrapolating a multiple of present-day observed discharge forward in time to 2100 (see Appendix E). For glaciers and ice caps, an increment of flux equal to 50 percent of the present-day discharge was added, equivalent to 162.4 GT yr^{-1}. For Greenland, the average speed of all outlet glaciers was increased by 2 km yr^{-1}, equivalent to a net discharge of 375.1 GT yr^{-1}. These values are consistent with the observed doubling of Greenland's mass balance deficit between ca. 1996 and 2000 (Rignot and Kanagaratnam, 2006). For Antarctica, the net outlet flux was doubled from its ca. 2006 value to 264 GT yr^{-1}. All values were increased linearly over 20 years and held constant thereafter. The exact choice of values for the individual components is less important than the net added flux after the 20-year increase, which is approximately 800 GT yr^{-1} (2.2 mm yr^{-1} sea-level equivalent).

Decreased ice discharge was simulated by reducing the projected output of the Greenland Ice Sheet by 25 percent from its projected base value. Currently, about 50 percent of Greenland's ice loss rate is caused by iceberg calving; a hypothetical 50 percent reduction in calving discharge yields a 25 percent reduction in the total ice loss rate. Other cryosphere terms were left unchanged. For Antarctica, systems likely to experience rapid change are concentrated on the Amundson Coast, and there are no known geographic features in the region that would likely serve as points of stabilization. Moreover, there is no reason to think that the dynamic slowdown seen recently in Greenland is likely to occur soon in Antarctica. Given the larger size of the Amundson Coast outlet glaciers, it is reasonable to hypothesize that any reversals will occur on longer timescales than the committee's projections. For glaciers and ice caps, future discharge was left unchanged from the base-rate projection in this experiment. The fraction of glacier and ice cap loss from calving discharge is unknown, but is probably less than 50 percent. Thus, the committee assumed that direct climatically-forced surface mass balance is the primary control on future changes in the loss rate of glaciers and ice caps.

The variations listed above were intended to capture the general magnitude of plausible changes in ice dynamics. Although these exact events may not occur, the calculations provide a means to develop a quantitative, albeit crude, estimate of the influence of rapid glacier dynamics on sea-level rise.

Results

The results of the extrapolation are presented in Table 5.2. All of the cryosphere extrapolations to 2100 are higher than the IPCC (2007) cryosphere projections. Among the most important reasons for this increase are the following:

1. Observed rates of loss from the ice sheets have accelerated significantly since the IPCC Fourth Assessment Report was finalized. Prior to 2004, published mass balances for the ice sheets were near zero or even negative, but subsequent work indicates that loss rates are rapidly accelerating (see Chapter 3). Thus, the present-day loss rates from the ice sheets constitute significantly different initial conditions than were applied in the IPCC (2007) model calculations.
2. The extrapolation method gives more weight to recent observations than to past observations (Appendix E). Thus the high present-day observed loss rates have a larger effect on extrapolations than on model calculations.
3. Rapid dynamic response was hypothesized as significant in the IPCC (2007) analyses, but was incorporated at only a rudimentary level in the projections. In the committee's analysis, added dynamics can account for 26 percent to 58 percent of total sea-level rise.

Even accounting for the possibility of slowing discharge in Greenland, the committee's cryosphere extrapolations are substantially higher than the IPCC (2007) cryosphere projections. A 25 percent reduction in the Greenland dynamic discharge lowers the committee's sea-level projections by 6 percent for 2100 (see Table E.4 in Appendix E). This result is not surprising, given the fraction of Greenland's contribution to global sea-level rise. If calving is responsible for 50 percent of Greenland's ice loss rate, or about 10 percent of total global sea-level rise, then halving the amount of calving should affect sea level at about the 5 percent level.

The committee's extrapolations also are higher than recent numerical model projections. For example, Price et al. (2011) simulated the net dynamic losses from the Jakobshavn, Helheim, and Kangerdlugssuaq glaciers, including their effects on the interior of the ice sheet, to 2100, and then scaled up that response to the entire Greenland Ice Sheet. They projected a dynamic sea-level contribution from Greenland of 5.8 ± 2.1 mm SLE by 2100, regarding it as a lower bound. By comparison, the committee's projection for Greenland is 20.1 ± 2.7 cm SLE, which includes both surface mass balance and dynamic contributions. If the dynamic response to climate change constitutes half of Greenland's recent ice loss rate, then the Greenland dynamic contribution to sea-level rise is about 10 cm SLE by 2100, more than an order of magnitude higher than the Price et al. (2011) projection. Even if dynamics constitutes only 13 percent of the Greenland's recent ice loss rate, as estimated by Price et al. (2011), the Greenland dynamic contribution projected by the committee would be 2.6 cm, greater than the Price et al. (2011) projection by a factor of 4. The two projections differ in part because of simplifications and uncertainties in both approaches, underscoring the need for more complete knowledge of processes and more complete information about initial and boundary conditions.

Uncertainty

The cryosphere projections presented here have two types of uncertainties: quantified uncertainty and unquantified uncertainty. The quantified uncertainty, which is calculated from the 5–95 percent projection intervals (Appendix E) then converted to 1 σ uncertainties, is a statistical product representing the uncertainty of the curve fitting process. The unquantified uncertainty is associated with the assumption that past system behavior is a good predictor of future system behavior. Rapid dynamic response may play a different role in future sea-level rise than it did during the period of observations, making that period potentially a poor predictor of future system behavior. However, deviations of actual sea-level rise from the simple extrapolation will take time to emerge. Extrapolation of unstable or unpredictable dynamics will thus be reliable initially, but the errors may increase dramatically as the timescale of the extrapolation exceeds the characteristic timescale of the dynamics.

In theory, the uncertainty of the extrapolations could be evaluated by determining the characteristic timescale of rapid dynamic response of vulnerable land-based ice. The timescale for dynamics of individual outlet glacier systems is thought to be decades, whereas the timescale of aggregate outlet glacier systems, such as the marine-ending glaciers draining the Greenland Ice Sheet, may be a century or longer. This timescale has not been established, however, and contributes uncertainties that are both quantifiable and unquantifiable. New work on the time-varying aspects of dynamic response of outlet glaciers, such as the modeling study of Price et al. (2011), may lead to constraints on the timescales of rapid dynamic response.

Discussion of Global Projections

The committee's projections of global sea-level rise are summarized in Table 5.2 and illustrated in Figures 5.4 and 5.5. For the three projections periods (2030, 2050, and 2100), the committee provides a projection and a range, which attempt to incorporate the various sources of uncertainty discussed above and to provide guidance on possible outcomes. The projection for the steric component is derived from the A1B emission scenario, which was used in the Pardaens et al. (2010) analysis, and the range is the corresponding value for the lowest emission scenario (B1) and the highest emission scenario (A1FI). Extrapolations are based on observations and thus take no account of emission scenarios. For the cryospheric component, the projection is the extrapolation from observed changes and the range includes a possible additional dynamic contribution. No formal probability analysis of the individual contributors of uncertainty was performed, so the projections are not necessarily the likeliest outcomes, and the ranges are not the highest or lowest possibilities.

The committee's projected contributions of the steric and cryospheric components of future sea-level rise are illustrated in Figure 5.4. The steric component, which the IPCC (2007) projected as substantially larger than the cryospheric component (Table 5.1), is roughly similar in magnitude to the cryospheric component for the first few decades. By mid-century, however, the cryospheric component greatly exceeds the steric component for all GCM simulations. The steric projection of the various models ranges from 15 cm to almost 40 cm in 2100, relative to 2000, with a model average of 24 cm. The cryospheric extrapolation ranges from about 50 cm to 67 cm in 2100, and ice dynamics would add 18 cm.

Figure 5.5 shows the range of projections of global sea-level rise. For the projection (middle line), the steric estimate for the all-model average was added to the central value of the cryospheric extrapolation. The low estimate was derived by subtracting twice the standard deviation of the steric values (shown in the first row of

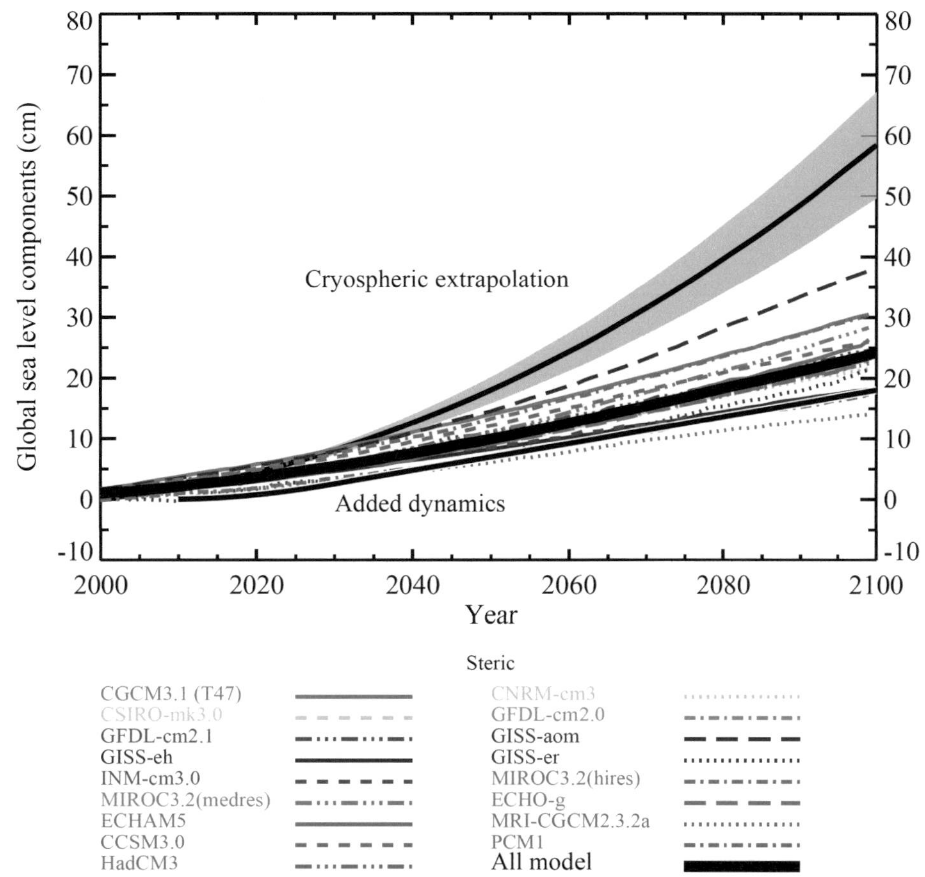

FIGURE 5.4 Committee projections of individual components of global sea-level rise. The colored lines are the steric contributions from various models with the A1B emission scenario, and the heavy black line (labeled "all model") is the model average. Gray shading is the cryospheric contribution, and the black line within the gray swath is the cryosphere average. The black line at the bottom is the added ice dynamics component.

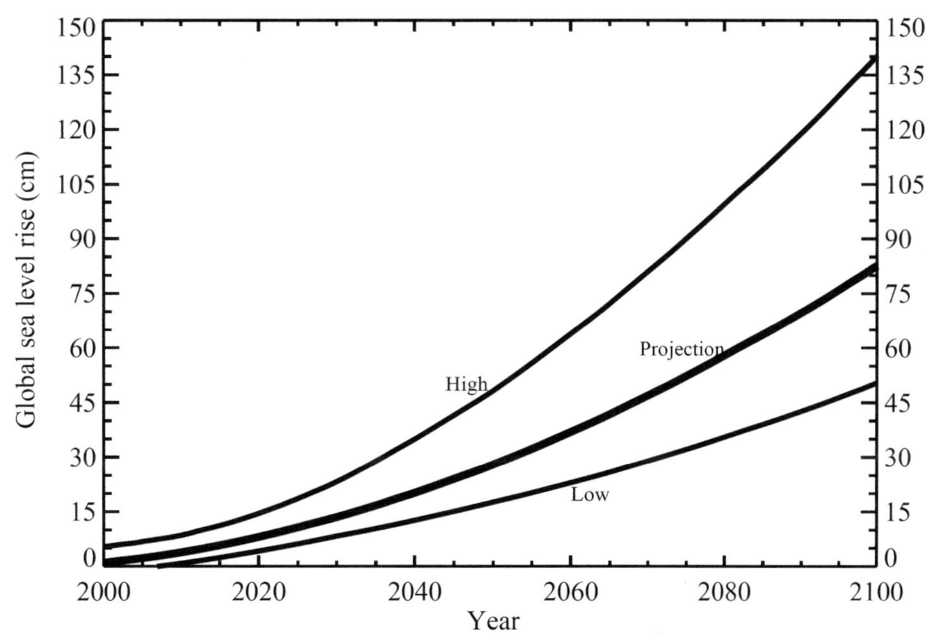

FIGURE 5.5 Range of committee projections for the sum of all individual components of global sea-level rise.

Table 5.2), interpolated between the years 2030, 2050, and 2100, and adding the lower fit (mean minus twice the standard deviation) of the extrapolated cryospheric estimate. For the high estimate, the steric curve plus twice the standard deviation of the steric values was added to the upper fit (mean plus twice the standard deviation) of the cryospheric extrapolation and the additional ice dynamics contribution.

All three curves in Figure 5.5 have a positive curvature, as do the semi-empirical projections. In the committee's projection, the acceleration originates primarily from the cryospheric extrapolation, although a small amount of acceleration comes from the steric term. In the semi-empirical estimates, the acceleration is built into the mathematical expression relating the rate of sea-level rise to the departure of temperature from some equilibrium value (see the first two equations in "Semi-Empirical Models" above). In this formulation, as long as temperature continues to rise, sea-level rise will accelerate.

Given the inherent thermodynamics of ice bodies in disequilibrium with their climatic environment, an accelerating cryospheric contribution is a reasonable supposition as long as the timescales of the climatic change are short relative to the timescales of the ice response. Although large glaciers may have response timescales of a few decades, the Greenland and Antarctic ice sheets, which dominate the cryospheric term, have response times of centuries to millennia.

Figure 5.6 compares the ranges of sea-level rise projected by the committee, IPCC (2007), and Vermeer and Rahmstorf (2009). The committee's projections for 2030 and 2050 are similar to the Vermeer and Rahmstorf (2009) projections, but have a wider range: 8–23 cm for 2030 and 18–48 cm for 2050. For 2100, when IPCC (2007) estimates are also available, the committee's projected range (50–140 cm) is substantially higher than the IPCC (2007) range (18–59 cm, with an additional 17 cm if rapid dynamical changes in ice flow are included) and lower than the Vermeer and

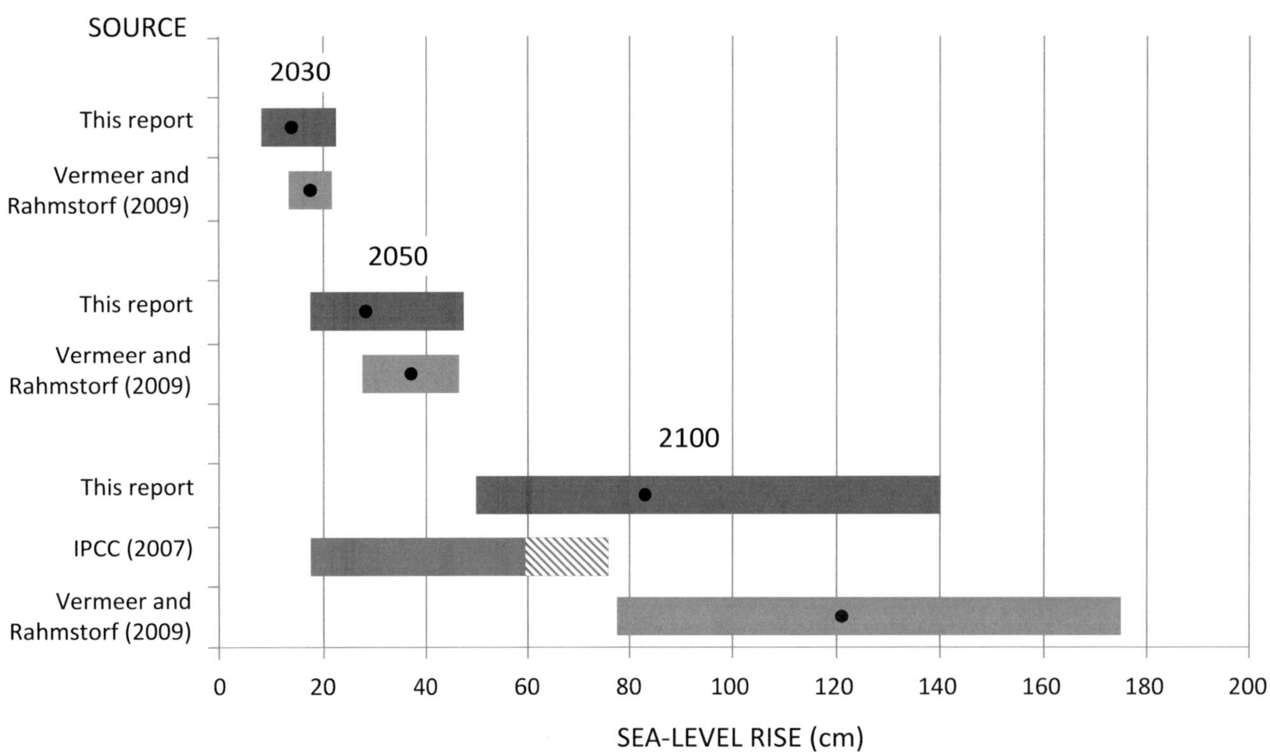

FIGURE 5.6 Global sea-level rise for 2030, 2050, and 2100 projected by this committee (red), Vermeer and Rahmstorf (2009; green), and IPCC (2007; blue). The dots are the projected values and the colored bars are the ranges. The IPCC value includes the sea-level projection (blue) plus the scaled-up ice sheet discharge component (blue diagonal lines).

Rahmstorf (2009) range (78–175 cm). The committee's results differ from the IPCC (2007) results because the committee considered more recent scientific observations and modeling and also used different methods to make projections. For example, although the steric contributions were drawn from the same global climate models used in IPCC (2007), the committee used the global climate model results directly, whereas IPCC (2007) used lower-order models to develop estimates for emission scenarios that were not simulated in global climate models (e.g., A1FI). In addition, the committee used extrapolation methods to project the cryosphere component of sea-level rise, whereas IPCC (2007) used climate models.

The global sea-level projections shown in Figures 5.4 and 5.5 do not include contributions from groundwater depletion and reservoir extraction. Estimates available at the time this report was being written (e.g., Milly et al., 2010) suggested that the sum of these contributions was near zero, within the stated uncertainties. Although some studies have pointed out that the number of new reservoirs has been declining over the past three decades (e.g., Chao et al., 2008), the committee had no firm basis for projecting a growing contribution to sea-level rise from groundwater extraction. A new paper published as this report was nearing release, however, projects that increasing groundwater extraction and decreasing reservoir impoundment will contribute about 1.5 ± 0.8 cm SLE to global sea level in 2030, 3.1 ± 1.1 cm SLE in 2050, and 7.5 ± 2.0 cm SLE in 2100 (Wada et al., 2012b). If confirmed by subsequent analyses, these results indicate that changes in the balance of groundwater depletion and reservoir impoundment could increase the magnitude of future sea-level change.

PREVIOUS PROJECTIONS OF U.S. WEST COAST SEA-LEVEL RISE

Only a few studies have attempted to project 21st century sea-level rise along the west coast of the United States. Methods varied, but each study used global climate models forced by the IPCC (2000) low and high greenhouse gas emission scenarios. The results were then downscaled, used in semi-empirical projections, or combined with local information, as discussed below. Each of the studies emphasized that the results represented a range of outcomes, not formal projections of sea-level rise.

The earliest of these studies, Hayhoe et al. (2004), used two global climate models, downscaled to a 150 km^2 grid, to simulate climate change in California. Projections of various aspects of climate change were averaged over the 2020–2049 and 2070–2099 periods, relative to the 1961–1990 period, following the approach taken in the IPCC Third Assessment Report. Hayhoe et al. (2004) estimated that sea level along the California coast would rise 8.7 cm to 12.7 cm for the 2020–2049 period, and 19.2 cm to 40.9 cm for the 2070–2099 period, depending on the model and emission scenario used.

Mote et al. (2008) estimated future sea-level rise off Washington for 2050 and 2100, dividing the coastline into three regions according to their vertical land motions. They used global climate models to calculate the thermal expansion and cryosphere contributions to sea-level rise. The rates of global sea-level rise were then adjusted for vertical land motions and for model-predicted seasonal and interannual wind-driven increases in sea level. Mote et al. (2008) projected low, medium, and high sea-level rise for the Puget Sound region of 16 cm, 34 cm, and 128 cm, respectively, by 2100. They also found that some parts of the Olympic Peninsula could experience tectonic uplift that would exceed the low end of projected rates of global sea-level rise, with the medium estimate for sea-level fall between 0 cm and -15 cm by 2050, depending on location, and 0 cm and -30 cm by 2100.

Cayan et al. (2009) projected sea-level rise off California using Rahmstorf's (2007) semi-empirical method with global average surface air temperature simulated from global models. Assuming that the rate of sea-level rise off the California coast will be the same as the global rate, Cayan et al. (2009) estimated sea-level rise of 30–45 cm by 2050, and 50–140 cm by 2100, relative to 2000.

Tebaldi et al. (2012) projected sea-level rise at 11 tide gage locations along California, Oregon, and Washington using the semi-empirical method of Vermeer and Rahmstorf (2009) to estimate global sea-level rise and 50 years (1959–2008) of tide gage records to estimate local rates and their deviations from global sea-level rise caused by local effects. Based on this information and output from an ensemble of GCM simulations, they obtained sea-level rise estimates of 3–12 cm by 2030 and 11–30 cm by 2050, relative to 2008, for the 11 locations.

COMMITTEE PROJECTIONS OF SEA-LEVEL RISE ALONG THE CALIFORNIA, OREGON, AND WASHINGTON COASTS

Sea-level rise along the west coast of the United States differs from global mean sea-level rise because of local steric (primarily thermosteric) contributions, dynamic height differences caused primarily by changes in winds, the gravitational and deformational effects of modern land ice melting, and vertical land motions along the coast (see Chapter 4). The committee projected the contributions of these components to sea-level rise off the California, Oregon, and Washington coasts for the years 2030, 2050, and 2100, relative to year 2000. The local steric and wind-driven contribution was estimated using GCMs; the land ice contribution, adjusted for gravitational and deformational effects, was extrapolated; and the contribution from vertical land motion was estimated using Global Positioning System (GPS) data. Values for the individual contributions are summarized in Table 5.3 and discussed below.

Steric and Dynamic Ocean Height Effects

The local steric and wind-driven components were determined from the same CMIP 3 global ocean models used to calculate the steric contribution to global sea-level rise. Thirteen of the CMIP 3 models examined in Pardaens et al. (2010) include global annual averages of the steric contribution and wind-driven dynamic ocean heights on a 1° latitude by 1° longitude grid. From this data set, the committee selected the ocean model grid points closest to the coastlines of California, Oregon, and Washington at each latitude, developing a time series for each model at each latitude. To obtain values

TABLE 5.3 Regional Sea-Level Rise Projections (in cm) Relative to Year 2000

	2030		2050		2100	
Component	Projection	Range	Projection	Range	Projection	Range
Steric and dynamic ocean[a]	3.6 ± 2.5	0.0–9.3 (B1–A1FI)	7.8 ± 3.7	2.2–16.1 (B1–A1FI)	20.9 ± 7.7	9.9–37.1 (B1–A1FI)
Non-Alaska glaciers and ice caps[b]	2.4 ± 0.2		4.4 ± 0.3		11.4 ± 1.0	
Alaska, Greenland, and Antarctica with sea-level fingerprint effect[c]						
Seattle, WA	7.1	5.4–9.5	16.0	11.1–22.1	52.7	32.7–74.9
Newport, OR	7.4	5.6–9.5	16.6	11.7–22.2	54.5	34.1–75.3
San Francisco, CA	7.8	6.1–9.6	17.6	12.7–22.3	57.6	37.3–76.1
Los Angeles, CA	8.0	6.3–9.6	17.9	13.0–22.3	58.5	38.6–76.4
Vertical land motion[d]						
North of Cape Mendocino	-3.0	-7.5–1.5	-5.0	-12.5–2.5	-10.0	-25.0–5.0
South of Cape Mendocino	4.5	0.6–8.4	7.5	1.0–14.0	15.0	2.0–28.0
Sum of all contributions						
Seattle	6.6 ± 5.6	-3.7–22.5	16.6 ± 10.5	-2.5–47.8	61.8 ± 29.3	10.0–143.0
Newport	6.8 ± 5.6	-3.5–22.7	17.2 ± 10.3	-2.1–48.1	63.3 ± 28.3	11.7–142.4
San Francisco	14.4 ± 5.0	4.3–29.7	28.0 ± 9.2	12.3–60.8	91.9 ± 25.5	42.4–166.4
Los Angeles	14.7 ± 5.0	4.6–30.0	28.4 ± 9.0	12.7–60.8	93.1 ± 24.9	44.2–166.5

[a] Projection indicates the mean and ± standard deviation computed for the Pacific coast from the gridded data presented in Pardaens et al. (2010) for the A1B scenario. Ranges are the means for B1 and A1Fl using the scaling in Table 10.7 of IPCC (2007; see also Table 5.1 of this report): (B1/A1B) = (0.1/0.13); (A1FI/A1B) = (0.17/0.13).

[b] Extrapolated based on ice loss rates for glaciers and ice caps except Alaska, Greenland, and Antarctica. No ranges are given because these sources are assumed to have a small or uniform effect on the gradient in sea-level change along the U.S. west coast (see "Sea-Level Fingerprints of Modern Land Ice Change" in Chapter 4).

[c] Extrapolation based on ice loss rates and gravitational attraction effects for Alaska, Greenland, and Antarctica. Ranges reflect uncertainty in ice loss rates.

[d] Assumes constant rates of vertical land motion of 1.0 ± 1.5 mm yr^{-1} for Cascadia and -1.5 ± 1.3 mm yr^{-1} for the San Andreas region. The signs were reversed to calculate relative sea level. Uncertainties are 1 standard deviation.

for 2030, 2050, and 2100 (relative to year 2000) that reflect the century-long change in sea level rather than year-to-year variations, quadratic fits to each time series were performed. In addition, the variance about the quadratic curve was computed for 21-year intervals centered on 2030, 2050, and for the last 20 years of the time series. Figure 5.7 shows the sum of the steric and dynamic ocean height terms for the 13 models. Although some models predict a north-south gradient in sea-level change, the average for all of the models predicts nearly uniform steric and dynamic ocean height contributions along the entire coast (see heavy black line in Figure 5.7). The average values were used in the projection (see first row of Table 5.3).

Cryosphere and Sea-Level Fingerprint Effects

The contribution of the cryosphere to sea-level rise along the west coast of the United States is influenced by gravitational and deformational effects associated with melting from the Alaska glaciers, which are nearby, and the Greenland and Antarctic ice sheets, which are large (see "Sea-Level Fingerprints of Modern Land Ice Change" in Chapter 4). To account for these effects in the projections, the committee subdivided ice loss rates into four categories: Greenland Ice Sheet, Antarctic Ice Sheet, Alaska glaciers, and all other glaciers and ice caps. The sea-level fingerprint scale factors were applied to the first three sources, then loss rates from all ice sources were extrapolated forward and converted to cumulative sea-level rise (see details in Appendix E). Because the gravitational and deformational effects associated with the sea-level fingerprints of the three ice sources varies strongly with latitude, the projections were made for four points along the coast: Seattle, Washington, Newport, Oregon, San Francisco, California, and Los Angeles, California (Table 5.3). A polynomial was then fit through the four points to estimate the influence of the sea-level fingerprints as a function of latitude.

To determine the magnitude of the gravitational and deformational effects of melting in Alaska, Greenland, and Antarctica on the projections, the committee compared the results of the above analysis with the global cryosphere extrapolations, which were done without an adjustment for the regional sea-level fingerprints. The difference between the adjusted and

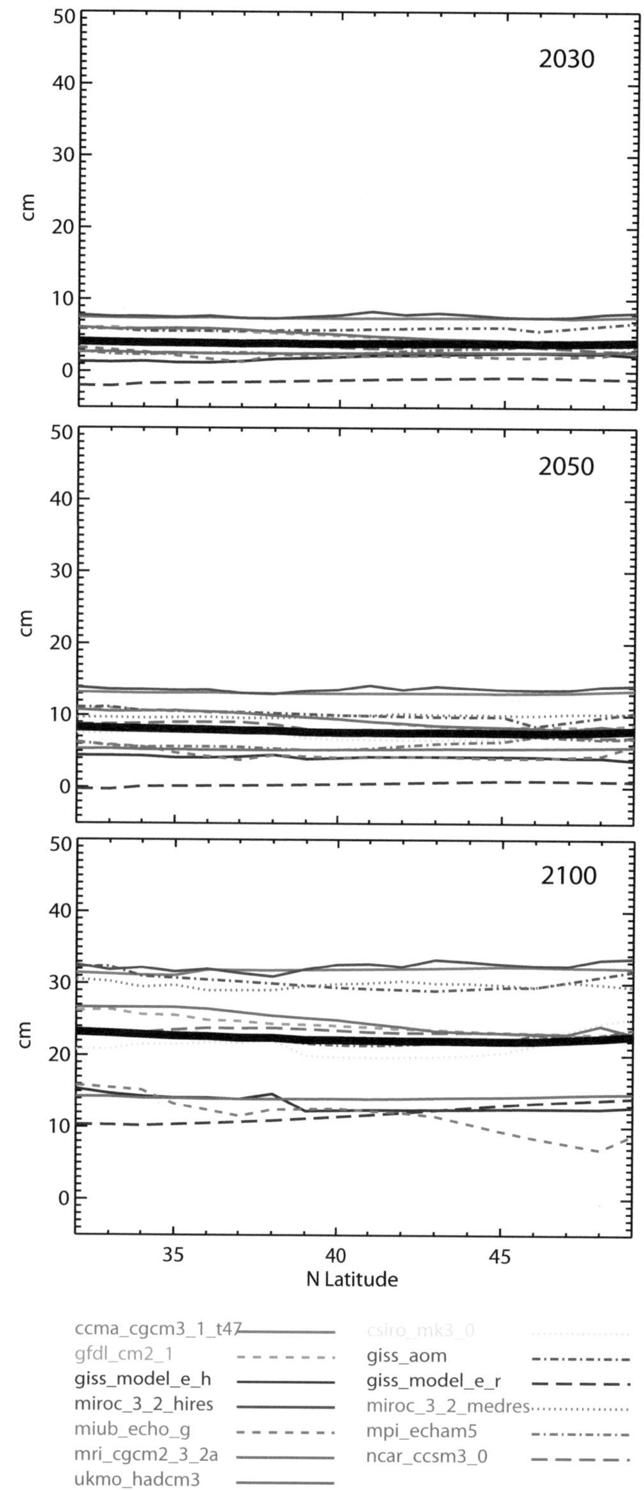

FIGURE 5.7 Combined thermosteric and ocean dynamic height sea-level change for 2030 (top), 2050 (middle), and 2100 (bottom), relative to 2000, from 13 GCMs as a function of latitude. The thick black line indicates the average of all the models.

unadjusted extrapolations is given in Table 5.4. The gravitational and deformational effects reduce the cryospheric contribution to relative sea-level rise projected for 2100 by 7–21 percent along the north coast and by 1–12 percent along the central coast. Along the south coast, these effects can increase the cryospheric contribution to sea-level rise by up to 2 percent or decrease it by up to 6 percent.

The projection assumes that the fingerprint scale factors remain constant for the small reduction in land ice volume expected by 2100, which is likely reasonable for the next several decades. Assuming that the sea-level fingerprint is correct, uncertainties in the calculation are associated with the ice loss rate. When uncertainties in the ice loss rate are factored in, the fingerprint-adjusted contribution of Alaska, Greenland, and Antarctica to relative sea-level rise ranges from 33–75 cm along the north coast and 39–76 cm along the south coast for 2100 (Table 5.3).

Vertical Land Motion

Major causes of vertical land motion along the west coast of the United States include tectonics, glacial isostatic adjustment (GIA), and subsidence as a result of sediment compaction and/or fluid withdrawal (see "Vertical Land Motion" in Chapter 4). The vertical land motion signal in Oregon, Washington, and northern California is dominated by regional tectonics associated with the Cascadia Subduction Zone. In California south of Cape Mendocino, vertical land motion depends on varying combinations of GIA, sediment compaction, fluid withdrawal or recharge, and local compressional tectonics that may or may not be related to the San Andreas Fault. Projections of the regional tectonic and GIA components of vertical land motion can be made using earthquake cycle deformation models and geophysical models, respectively. The total vertical land motion, including tectonics, GIA, compaction, and fluid withdrawal and recharge, can be projected using continuous GPS (CGPS), assuming that the vertical land motion is predominantly secular within ~100 years. Results from these three projection methods are summarized in Table 5.5 and discussed below.

Tectonic Projections

Tectonics causes significant vertical land motion along the coast above the Cascadia Subduction Zone. The tectonic component of vertical land motion in this area can be projected using earthquake cycle deformation models. The CAS3D-2 model (He et al., 2003; Wang et al., 2003; Wang, 2007), which is the most sophisticated and complete model of earthquake deformation along the Cascadia Subduction Zone, is the only model that has been used to make forward projections. Like other tectonic models, it is limited by incomplete knowledge of the temporal behavior of the earthquake process, such as the degree of periodicity of the Cascadia earthquake cycle, which adds uncertainties that are difficult to quantify. The model excludes vertical land motion from glacial isostatic adjustment.

The projected rates of interseismic deformation for the Cascadia Subduction Zone from the CAS3D-2 model are given in Table 5.5. The projections suggest that coastal sites, which are closest to the offshore subduction boundary, will undergo uplift, whereas more inland locations (Anacortes and Seattle) will undergo subsidence. Projected vertical land motions for the coastal sites range from -1.0 mm yr^{-1} (subsidence) to +3 mm yr^{-1} (uplift), with most rates varying by less than 0.2 mm yr^{-1} over the 21st century. The vertical land motions projected using the CAS3D-2 model

TABLE 5.4 Effect of Sea-Level Fingerprints of Alaska, Greenland, and Antarctica Ice Masses Expressed as Percentage Differences from Cryosphere Projections with No Fingerprint Effect[a]

	North Coast (Neah Bay)			Central Coast (Eureka)			South Coast (Santa Barbara)		
Year	Low	Central	High	Low	Central	High	Low	Central	High
2030	-19.9%	-13.4%	-9.6%	-10.4%	-5.8%	-3.1%	-4.9%	-1.6%	0.4%
2050	-20.3%	-12.2%	-8.6%	-10.6%	-4.8%	-2.2%	-5.1%	-0.8%	1.2%
2100	-21.1%	-10.8%	-7.4%	-12.3%	-3.7%	-1.2%	-5.5%	0.2%	2.1%

[a] Uncertainties, expressed as low-high values, were derived from the spread of projected contributions from Alaska, Greenland, and Antarctica.

TABLE 5.5 Vertical Land Motion Projections for the Two Tectonic Regimes

			Tectonic Component (mm yr^{-1})[a]			GIA Component (mm yr^{-1})[b]	Total Vertical Land Motion (mm yr^{-1})[c]
Location	Latitude	Longitude	2010–2030	2030–2050	2050–2100	2010–2100	2010–2100
Cascadia Subduction Zone							
Cherry Point, WA	48.87	-122.75				0.2 ± 0.4	1.0 ± 1.5
Anacortes, WA	48.56	-122.64	-0.9	-0.9	-1.0	-0.1 ± 0.5	1.0 ± 1.5
Seattle, WA	47.85	-122.73	-0.6	-0.6	-0.6	-0.5 ± 0.4	1.0 ± 1.5
Long Beach, WA	46.58	-123.83	1.9	1.8	1.7	-1.1 ± 0.5	1.0 ± 1.5
Pacific City, OR	45.38	-123.94	1.7	1.6	1.5	-1.0 ± 0.4	1.0 ± 1.5
Waldport, OR	44.42	-124.02	1.7	1.6	1.5	-0.9 ± 0.3	1.0 ± 1.5
Coos Bay, OR	43.36	-124.30	2.3	2.2	2.1	-0.9 ± 0.3	1.0 ± 1.5
Eureka, CA	40.87	-124.15	3.0	2.8	2.6	-0.7 ± 0.3	1.0 ± 1.5
San Andreas Fault Zone							
Point Reyes, CA	38.00	-122.98				-0.5 ± 0.3	-1.5 ± 1.3
San Francisco, CA	37.80	-122.47				-0.5 ± 0.3	-1.5 ± 1.3
Monterey, CA	36.60	-121.88				-0.5 ± 0.3	-1.5 ± 1.3
Port San Luis, CA	35.17	-120.75				-0.5 ± 0.3	-1.5 ± 1.3
Santa Monica, CA	34.02	-118.50				-0.3 ± 0.3	-1.5 ± 1.3
Los Angeles, CA	33.72	-118.27				-0.4 ± 0.3	-1.5 ± 1.3
San Diego, CA	32.72	-117.17				-0.4 ± 0.3	-1.5 ± 1.3

NOTE: Positive rates denote uplift and negative rates denote subsidence.
[a] Rates provided by Kelin Wang, Geological Survey of Canada, using the CAS3D-2 model described in Chapter 4.
[b] Rates were averaged from an ensemble of 16 GIA models (see Table 4.3) and are represented so positive GIA means falling relative sea level.
[c] Rates (± 1 standard deviation) were determined from CGPS data from the Scripps Orbit and Permanent Array Center taken within 15 km of the coast; see Table A.1.

generally agree with rates determined from leveling (Burgette et al., 2009) and from GPS (Mazzotti et al., 2008; this report).

GIA Projections

Projections of the GIA component of vertical land motion were made using an ensemble of 16 models. The projections show subsidence at all locations except for northernmost Washington, which shows negligible uplift (Table 5.5). Mean GIA model predicted rates of vertical land motion range from +0.2 mm yr^{-1} in northernmost Washington to -1.0 mm yr^{-1} in southern Washington and northern Oregon. This strong latitudinal gradient illustrates the importance of GIA in regions underneath or at the margins of the extinct Laurentide ice sheet. In southern Oregon and California, mean rates are generally between -0.4 mm yr^{-1} and -1.0 mm yr^{-1}. Given the slow pace of glacial isostatic adjustment, these rates are assumed constant for the three projection periods (2030, 2050, and 2100).

GPS Projections

The total vertical land motion, including signals from tectonics, GIA, sediment compaction, and/or fluid withdrawal or recharge, is recorded in GPS data. Consequently, the committee used CGPS data in its projections of sea-level rise for 2030, 2050, and 2100. It would be attractive to use the relatively densely-spaced CGPS vertical land motion data to make projections at high spatial resolution along the coast. However, vertical land motions can vary at length scales that are considerably smaller than the CGPS station spacing, so interpolation using the CGPS data carries substantial risk of spatial aliasing. Moreover, the data exhibit significant scatter because of local sediment compaction and/or fluid withdrawal (see Figure 4.14b and associated discussion). Consequently, the committee chose the conservative approach of projecting vertical land motion for the two tectonic regions—Cascadia and the San Andreas region—and characterizing them using simple statistics (mean and 1 standard deviation). With obvious outliers removed, the current rates of vertical land motion for these regions are 1.0 ± 1.5 mm yr^{-1} for Cascadia and -1.5 ± 1.3 mm yr^{-1} for the San Andreas area.

In using the current rates of vertical land motion in its projections, the committee assumed that the CGPS spatial pattern and rates in the two tectonic regions would remain constant for 2030, 2050, and 2100. This assumption is supported by leveling data in California (Appendix D) and in Washington and Oregon (Burgette et al., 2009). In addition, the CAS3D-2 tectonic model suggests that, in the absence of a great earthquake, the general vertical land motion pattern or trend in Cascadia will not change significantly in the coming century. The projected rates of vertical land motion are given in Table 5.3.

Discussion of Regional Projections

The projections of sea-level rise off California, Oregon, and Washington were made by summing the cryosphere component, adjusted for the effects of the sea-level fingerprints of Alaska, Greenland, and Antarctica; the local steric and dynamical ocean component; and the vertical land motion component. The values used for the projections appear in Table 5.3 and are illustrated in Figure 5.8. The cryosphere is the only component with pronounced upward curvature (acceleration) over the 21st century. Ice mass loss rates for Alaska, Greenland, and Antarctica were adjusted for gravitational and deformational effects and then added to loss rates from other glaciers and ice caps. The sum was then extrapolated forward. The steric and dynamical ocean components (blue swath in Figure 5.8) were extracted from the ocean data provided by Pardaens et al. (2010), averaged for the west coast, then smoothed for plotting using locally weighted regression. The vertical land motion components and their uncertainties for the northern and southern part of the coast are shown in the shaded areas; the bars on the right margin indicate the range for 2100. North of Cape Mendocino, the coast is experiencing mean uplift, so vertical land motion contributes negatively to relative sea-level rise (although uncertainties are large and include positive contributions), whereas the coast south

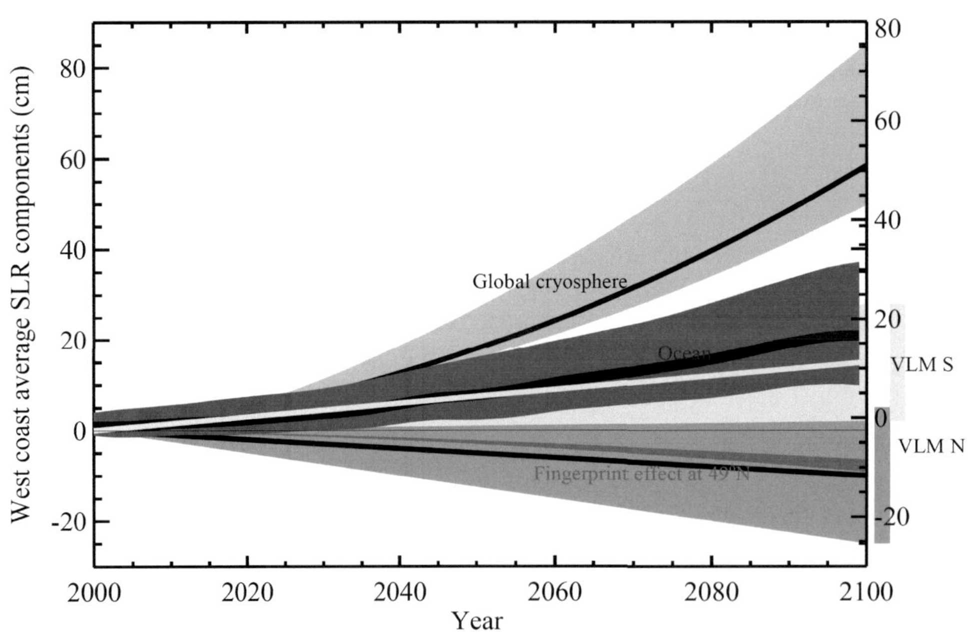

FIGURE 5.8 Committee projections of components of sea-level rise off California, Oregon, and Washington. The blue band represents the model results for combined global steric and local dynamical sea-level change, averaged between 32° and 49° latitude, from 13 GCMs. Light gray shading in the middle of the figure shows estimated effects of vertical land movement in the San Andreas region (VLM S), and dark gray shading at the bottom of the figure shows the vertical land movements for Cascadia (VLM N). Light gray shading at the top of the figure shows the global cryosphere, including added ice dynamics. The red line is the effect of the sea-level fingerprint of ice melt from the Alaska, Greenland, and Antarctica sources, shown for the north coast (49°N). The fingerprint effect is subtracted from the global cryosphere.

of Cape Mendocino is experiencing mean subsidence, so vertical land motion contributes positively to relative sea-level rise.

Figure 5.9 shows the total regional sea level projected for the years 2030, 2050, and 2100, relative to year 2000, for a transect along the west coast. The shape of the curve is dominated by the change in vertical land motion at about 40° latitude from uplift in the north to subsidence in the south. The sea-level fingerprint effect reduces the projected sea levels along the entire coast and is most pronounced in Washington. The fingerprint effect has not been included in previous studies and projections of sea level for the west coast (e.g., Mote et al., 2008; Cayan et al., 2009; Tebaldi et al., 2012). The ocean components have little effect on the north-south gradient in projected sea-level change.

The committee's projections for the west coast of the United States are significantly different from global projections (Figure 5.10). The difference is largest off the Washington coast, where sea-level fingerprint effects lower the height of the ocean surface and regional tectonics raises the height of the land surface, resulting in rates of relative sea-level rise that are substantially lower than the global mean. Off the California coast, where subsidence is lowering the land surface, the projected relative sea-level rise is slightly higher than the global mean. The committee's projected values for California are somewhat lower than the Vermeer and Rahmstorf (2009) projections, which are being used by California state agencies on an interim basis for coastal planning (CO-CAT, 2010). For California and Washington, the committee's projections fall within the range presented in Cayan et al. (2009) and Mote et al. (2008), respectively. The committee's projected values for 2030 and 2050 also are comparable to those of Tebaldi et al., (2012), although the committee found a larger north-south difference in the magnitude of sea-level rise.

UNCERTAINTY

Projections of future sea-level rise carry numerous sources of uncertainty. This uncertainty arises from an incomplete understanding of the global climate system, the inability of global climate models to accurately represent all important components of the climate system at global or regional scales, a shortage of data at the temporal and spatial scales necessary to constrain the models, and the need to make assumptions about future conditions (e.g., population growth, technological developments, large volcanic eruptions) that drive the climate system. Although a systematic analysis of these uncertainties was beyond the ability of the committee, this report attempts to describe and combine the most important uncertainties. For the committee's global sea-level rise projections, important uncertainties are associated with assumptions about the growth of concentrations of greenhouse gases and sulfate aerosol, which affect the steric contribution, and future ice loss rates and the effect of rapid dynamic response, which affect the land ice contribution. Additional, unquantified uncertainties arise from neglecting the terrestrial water component in the projections and from combining model-projected steric contributions with extrapolation-projected land ice contributions (e.g., model projections account for future emissions whereas extrapolations do not).

Regional projections carry additional uncertainties because more components are included and some components are estimated from global scale analyses. The uncertainties are larger for the committee's projections for California, Oregon, and Washington than they are for the global projections, primarily because uncertainties in the steric component are larger at smaller spatial scales and because some of the additional components (e.g., vertical land motion) have relatively large uncertainties.

For both global and regional projections of sea-level rise, uncertainties grow as the projection period increases because the chances of the observations and models deviating from actual climate changes increases. Currently, all projection methods—including process-based numerical models, extrapolations, and semi-empirical methods—have large uncertainties at 2100. Although the actual value of sea-level rise will almost surely fall somewhere within these wide uncertainty bounds, confidence in specifying the exact value is relatively low. At short timescales, the models more closely represent the future climate system, so uncertainties are smaller and confidence is higher. Confidence in the committee's projections is likely to be highest in 2030 and perhaps 2050, which are likely of greatest interest to coastal planners, engineers, and other decision makers tasked with planning for sea-level rise along the west coast of the United States.

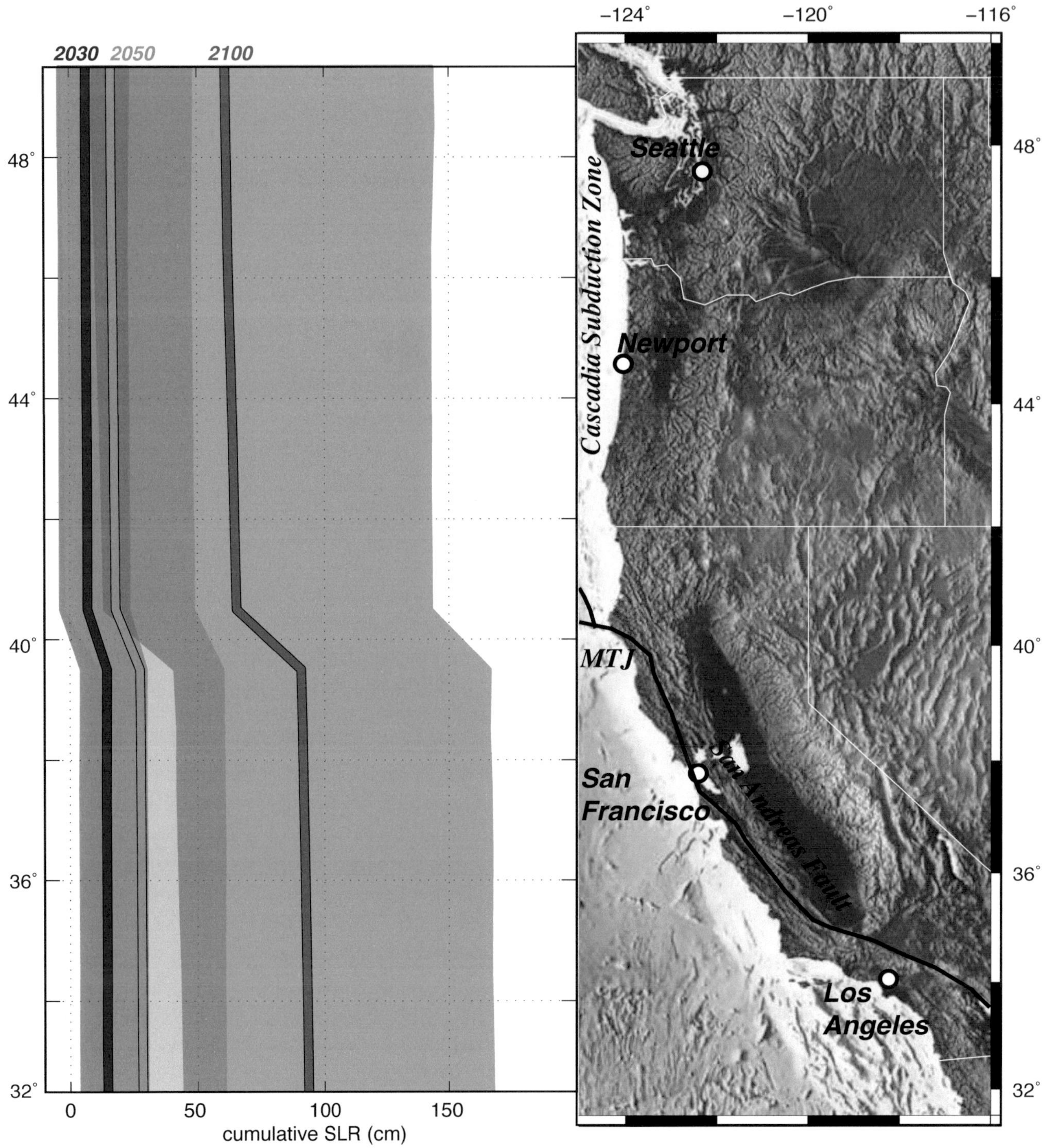

FIGURE 5.9 Projected sea-level rise off California, Oregon, and Washington for 2030 (blue), 2050 (green), and 2100 (pink), relative to 2000, as a function of latitude. Solid lines are the projections and shaded areas are the ranges. Ranges overlap, as indicated by the brown shading (low end of 2100 range and high end of 2050 range) and blue-green shading (low end of 2050 range and high end of 2030 range). MTJ = Mendocino Triple Junction, where the San Andreas Fault meets the Cascadia Subduction Zone.

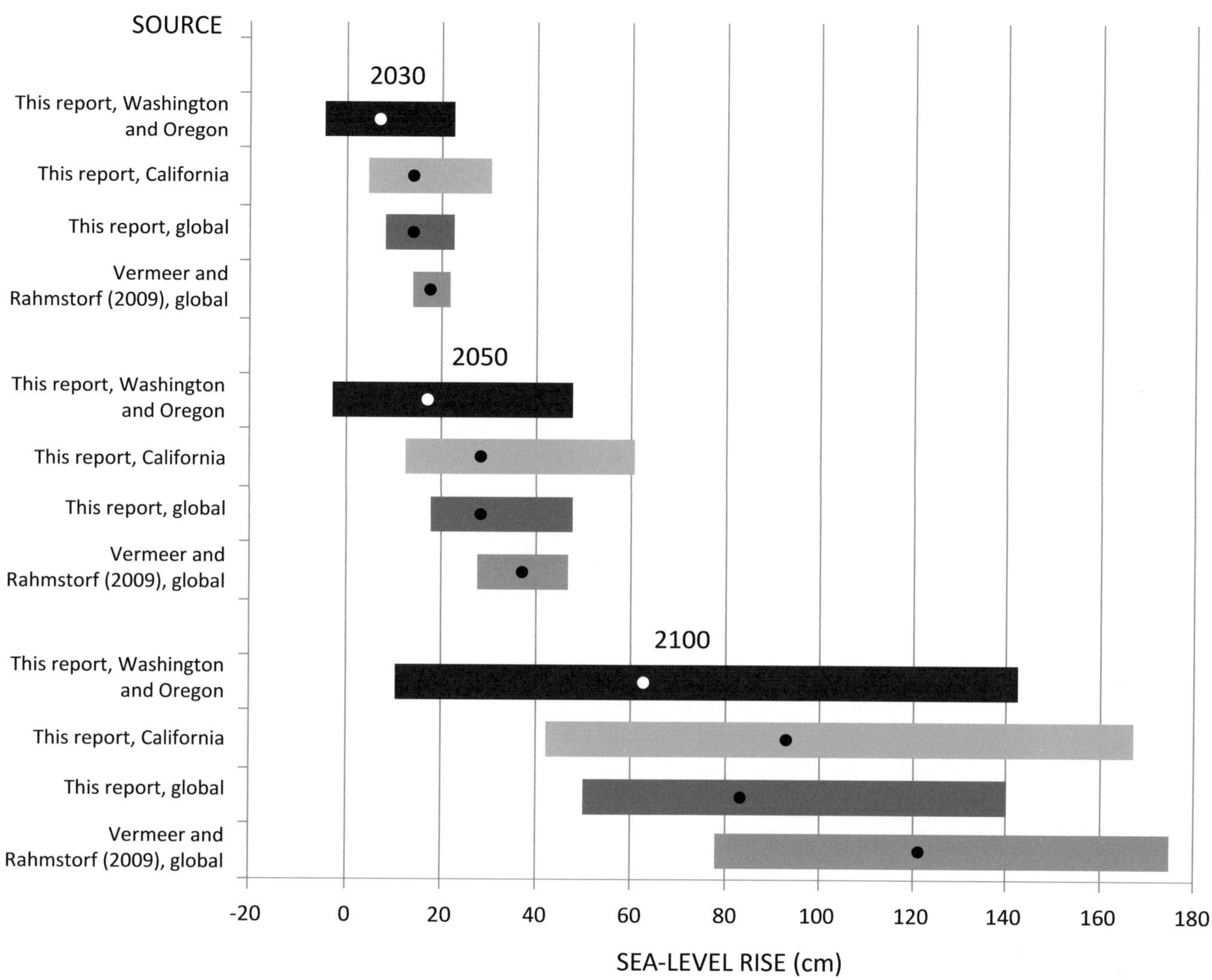

FIGURE 5.10 Committee's projected sea-level rise for California, Oregon, and Washington compared with global projections. The dots are the projected values and the colored bars are the ranges. Washington and Oregon = coastal areas north of Cape Mendocino; California = coastal areas south of Cape Mendocino.

RARE EXTREME EVENTS

Extreme events can raise sea level much faster than projected above. The rapid rise in sea level could be temporary, as in the case of a severe storm, or permanent, as in the case of a great subduction zone earthquake. The potential contribution of such extreme events to future sea-level rise is described below.

Extreme Sea Level

In the first 3 months of 1983, the west coast of the United States experienced a sequence of strong storms, with the coincidence of El Niño conditions, high astronomical tides, and large waves producing record sea levels along virtually the entire coast (see "Changes in Ocean Circulation" in Chapter 4). Damage was extensive (e.g., Figure 5.11), with losses totaling $215 million (in 2010 dollars; Griggs et al., 2005). Some models predict that such extreme events will become more common and that heightened sea level will persist longer as sea level rises, increasing the potential for damage (Cayan et al., 2008; Cloern et al., 2011).

Cloern et al. (2011) used a GCM forced by the IPCC (2000) B1 emission scenario to assess possible climate change impacts in the San Francisco Bay and delta. As part of the analysis, they used a local sea-level model, introduced by Cayan et al. (2008), to investi-

FIGURE 5.11 Rio Del Mar on northern Monterey Bay was damaged during the El Niño winter of 1983 by large waves arriving simultaneously with high tides and elevated sea levels. SOURCE: Courtesy of Gary Griggs, University of California, Santa Cruz.

gate sea-level extremes that occur in conjunction with broad-scale sea-level rise. Historical (1961–1999) and projected (2000–2100) hourly sea level was simulated using predicted tides, simulated weather and El Niño-Southern Oscillation conditions, and long-term rates of sea-level rise from Vermeer and Rahmstorf (2009). Wind, surface atmospheric pressure, and tropical Pacific sea surface temperature were obtained from the National Center for Atmospheric Research PCM1 climate model simulation.

The committee reproduced the Cloern et al. (2011) analysis using its own sea-level projection for the San Francisco area and the Geophysical Fluid Dynamics Laboratory CM2.1 model. This exercise showed that as mean sea level rises, the incidence of extreme high-sea-level events becomes increasingly common (Figure 5.12). According to the model, the incidence of extreme water heights that exceed the 99.99th percentile level (1.41 m above historical mean sea level) increases from the historical rate of approximately 9 hours per decade to more than 250 hours per decade by mid-century, and to more than 12,000 hours per decade by the end of the century. The model also shows that the duration of these extremes would lengthen from a maximum of 1 or 2 hours for the recent historical period to 6 or more hours by 2100, increasing the exposure of the coast to waves.

The marked rise in the occurrence of extreme sea levels is qualitatively similar for different sea-level rise scenarios, but the duration of extremes can differ substantially. For example, for the low end of the Vermeer and Rahmstorf (2009) sea-level projection (78 cm by 2100), extreme water heights (exceeding the 99.99th percentile) are predicted to occur more than 300 hours per decade by 2050 and more than 7,500 hours per decade by 2100.

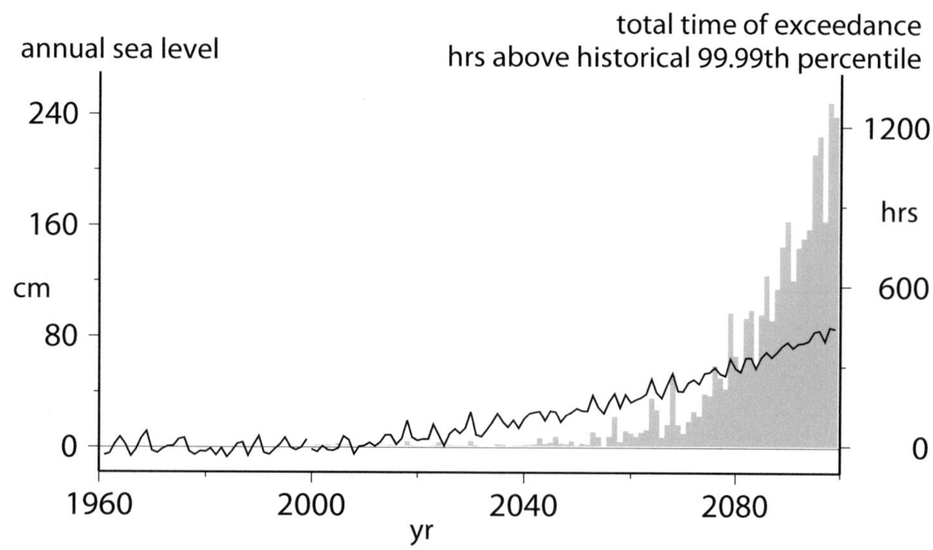

FIGURE 5.12 Projected number of hours (blue bars) of extremely high sea level off San Francisco under an assumed sea-level rise and climate change scenario. In this exercise, a sea-level event registers as an exceedance when San Francisco's projected sea level exceeds its recent (1970–2000) 99.99th percentile level, 1.4 m above historical mean sea level. In the recent historical period, sea level has exceeded this threshold about one time (1 hour) every 14 months. Sea-level rise (black line) during 1960–1999 was arbitrarily set to zero, then increased to the committee's projected level for the San Francisco area over the 21st century (92 cm). SOURCE: Adapted from Cloern et al. (2011).

Great Earthquakes Along the Cascadia Subduction Zone

Measurements of current deformation and geologic records (e.g., Savage et al., 1981; Atwater, 1987; Nelson et al., 1996; Atwater and Hemphill-Haley, 1997) establish the potential for great (magnitude greater than 8) megathrust earthquakes and catastrophic tsunamis along the Cascadia Subduction Zone. In Washington and Oregon, a great earthquake would cause some areas to immediately subside and sea level to suddenly rise perhaps by more than 1 m. This earthquake-induced rise in sea level would be in addition to the relative sea-level rise projected above. A great earthquake also would produce large postseismic vertical land motions in the area for years to decades.

Sudden subsidence during great earthquakes is revealed in the geological record as abrupt changes in sedimentary sequences (Nelson, 2007). When a great earthquake occurs, salt marsh or terrestrial soils are lowered into the intertidal zone, killing the vegetation (e.g., Figure 5.13). These peaty soils are quickly covered by tsunami-deposited sand or muddy tidal sediments. In the decades after an earthquake, the coast slowly rises, producing a gradual transition back to a salt marsh or terrestrial soil (e.g., Nelson et al., 1996; Leonard et al., 2010).

Cycles of buried peat-mud couplets beneath coastal marshes (Figure 5.14) suggest that 6 to 12 great earthquakes have occurred at irregular intervals ranging from a few hundred years to 1,000 years along the central Cascadia margin over the past 6,000 years (Long and Shennan, 1998). Geologic evidence also has been found for six great earthquakes along the northern Oregon coast in the past 3,000 years (Darienzo and Peterson, 1995), 11 or 12 great earthquakes in southern Oregon in the past 7,000 years (Kelsey et al., 2002; Witter et al., 2003), and seven great earthquakes in southwest Washington in the past 3,500 years (Atwater and Hemphill-Haley, 1997). Turbidite deposits identified in marine cores suggest that 18 great earthquakes ruptured at least the northern two-thirds of the Cascadia margin during the Holocene (Goldfinger et al., 2003, 2008).

The last great earthquake on the Cascadia megathrust occurred on January 26, 1700 (Satake et al., 1996, 2003). The date of the earthquake was determined by radiocarbon dating of suddenly buried marsh herbs, tree-ring records of trees stressed by coastal flooding

FIGURE 5.13 Ghost forests, such as this grove of weather-beaten cedar trunks near Copalis River, Washington, are evidence of sudden subsidence. SOURCE: Courtesy of Brian Atwater, U.S. Geological Survey.

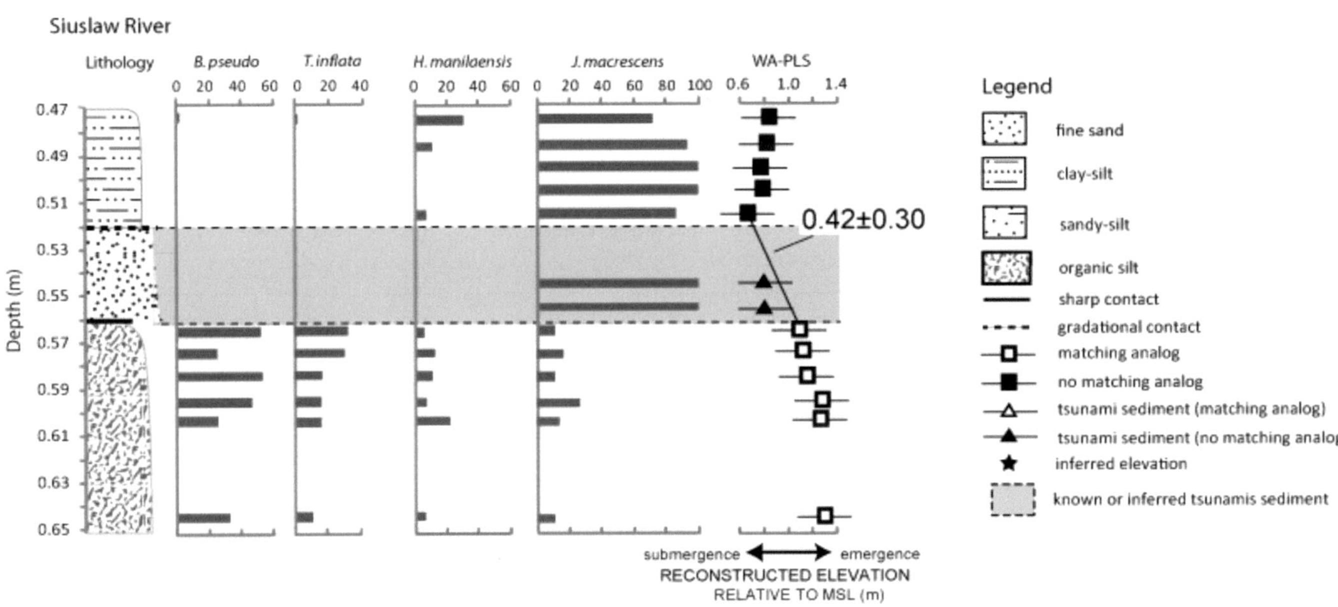

FIGURE 5.14 Stratigraphy and abundance of foraminifera in the sediment sequence recording the 1700 earthquake at Siuslaw River, Oregon. Also shown is a reconstruction of elevation during this interval (WA-PLS column). Sediment likely deposited by tsunamis is shaded in gray. SOURCE: Modified from Hawkes et al. (2011).

during subsidence (e.g., Yamaguchi et al., 1997), and Japanese historical records of a tsunami from a distant source. Modeling of the tsunami waveform (Satake et al., 1996) and estimates of coastal subsidence based on detailed microfossil studies (Hawkes et al., 2011) suggest an earthquake magnitude of 8.8 to 9.2. The coastal subsidence and associated sea-level rise were spatially variable, with the largest rise in sea level (1–2 m) occurring in northern Oregon and southern Washington, where the plate boundary forms a wide, shallow arch (Leonard et al., 2004, 2010; Hawkes et al., 2011). Other sections of the margin subsided <1 m and the southernmost part of the subduction zone was uplifted (Leonard et al., 2004, 2010; Hawkes et al., 2011).

Discussion

Changes in regional meteorological and climate patterns, including El Niños, coupled with rising sea level, are predicted to result in increasing extremes in sea levels. Models suggest that sea-level extremes will become more common by the end of the 21st century. Waves riding on these higher water levels will cause increased coastal damage and erosion—more than that expected by sea-level rise alone.

The biggest game changer for future sea level along the west coast of the United States is a great Cascadia earthquake. The related coastal subsidence of such an earthquake would, in a matter of minutes, produce significantly higher sea levels off the Cascadia coast than 100 years of climate-driven sea-level rise. A great earthquake could cause 1–2 m of sea-level rise in some areas, which is significantly higher than the committee's projection for Cascadia in 2100 (0.6 m). Further, the earthquake-induced sea-level rise would be an addition to the expected global warming-related sea-level rise.

CONCLUSIONS

Global projections are commonly made using ocean-atmosphere GCMs, which provide a reasonable representation of the steric contribution to global sea-level rise, but do not yet fully capture the cryospheric contribution. The IPCC (2007) projections made using this method are likely too low, even with an added ice dynamic component. Some studies project the cryospheric contribution by extrapolating current observations into the future, but the results depend on assumptions about the future behavior of the system. Semi-empirical methods avoid these difficulties by projecting global sea-level rise based on the observed relationship between sea-level change and global temperature. However, the highest projections made using this method (e.g., Grinsted et al., 2009) require unrealistically rapid acceleration of glaciological processes.

Given the strengths and weaknesses of the different projection approaches and the resource constraints of an NRC study, the committee chose to use GCMs developed for the IPCC Fourth Assessment Report to estimate the steric contribution and extrapolation techniques to estimate the cryospheric contribution. The contributions were then summed. The land hydrology component was assumed to be near zero and was not factored into the projection. The committee's global projections for 2100 are substantially higher than the IPCC's (2007) projection, mainly because of a faster growing cryosphere component, and are somewhat lower than the Vermeer and Rahmstorf (2009) projections. The committee estimates that global sea level will rise 8–23 cm by 2030, 18–48 cm by 2050, and 50–140 cm by 2100, relative to 2000 levels. As the projection horizon lengthens, the uncertainties grow, and hence the ranges widen. The major sources of uncertainty in the global projection are related to assumptions about the increase in rapid ice dynamics and the growth of future greenhouse gas emissions.

Formal projections of future sea-level rise along the west coast of the United States have not been made, although a few studies have presented ranges of possible outcomes for California and Washington. Methods vary but usually involve a combination of global models and local information. The committee's projections account for factors that affect sea level in the area, including local steric variations; wind-driven differences in ocean heights; the gravitational and deformational effects associated with melting of Alaska, Greenland, and Antarctic glaciers; and vertical land motions along the coast. The local steric and wind-driven components were estimated by extracting northeast Pacific data from the same ocean models used for the global projections. The cryosphere component was adjusted for gravitational and deformational effects and then extrapolated forward. Vertical land motion was estimated using continuous GPS measurements.

The projected values vary by latitude, with the highest sea levels expected off the coast south of Cape Mendocino (4–30 cm for 2030, 12–61 cm for 2050, and 42–167 cm for 2100, relative to 2000) and the lowest sea levels expected off the coast north of Cape Mendocino (-4–23 cm for 2030, -3–48 cm for 2050, and 10–143 cm for 2100). The lower sea levels projected for Washington, Oregon, and northernmost California reflect coastal uplift and gravitational and deformational effects, which lower the relative sea level. Major sources of uncertainties in the regional projections are related to assumptions about the rate of future ice losses and the constant rate of vertical land motion over the projection period. Uncertainties are larger for the regional projections than for the global projections because more components are considered and because uncertainties in the steric and ocean dynamic components are larger at a regional scale than at a global scale.

Extreme events can raise sea level much faster than the rates projected by the committee. For example, unusually high sea levels may occur temporarily when major storms coincide with high astronomical tides, and especially during years when regional sea levels are anomalously heightened during El Niño events. As mean sea level continues to rise, the number of extreme high water events and their duration are expected to increase. A simulation based on predicted tides, projected weather and El Niño conditions under a mid-range greenhouse gas emission scenario, and the committee's projections of sea-level rise suggests that the incidence of extreme water heights in the San Francisco Bay area would increase from about 9 hours per decade for the recent historical period (1961–1999) to hundreds of hours per decade by 2050 and several thousand hours per decade by 2100. In addition, the duration of these extremes would lengthen from 1 or 2 hours in the historical period to about 6 hours by 2100.

The biggest game changer for future sea-level rise along the U.S. west coast would be a great earthquake (magnitude greater than 8) along the Cascadia Subduction Zone. Such earthquakes have occurred every several hundred to 1,000 years, with the most recent occurring in 1700. During a great earthquake, some land areas would immediately subside and relative sea level would suddenly rise, perhaps by 1 m or more. This earthquake-induced rise in sea level would be added to the projected rise in relative sea level (about 60 cm by 2100).

6

Responses of the Natural Shoreline to Sea-Level Rise

Sea-level rise affects the natural shoreline in several ways. Higher water levels erode beaches, dunes, and cliffs; inundate wetlands and other low-lying areas; and increase the salinity of estuarine systems, displacing existing coastal plant and animal communities. These coastal environments provide a protective buffer to areas further inland, as wetlands can reduce flooding and cliffs, beaches, and dunes protect coastal property from storm waves.

The distribution and character of coastal habitats and geomorphic environments varies along the California, Oregon, and Washington coasts, as does their response to sea-level rise. The coast of California is dominated by uplifted terraces fronted by low cliffs, but also includes steep coastal mountains and areas of coastal lowlands and dunes. Oregon's coast is similar and is characterized by rugged volcanic headlands separating areas of uplifted marine terraces and river mouth estuaries, dunes, and beaches. The southern coast of Washington is dominated by low relief sand spits, occasionally backed by bays. The northern coast and Olympic Peninsula are rocky and rugged, whereas Puget Sound retains the signature of Ice Age glaciation—a crenulated coastline with islands, embayments, and typically sandy bluffs.

This chapter summarizes what is known about (1) the responses of coastal habitats and geomorphic environments—including coastal cliffs and bluffs, beaches, dunes, estuaries, and marshes—to future sea-level rise and storminess along the west coast of the United States and (2) the role of coastal habitats (including benthic habitats), natural environments, and restored tidal wetlands in providing protection from future inundation and the impact of waves. The objective was to summarize existing knowledge, not to predict specific future shoreline responses or to assess coastal impacts of sea-level rise and storminess (see Box 1.1).

COASTAL CLIFFS AND BLUFFS

Cliffs and bluffs are dominant features along the west coast of the United States, and they have been retreating for thousands of years. The rate of coastal cliff and bluff retreat is controlled by the properties of the rock materials and the physical forces acting on the cliffs. Important rock properties include the hardness or degree of consolidation or cementation, the presence of internal weaknesses (e.g., fractures, joints, faults), and the degree of weathering. Rates of cliff retreat are generally well documented along the California coast (Dare, 2005; Hapke and Reid, 2007), and range from a few cm per year in granitic or volcanic rock to tens of cm per year or more in sedimentary rocks or unconsolidated materials (Griggs, 1994; Griggs et al., 2005). Moore et al. (1999) found cliff and bluff erosion rates of 2–20 cm yr^{-1} for 1932–1994 in San Diego County, and 6–14 cm yr^{-1} for 1953–1994 in Santa Cruz County. In California, cliffs and bluffs made of sedimentary rocks typically erode at rates of 15–30 cm yr^{-1} (Griggs and Patsch, 2004).

Fewer bluff retreat rates are available for the Oregon and Washington coasts. Komar and Shih (1991) and Komar (1997) described the temporal and spatial variability in cliff and bluff erosion along the Oregon coast, noting that cliff erosion is slower where uplift rates are highest and the base of the cliff has been raised to an elevation seldom reached by wave runup. Priest (1999) found that cliffs and bluffs in Lincoln County, Oregon, generally retreated at rates less than 19 cm yr^{-1} for 1939–1991. In landslide areas, bluff retreat rates were somewhat higher, ranging from 11–50 cm yr^{-1}.

The physical forces driving cliff and bluff erosion include marine processes—primarily wave energy and impact, but also tidal range or sea-level variations—and terrestrial processes, such as rainfall and runoff, groundwater seepage, and mass movements such as landslides and rockfalls. As discussed in Chapter 4 ("Short-Term Sea-Level Rise, Storm Surges, and Surface Waves"), waves may be getting higher (e.g., Figure 6.1). Increased wave heights mean that more wave energy is available to erode the coastline. Rising sea level would exacerbate this effect because waves will break closer to the coastline and will reach the base of the cliff or bluff more frequently, thereby increasing the rate of cliff retreat.

Cliff and bluff retreat is an episodic process whereby large blocks fail suddenly under conditions of heavy rainfall, large waves at times of elevated sea levels or high tides, or earthquakes, followed by periods of little or no failure. In steep, mountainous areas, failure is often through large landslides or rock falls (Figure 6.2), usually driven by excess or prolonged rainfall during the winter months. With very large landslides, such as the Portuguese Bend slide on the Palos Verdes Peninsula, the shoreline may actually be extended seaward for a decade or more before basal wave action removes the protrusion (Orme, 1991). The episodic nature of cliff retreat, combined with the frequent absence of an identifiable edge or reference feature, makes it difficult to quantify or verify cliff erosion rates in mountainous areas over short time intervals, such as a few decades, or to project future erosion rates (Priest, 1999).

FIGURE 6.1 Boiler Bay, Oregon. Some evidence suggests that waves have been increasing in height off the west coast. SOURCE: Courtesy of Erica Harris, Oregon State University.

FIGURE 6.2 Large-scale landsliding along the Humboldt County, California, coast at Centerville. SOURCE: Copyright 2002–2012 Kenneth & Gabrielle Adelman, California Coastal Records Project, <www.Californiacoastline.org>.

Cliff and bluff erosion is not reversible. The most common human response has been to armor the cliff base with rock revetments (Figure 6.3) or seawalls (Figure 6.4). Ten percent of the California coastline has now been armored, including 33 percent of the coastline of the four most developed southern California counties (Ventura, Los Angeles, Orange, and San Diego; Griggs, 1999). Shoreline armoring also has increased over the past several decades in Oregon and Washington. Approximately one-third of the Puget Sound shoreline is now armored (Shipman et al., 2010). Despite this protection, coastal storm damage has increased over the past several decades because of intense development and the occurrence of a number of severe El Niño events, raising questions about the long-term efficacy of existing coastal protection structures (Griggs, 2005; Shipman et al., 2010). Moreover, while seawalls and revetments may provide current protection for oceanfront development and infrastructure, they are usually designed for a particular set of wave and sea-level conditions. If sea level increases substantially and wave heights continue to increase, the original freeboard will be gradually exceeded and overtopping will become more frequent.

BEACHES

Beaches respond quickly to the forces acting on them as waves and littoral currents easily move the sand. Along the west coast, beaches change seasonally in response to the different winter and summer wave climates. These fluctuations in beach width are predictable and temporary, and the losses of sand experienced each winter are normally recovered the following summer. Longer-term fluctuations in beach widths associated with the El Niño-Southern Oscillation and the Pacific Decadal Oscillation (PDO) also have been documented in southern California (Orme et al., 2011).

FIGURE 6.3 Erosion of poorly consolidated sedimentary cliffs at Pacifica, south of San Francisco, is threatening these apartments, and residents have had to move out. Riprap protection has been placed at the toe of the bluff in an attempt to slow the erosion. SOURCE: Hawkeye Photography.

FIGURE 6.4 Seawalls and revetments fronting coastal cliffs and bluffs in California and Oregon. (*Left*) Concrete and timber seawalls protecting cliff top homes in Solana Beach, California. SOURCE: Copyright 2002–2012 Kenneth & Gabrielle Adelman, California Coastal Records Project, <www.Californiacoastline.org>. (*Right*) Rip rap protecting bluff top housing along the central Oregon coast. SOURCE: Courtesy of Gary Griggs, University of California, Santa Cruz.

Periodic El Niño events both enhance storm wave activity, leading to severe beach erosion, and increase rainfall and runoff, increasing sand delivery to the shoreline and thus sometimes leading to wider beaches in subsequent months. More frequent storms during warmer PDO cycles can lead to extended periods when beach widths are narrower than average. Over the long term, rising sea level will cause landward migration or retreat of beaches. The retreat is caused partly by inundation of the beach by the rising sea and partly by offshore transport of sand to maintain the beach profile. Because the berm or back beach is essentially a horizontal surface, even a small rise in sea level may lead to a horizontal retreat that is considerably larger than the sea-level rise (Edelman, 1972).

Beaches also can undergo erosion or long-term retreat in response to a reduction of sand supply. Coastal rivers and streams—many of which have been dammed for water supply, flood control, hydroelectric power, or recreation—provide most beach sand along the west coast. Willis and Griggs (2003) determined that more than 500 dams have reduced the average annual sand and gravel flux to California's coastal watersheds by 25 percent. Sherman et al. (2002) calculated that 28 dams and more than 150 debris basins in the watersheds of eight major rivers in southern California have impounded more than 4 million m^3 yr^{-1} of sand. Statewide, approximately 152 million m^3 of sand that would have been delivered to the shoreline to nourish beaches since 1885 has been trapped by coastal dams (Slagel and Griggs, 2008). The long-term effect of declining sand supply works in concert with rising sea level to progressively narrow beaches.

Barrier spits or other sandy peninsulas, which are common along the northern Oregon and southern Washington coastlines (Figure 6.5), will tend to erode

FIGURE 6.5 Oregon's Cape Lookout State Park on Netarts Spit, which is backed by Netarts Bay. Long sand spits commonly form at the mouths of estuaries along the central and northern Oregon coasts and Washington coast. SOURCE: Courtesy of Erica Harris, Oregon State University.

or migrate under elevated sea levels and large storm waves. Erosion or landward migration of sand spits or barrier bars will occur more frequently with sea-level rise (Pilkey and Davis, 1987).

Back-beach barriers can slow or halt the natural inland migration of beaches because of rising sea level. Where a seawall, revetment, or structure exists, the shoreline cannot advance landward and the beach is progressively inundated (Figure 6.6). This process, known as coastal squeeze or passive erosion, has been documented in a number of locations along the west coast. Similarly, barrier spits that have been developed and then protected with revetments cannot migrate with sea-level rise (Figure 6.7). Depending on the rate of sea-level rise, all west coast beaches with hardened or constrained back beach edges will gradually be inundated.

Only a few studies have quantified rates of change along the sandy shoreline of the U.S. west coast. Kaminsky et al. (1999) found widely varying rates of change for the sandy shoreline of Pacific County, Washington, ranging from +0.8 to +14.2 m yr^{-1} for 1870–1926, -13.6 to +8.8 m yr^{-1} for 1926–1950, and -7.0 to +4.2 m yr^{-1} for 1950–1995. Sand spits eroded or accreted, depending on sand supply, wave energy, and relative sea level. Coastal land change along the sandy shoreline of California was assessed as part of the U.S. Geological Survey's National Assessment of Shoreline Change program (Hapke et al., 2006). Maps, aerial photographs, and, more recently, lidar (light detection and ranging) were used to determine both long-term (1800s to 1998–2002) and short-term (1950s–1970s to 1998–2002) rates of shoreline or beach change. More than 16,000 transects revealed that the shoreline eroded 0.2 ± 0.4 m yr^{-1} over the short term. The average rate of long-term change was 0.2 ± 0.1 m yr^{-1}, an accretional trend, although 40 percent of the transects showed net erosion. This net accretional trend was attributed to the large volumes of sediment that were added to the system from large rivers and to the impact of coastal engineering and beach nourishment projects (Hapke et al., 2006). A similar assessment effort is planned for the Oregon and Washington coasts.

COASTAL DUNES

Cooper (1958, 1967) mapped and described the coastal dunes of Washington, Oregon, and California, and found that extensive coastal sand dunes accumulate when the following conditions are met: (1) a large supply of fine-grained sand, (2) a barrier such as a headland to trap littoral drift and accumulate sand, (3) a low-relief area landward of the beach where sand can accumulate, and (4) a dominant or persistent onshore wind. Large dune fields are best preserved in areas that have undergone either net subsidence or limited uplift during the Quaternary (Orme, 1992). Dunes back about 45 percent of the Oregon coast and 31 percent of the

FIGURE 6.6 (*Left*) Passive erosion in front of a revetment, illustrating the loss of beach where the structure restricts the shoreline from migrating landward. The beach continues to migrate inland on either side of the revetment. (*Right*) Recovery of the beach following removal of the revetment and bluff top structure. SOURCE: Copyright 2002–2012 Kenneth & Gabrielle Adelman, California Coastal Records Project, <www.Californiacoastline.org>.

FIGURE 6.7 Developed sand spit at Stinson Beach in Marin County, California, where a revetment has been constructed in an effort to protect the homes. This spit cannot migrate with sea-level rise. SOURCE: Copyright 2002–2012 Kenneth & Gabrielle Adelman, California Coastal Records Project, <www.Californiacoastline.org>.

Washington coast (Komar, 1997). Many of the dune areas exposed along and inland from the west coast shoreline today formed during the lower sea levels of the past. At the end of the last ice age, when sea levels were about 120 m lower than today, the entire continental shelf was exposed. Sand from rivers and streams was deposited across this extensive plain, and onshore winds produced large dune fields, such as those in the Coos Bay area of central Oregon, which extend along the coast for nearly 240 km and are encroaching into some developed areas (Figure 6.8; Komar, 1997). As sea level rose, many of the dunes were cut off from their vast reservoir of offshore sand. Dunes still form and are active today along the shorelines of all three states, but they have a lower supply of sediment and are much less extensive than those that formed in the past.

Decades of observations of coastal dunes around the world have shown that the frontal dune, which is closest to the beach, is an ephemeral and unstable feature (e.g., McHarg, 1969). Sand dunes typically accrete or expand under the force of onshore winds and an ample supply of sand, but they can erode quickly under severe wave attack at times of high tide or elevated sea level. The hazards of building on the frontal dune have been known for centuries (McHarg, 1969). Nevertheless, many housing developments in California, Oregon, and Washington have been constructed on dunes and are periodically threatened or damaged (Figure 6.9). Dunes, whether modern or Pleistocene, can be expected to retreat quickly under rising sea levels and larger waves.

RETREAT OF CLIFFS AND BEACHES UNDER SEA-LEVEL RISE

Coastlines have been retreating globally since sea level began rising at the end of the last ice age, ap-

FIGURE 6.8 Dunes along the central Oregon coast at Florence are encroaching into development. SOURCE: Courtesy of Phoebe Zarnetske, Oregon State University.

FIGURE 6.9 Construction of private homes on the frontal dunes. (*Left*) Homes in central Monterey Bay were threatened by erosion during the high tides, elevated sea levels, and large storm wave of the 1983 El Niño. SOURCE: Courtesy of Gary Griggs, University of California, Santa Cruz. (*Right*) Placement of riprap during storm conditions to protect development on dunes in Neskowin, Oregon. SOURCE: Courtesy of Armand Thibault.

proximately 21,000 years ago. At that time, the western shoreline of North America was located at the edge of the continental shelf (Shepard, 1963; Nummedal et al., 1987), which for Oregon and Washington is typically 25–50 km offshore (Komar, 1997). Off the California coast, the shelf width varies, averaging 15–30 km, but narrowing to 5 km or less off Big Sur and parts of southern California, and widening to 40 km off San Francisco (Figure 6.10). The average rate of coastline retreat over the post-glacial period of sea-level rise can be estimated by dividing the width of the continental shelf at a specific location by 21,000 years. For example, a shelf width of 5 km corresponds to an average retreat rate of 23.8 cm yr^{-1}, and a 40 km wide shelf corresponds to an average rate of 190 cm yr^{-1}. Of course, the actual rate at any given time and place may be significantly higher or lower, depending on variations in the rate of sea-level rise over the 21,000-year period as well as geographic variations in coastal geology, regional wave climate, offshore bathymetry, and the degree of coastal armoring.

Few studies have projected future shoreline and sea cliff retreat rates under rising sea level. For example, a Federal Emergency Management Administration-sponsored effort to assess future coastal erosion hazards (Crowell et al., 1999) simply projected historic erosion rates without considering changes in rates of sea-level rise or wave climate. Where data are available, projections for future coastal retreat could be made by extrapolating existing erosion trends (e.g., Box 6.1) and adding an appropriate safety factor to accommodate expected future sea-level rise and potential increases in storm wave heights. Because projected rates of sea-level rise are moderate in the near term (Chapter 5), extrapolation of current erosion rates is likely reasonable to at least 2030.

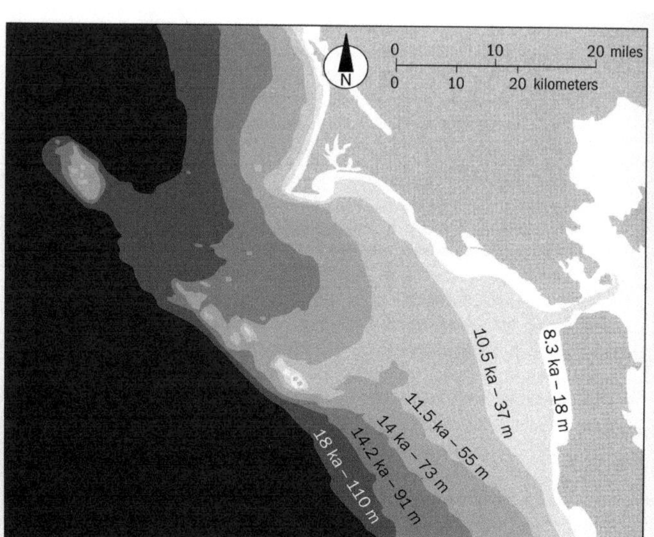

FIGURE 6.10 Sea-level rise has moved the San Francisco shoreline eastward by about 40 km since the last Ice Age ended. SOURCE: Griggs (2010).

An alternative approach to projections, developed by PWA (2009), relates rates of shoreline change to the coastal geology, then applies changes in total water level at the shoreline in exceedance of the elevation of the base of the bluff or cliff to predict erosion (Figure 6.11). Based on this approach, the central and northern California coast is projected to lose 81 km^2 of land by 2100 relative to 2000 for 1 m of sea-level rise and 99 km^2 of land for 1.4 m of sea-level rise (Table 6.1; Heberger et al., 2009; PWA, 2009). Due to their differing resistance to erosion, dunes and cliffs will respond differently to rising sea levels. Under the scenario of 1.4 m of sea-level rise by 2100, Revell et al. (2011) predicted that cliffs would erode an average distance of 33–60 m, depending on assumptions about geologic variability, and that dunes would erode an average distance of 170 m in the 11 counties studied. However, projected land losses vary significantly within each county and along the coast. In Del Norte County, for example, the average distance cliffs are projected to erode is 85 m by 2100 and the maximum distance is 400 m (Revell et al., 2011). The variability in how far cliffs are expected to erode under sea-level rise is illustrated in Figure 6.12. Such uncertainties in land losses, combined with uncertainties in exactly how sandy shorelines with back beach barriers or armor will respond to sea-level rise and with uncertainties in rates of future sea-level rise, make precise projections of future beach retreat or erosion in these areas problematic.

Wave Energy and Coastal Erosion

Wave-induced cliff and shoreline erosion is a significant problem along the west coast of the United States, and an increase in wave energy will only increase the rates of retreat. The amount of wave energy expended at any position on the coast is determined by the effects of wave height, tidal elevation or sea level,

BOX 6.1
Technology, Tools, and Resources for Evaluating Sea-Level Rise and Coastal Change

Most historic assessments of coastal change have relied on stereo vertical aerial photographs, which can be used to measure coastal erosion or retreat over time. However, most vertical photographs are in university libraries or must be obtained at considerable cost and time from aerial photographic companies or state or federal agencies. California has an online resource of oblique aerial photographs[a] as well as a selection of vertical photos. Ken and Gabrielle Adelman began flying and photographing the entire coast of California in 2002 and have rephotographed the coastline in 2004, 2005, 2006, 2008, and 2010. Three additional sets of oblique color slides taken in 1972, 1979, and 1987 by state agencies and some vertical aerial photographs have been scanned and added to the site. More than 90,000 high-resolution color photographs, covering every kilometer of the California coast, are available from the website. Using a time comparison option on the site allows users to immediately access photographs spanning 40 years of coastal change in California.

Lidar systems use a laser to precisely measure ground surface elevations or topography. Airborne scanning lidar can be used to estimate elevation every few square meters over tens to hundreds of kilometers of coast, allowing precise assessments of the spatial variability of beach and sea cliff changes (Sallenger et al., 2002). The first lidar topographic survey of the California coast was flown in October 1997 as a large El Niño event was approaching the west coast. A second survey of the same areas was flown in April 1998, after sea levels and storm waves returned to their normal state. The two surveys provided the first accurate comparison of the coastline before and after a severe event, and documented how much erosion or beach scour occurred (Figure).[b]

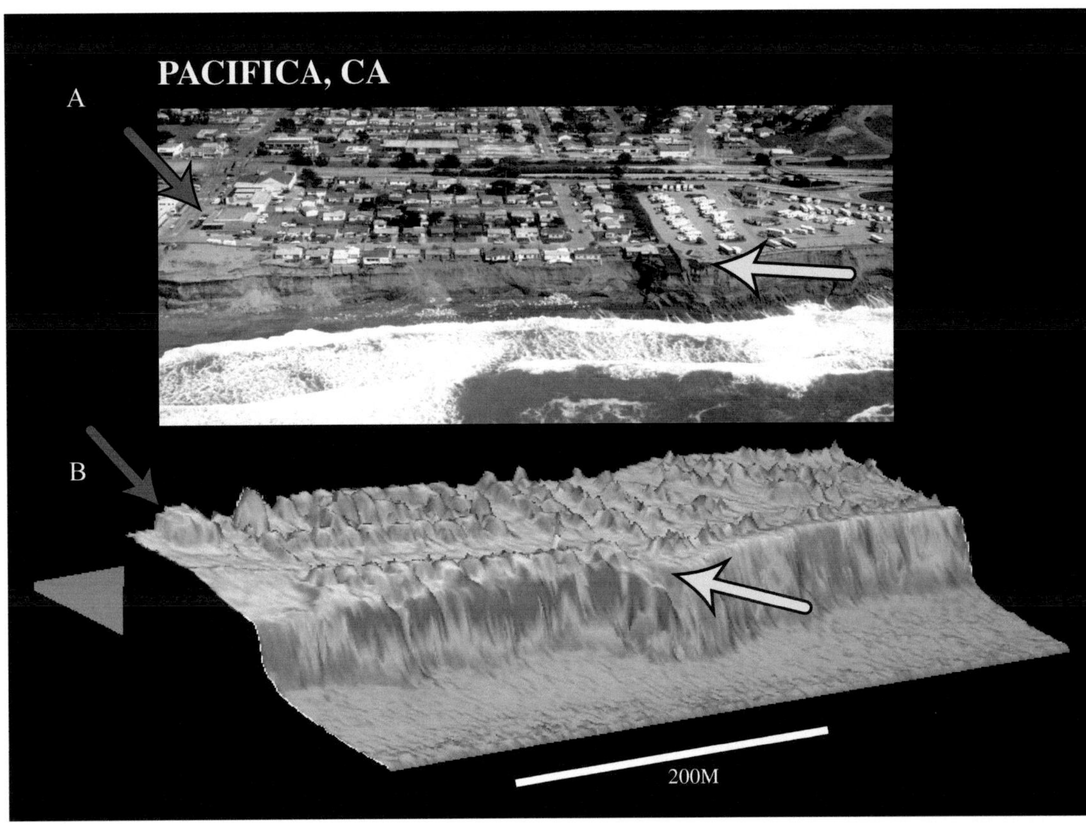

FIGURE (*A*) Photograph of the Pacifica region where extensive sea-cliff erosion occurred during the El Niño winter showing threatened houses at the top of the cliff. (*B*) Three-dimensional view using lidar data acquired prior to the El Niño winter of the area shown in (A). Note that the buildings are clearly shown. Superimposed on the topography is vertical change with warm (red) colors indicating loss over the El Niño winter. SOURCE: Sallenger et al. (2002).

[a] See <http://www.Californiacoastline.org>.
[b] See <http://coastal.er.usgs.gov/lidar/AGU_fall98/>.

FIGURE 6.11 Example of projected sea-level rise hazard zones, defined as the historic erosion rate times the percent increase in total water level, in map view. SOURCE: PWA (2009).

TABLE 6.1 Projected Land Loss for 11 Central and Northern California Counties Under 1.0 m and 1.4 m of Sea-Level Rise

Year	Cliff Land Loss[a] (km^2)	Dune Land Loss[b] (km^2)	Total Land Loss (km^2)
2025	5	21	27
2050	21	22–25	43–46
2100	53–61	27–38	81–99

SOURCE: Adapted from PWA (2009).
NOTE: Low end of the range is for 1.0 m of sea-level rise, and the high end of the range is for 1.4 m of sea-level rise.
[a] Includes 2 standard deviations of the historic shoreline change rates.
[b] Includes erosion associated with a 100-year storm event.

offshore and beach profile/slope, and beach width/height. Combined, these factors may significantly influence wave run-up and thus exert a major control on the hydraulic forces applied to the cliff, bluff, dune, or beach face (Benumof and Griggs, 1999). Conventional wisdom is that waves are the primary agent for seacliff erosion at the base of the cliff (Sunamura, 1992; Shih and Komar, 1994). Large storm waves occurring during high tide or times of elevated sea level are particularly effective in causing basal cliff erosion (Lee et al., 1976; Kuhn and Shepard, 1984; Griggs and Trenhaile, 1994; Benumof and Griggs, 1999). Any significant increase in wave heights or the amount of wave energy reaching the cliff will, therefore, lead to an increase in the erosive forces and the erosion rate.

A detailed investigation of cliff erosion in San Diego County, California, found significant variation in the rate of erosion, as well as in intrinsic proper-

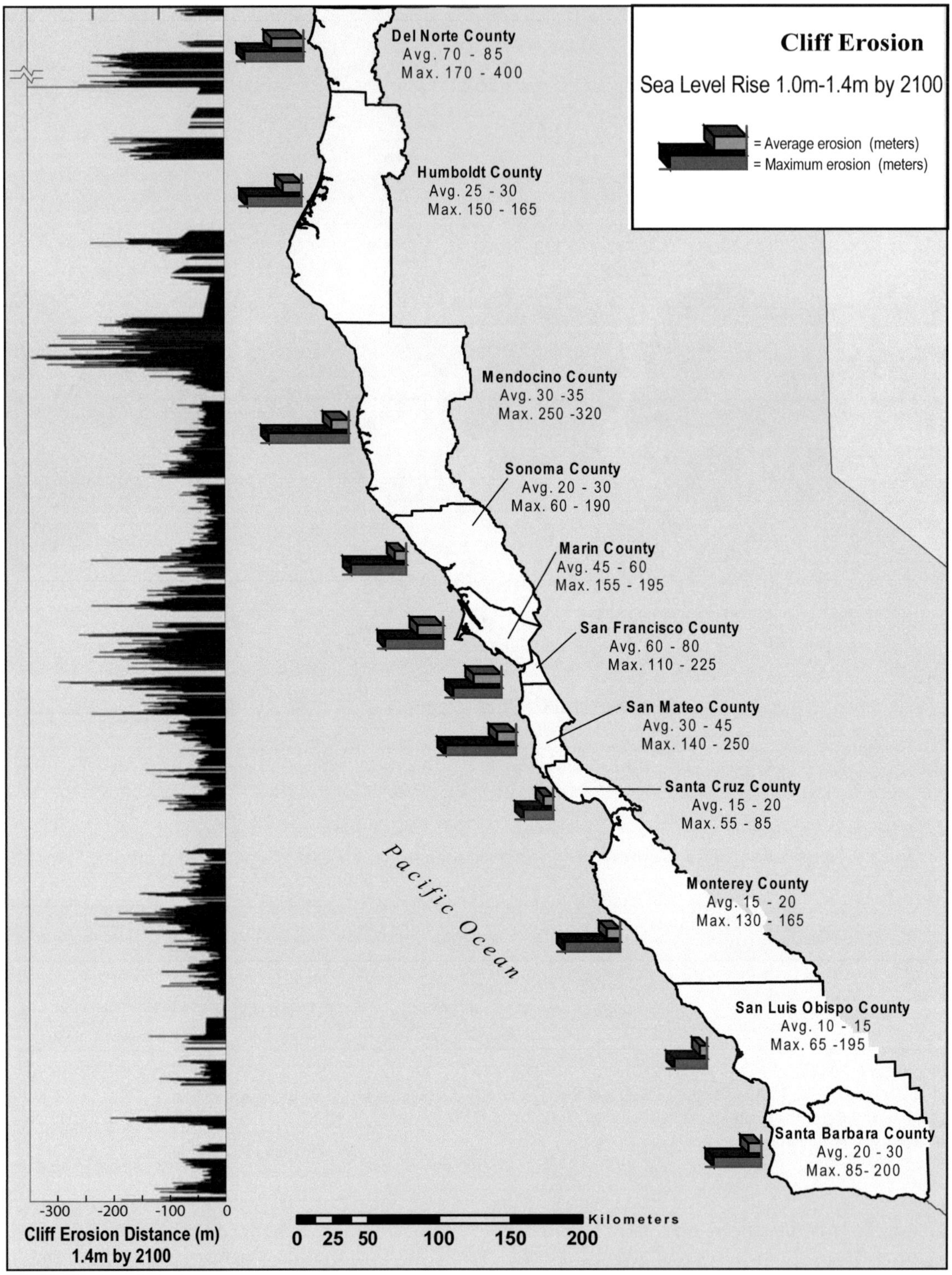

FIGURE 6.12 Variability in the distances coastal cliffs along the central and northern California coast are projected to erode under sea-level rise of 1.0 and 1.4 m by 2100 relative to 2000. Bar charts along the coast show average and maximum erosion distances between the 1.0 and 1.4 m sea-level rise scenarios. Chart on the left shows the projected erosion distance for each 500-m block for 1.4 m of sea-level rise. SOURCE: Revell et al. (2011).

ties of the cliff materials (e.g., lithology, structural weaknesses, rock strength, weathering) and extrinsic factors impacting the cliffs (e.g., wave energy, offshore bathymetry, rainfall). Although waves are the primary cause of seacliff erosion, the physical properties of the cliff materials in the San Diego study area strongly affect the erosion rates (Benumof and Griggs, 1999; Benumof et al., 2000).

Coastal Hazard Assessments

The U.S. Geological Survey developed an index of coastal vulnerability to sea-level rise in 2000 (Thieler and Hammar-Klose, 2000). The relative vulnerabilities of different coastal environments along the U.S. west coast to long-term sea-level rise were quantified based on variables including coastal geomorphology, regional coastal slope, rate of sea-level rise, wave and tide characteristics, and historical shoreline change rates. The rankings for each of the six variables at any particular location can be averaged to produce an overall coastal vulnerability index from 1 (very low) to 5 (very high; Table 6.2). This index provides a broad overview of how different regions of the west coast are likely to change in response to sea-level rise. Two specific regions (southwestern Washington/northwestern Oregon and San Francisco to Monterey, California) are covered in more detail, with maps delineating the distribution of various risk factors and an overall ranking of risk (Figure 6.13).

Living with the Changing California Coast (Griggs et al., 2005) provides a different approach for assessing coastal hazards in California and includes mile-by-mile maps of the entire coastline. Information on the maps include shoreline environment, erosion rates where published or known, presence and type of armoring, notes or comments on individual coastal areas and specific issues or problems, and a hazard ranking ranging from stable/low risk to hazard/high risk. An example is shown in Figure 6.14.

The high spatial variability portrayed in these maps underscores the difficulty of generalizing the response of coastal cliffs and bluffs, beaches, and dune to sea-level rise along the west coast of the United States.

ESTUARIES AND TIDAL MARSHES

Estuaries and tidal marshes are valuable ecosystems, providing a variety of services as well as the economic livelihoods of many communities (Mitsch and Gosselink, 2000; MA, 2005). Open waters, mudflats, and marshes offer refuge and forage for wildlife, fishes, and invertebrates. Shallow ponds and seed-producing vegetation provide overwintering habitat for millions of migratory waterfowl. Wetlands help absorb nutrients and reduce loading to the coastal ocean. They also help protect local communities from flooding, either by storing riverine floodwaters or by damping storm surges from the ocean.

Estuaries are bodies of water formed at the coastline where fresh water from rivers and streams flows into the ocean. The largest estuaries along the west coast of the United States include Puget Sound, the Columbia River Estuary, and the San Francisco Bay-Delta. Tidal marshes—herbaceous wetlands frequently or continu-

TABLE 6.2 Ranking of Variables Determining the Coastal Vulnerability Index

	Ranking of Coastal Vulnerability Index				
	Very low	Low	Moderate	High	Very high
Variable	1	2	3	4	5
Geomorphology	Rocky, cliffed coasts; fiords; fiards	Medium cliffs, indented coasts	Low cliffs, glacial drift, alluvial plains	Cobble beaches, estuary, lagoon	Barrier beaches, sand beaches, salt marsh, mudflats, deltas, mangrove, coral reefs
Coastal slope	> 1.9	1.3–1.9	0.9–1.3	0.6–0.9	< 0.6
Relative sea-level change (mm yr^{-1})	< -1.21	-1.21–0.1	0.1–1.24	1.24–1.36	> 1.36
Shoreline erosion or accretion (m yr^{-1})	> 2.0 Accretion	1.0–2.0	-1.0–1.0 Stable	-1.1– -2.0	< -2.0 Erosion
Mean tide range (m)	> 6.0	4.1–6.0	2.0–4.0	1.0–1.9	< 1.0
Mean wave height (m)	< 1.1	1.1–2.0	2.0–2.25	2.25–2.60	> 2.60

SOURCE: Thieler and Hammar-Klose (2000).

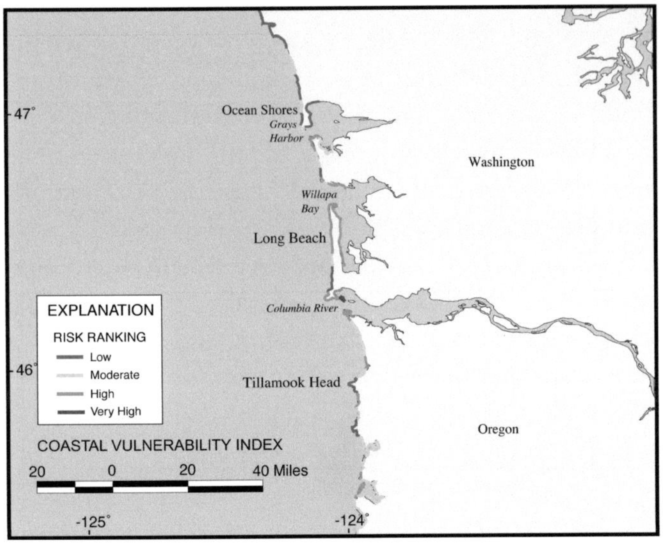

FIGURE 6.13 Coastal vulnerability index for southwestern Washington and northwestern Oregon. SOURCE: Thieler and Hammar-Klose (2000).

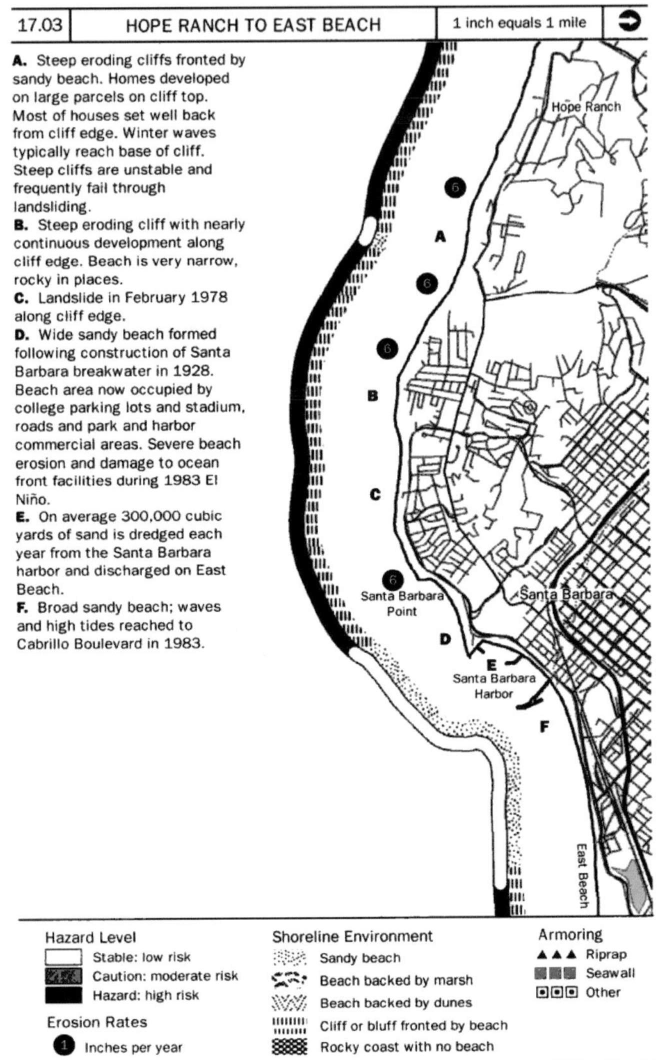

FIGURE 6.14 Map of portion of the Santa Barbara County coastline, delineating specific characteristics and an overall hazard rating. SOURCE: Griggs et al. (2005).

ally inundated with fresh, brackish, or saline water—are found within estuarine embayments or along protected coastlines (Figure 6.15). In California, extensive tidal marshes occur in Elkhorn Slough off Monterey Bay, San Francisco Bay, and the Sacramento-San Joaquin Delta, although smaller areas of marsh exist along the coast from San Diego to Humbolt Bay. In the Pacific Northwest, tidal marshes are common along the margins of rivers that flow directly into the ocean, such as the Salmon and Columbia rivers, and within bar-built estuaries such as the Tillamook Estuary in Oregon and Willapa Bay in Washington (Seliskar and Gallagher, 1983). Extensive tidal marshes also existed historically within the deltas of major rivers flowing into Puget Sound, such as the Nisqually and Skagit rivers.

Estuaries comprise subtidal (permanently flooded) areas, intertidal flats (unvegetated area regularly exposed by falling tides), and vegetated marshes (Figure 6.16). The transition between these environments depends on the interaction of tides with the local topography on timescales ranging from weeks to millennia (Figure 6.17). Changes in relative sea level may change the tidal dynamics within the estuary, including the tidal range. Changes in tidal dynamics affect saltwater penetration, the duration of flooding or exposure of intertidal flats and marshes, and the depth of flooding, which in turn influences wave activity, the potential for erosion, and a host of biological processes.

Vegetation plays a critical role in determining both the character of estuarine environments and their response to sea-level rise. The spread of emergent vegetation provides an effective trap for suspended sediment, stabilizing intertidal flats (e.g., Steers, 1948). The colonization of intertidal flats by vegetation depends on elevation (Williams and Orr, 2002), soil drainage, local dispersal mechanisms (e.g., Wolters et al., 2005), and/or exposure times, which determines whether propagules will survive or seeds will germinate. These conditions can change rapidly. Ward et al. (2003) noted

FIGURE 6.15 (*Left*) Marshes occur at the margins of river estuaries along parts of the west coast with steep gradients close to the ocean, such as the Alsea River Estuary near Waldport, Oregon. SOURCE: Courtesy of Laura Brophy, Green Point Consulting. (*Right*). More extensive coastal marshes occur in deltaic areas like the Skagit River Estuary, shown here near Milltown, Washington. Plant communities can vary with distance from the main channel as well as along the salinity gradient from river to sea. SOURCE: Courtesy of Greg Hood, Skagit River System Cooperative, <http://pers-erf.org/Gallery/>.

FIGURE 6.16 (*Left*) Tidal fluctuations periodically expose mudflats, such as this tidal channel and brackish marsh in the Umpqua River Estuary in Oregon. In areas of the estuary that are relatively remote from ocean influences, brackish water supports diverse vegetative communities. (*Right*) Low marsh in Alsea River Estuary, Oregon, showing that even small areas of tidal marsh can support well-developed tidal creek systems. SOURCE: Courtesy of Laura Brophy, Green Point Consulting.

that a single storm created the soil and elevation conditions necessary to allow the native *Spartina foliosa* to colonize a mudflat in the Tijuana Estuary in southern California.

Intertidal flats give way to marshes when the land surface reaches an elevation that supports salt- and/or flood-tolerant emergent vegetation (Pestrong, 1965). Important west coast marsh species in high-salinity habitats include *Spartina foliosa* (from Bodega Bay south), *Spartina densiflora*, *Salicornia virginica*, *Scirpus* spp., *Distichlis* spp., and *Jaumea* spp. Dominant species in low-salinity habitats include *Carex lyngbyei*, *Scirpus californicus*, *Juncus balticus*, *Potentilla pacifica*, and *Typha* spp. (Seliskar and Gallagher, 1983; Barnhart et al., 1992). The transition from intertidal flats to marshes is especially sensitive to changes in sea level. If tolerance limits of the vegetation are exceeded, abrupt transitions could occur.

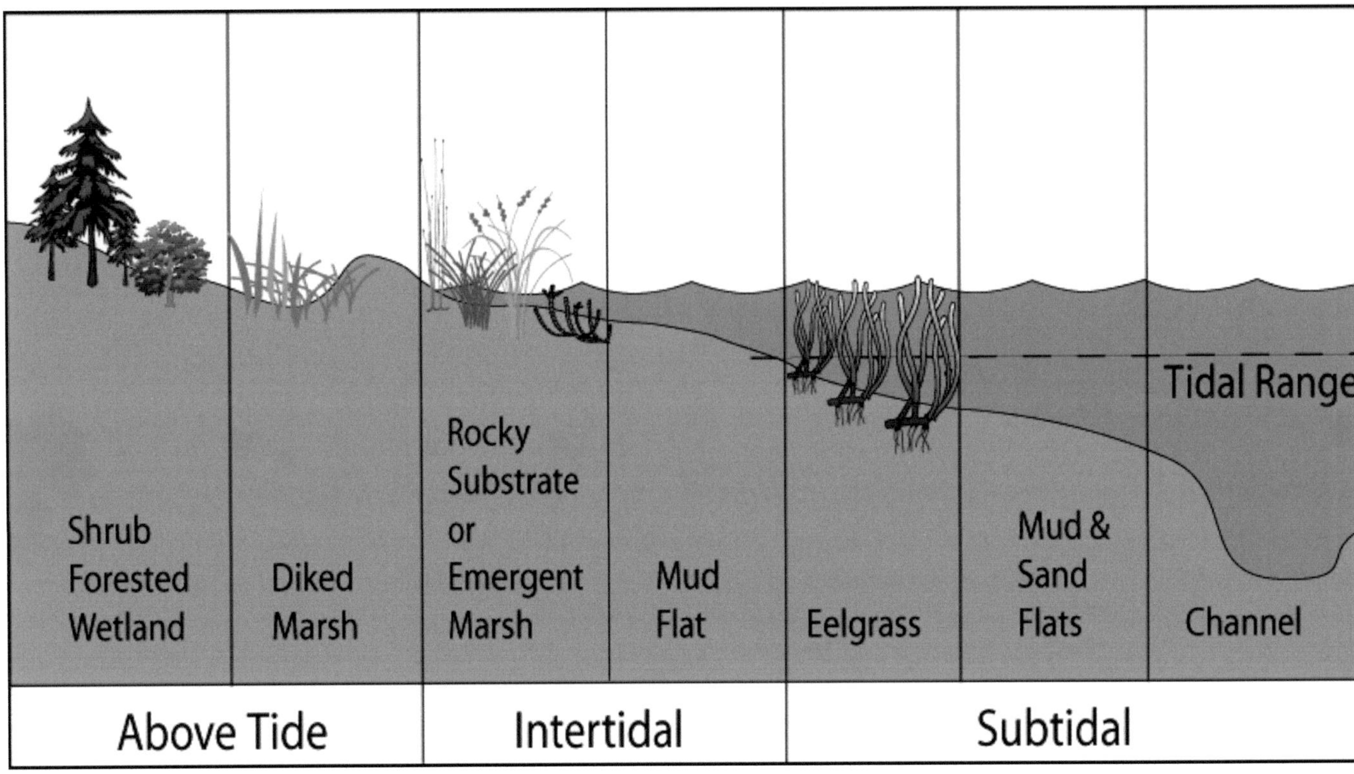

FIGURE 6.17 Relationship between elevation and types of habitat in the Columbia River Estuary. SOURCE: Thom et al. (2004).

The rate of transition from intertidal flat to emergent marsh depends on the vigor of the vegetative growth. Pestrong (1965) observed *Spartina* spp. colonizing tidal flats in San Francisco Bay and described luxuriant growth of dense stands. The efficacy of some *Spartina* species in trapping suspended sediments has been demonstrated in many areas, especially where invasive species have quickly covered large intertidal areas and raised elevations. For example, Feist and Simenstad (2000) noted that new colonies of invasive *Spartina alterniflora* expanded at rates of almost 80 cm yr^{-1} in Willapa Bay, Washington.

Mature marsh systems include a number of subsystems—vegetated plains, tidal courses, pans and ponds—as well as the adjacent intertidal zone (see Perillo, 2008, for a summary of the dynamics and interdependence of these subsystems). The biophysical characteristics of these environments influence the ability of estuaries to attenuate the effects of sea-level rise and storm waves on adjacent natural and human environments.

Historic and Current Patterns of Estuary Change on the West Coast

Many of the estuarine habitats along the U.S. west coast are a product of their sea level and tectonic histories, which control the position of the sea relative to valleys and coastal embayments and influence sediment delivery from adjacent steep watersheds. San Francisco Bay, for example, began to form 10,000 years ago as sea level rose through the Golden Gate (Atwater et al., 1977, 1979). When sea-level rise slowed, vegetation began to colonize and persist on tidal mudflats along the estuarine margins (Atwater et al., 1979; Collins and Grossinger, 2004). In the estuary's marine embayments, the high availability of reworked sediments and the low rates of sea-level rise enabled the formation of extensive marsh plains capable of accreting with rising sea level (Orr et al., 2003). Sediment was delivered to the nearby freshwater delta during flood flows of the Sacramento and San Joaquin rivers. Far from the delta, organic-rich marshes began to accumulate (Atwater, 1982).

More recently, human activities have become a powerful force on estuarine wetlands. The 1850 Swamp and Overflow Land Act transferred ownership of all swamp and overflow land, including estuarine marshes like those in the Sacramento-San Joaquin Delta, from the federal government to the states. By 1871, most of California's "swampland" was privately owned, and much of it was being converted to other uses (The Bay Institute, 1998). In the Puget Sound watershed, where approximately 70 percent of the Washington state population lives, the loss of historical nearshore ecosystems through development has been profound (see "Case Study on the Puget Sound" below).

Processes Determining Future Changes in Estuaries

The primary physical factors that influence coastal marsh development and survival are the fine sediment regime, tidal conditions, coastal configuration, and local sea-level history (French and Reed, 2001). Changes in soil elevation also may be important (Cahoon et al., 2002). If compaction rates exceed vertical accretion, the plant species dominating any particular tidal marsh ecosystem may cease to function physiologically (Kirwan and Murray, 2007). The response of flats and marshes to sea-level rise depends on the balance between submergence, erosive forces, and sediment supply, and is mediated by climatic influences on biotic processes (Reed, 1995). The cross-profile shape of intertidal flats in southern San Francisco Bay, an important determinant of their role in wave attenuation, is influenced by sediment deposition, tidal range, fetch length, sediment grain size, and tidal flat width (Bearman et al., 2010).

The resistance of tidal flats to erosion by waves and tides is heavily dependent on the biota. Widdows et al. (2004) estimated that as much as 50 percent of the sediment accumulation on a tidal flat in the Westerschelde Estuary (Netherlands) was due to biostabilization. The timescale on which various biotic factors influence tidal flat stability was conceptualized by Widdows and Brinsley (2002), who identified longer-term changes associated with the presence of persistent biota and shorter-term cyclic changes in the balance between microphytobenthos and sediment destabilizers, such as burrowing clams. Few studies have examined the interactions of biotic agents with intertidal morphology, although there is some evidence that invasive species can have dramatic effects (e.g., Hosack et al., 2006, studies in Willapa Bay) and that biotic agents can play a key role in wave attenuation within estuaries (Lacy and Wyllie-Echeverria, 2011; see also "Puget Sound Case Study" below).

The redistribution of sediments within and between tidal flats and marshes is strongly influenced by vegetation, which traps sediment, and by storm surges and waves (e.g., Reed et al., 2009; Williams, 2010). Storm surges can introduce sediment into marshes and redistribute sediment as portions of the marsh are eroded, resulting in substantial sediment accretion and geomorphic changes (Cahoon, 2006). The total amount of sediment deposition in coastal marshes depends on the prevailing meteorological and hydrographical conditions as well as on the number and magnitude of storm events (Bartholdy, 2001). The source of storm sediment deposited in marshes is rarely clear (e.g., Burkett et al., 2007), and erosion or deposition associated with individual storms is difficult to observe directly or infer from satellite images. Detailed geomorphologic surveys carried out before and after the storms are required to elucidate how these systems are linked during extreme events.

On the west coast of the United States, the role of storms in marsh sedimentation is mediated by precipitation and local runoff. In California's Elkhorn Slough Watershed, the relationship between estuarine marshes and adjacent uplands varies according to the changing nature of watershed inputs (Byrd and Kelly, 2006). Along the coast, these inputs are controlled by watershed size and runoff (Figure 6.18) and by the sensitivity of sediment delivery to land-cover change caused by anthropogenic and climatic factors, including changes in storminess. Short, steep drainage areas, such as commonly found on the Oregon coast, are likely more sensitive to changes in coastal storm precipitation. Sediment runoff from rivers not bounded by the Coast Range (e.g., Sacramento and Columbia rivers) is influenced by a wider set of climatic factors. Thus, the effect of individual storms on water and sediment delivery to estuaries in these areas is more buffered. However, sediment delivery to the coast is highly variable, even from the larger drainage areas. McKee et al. (2006) examined almost 20 years of data on suspended sediment delivery from the Sacramento and

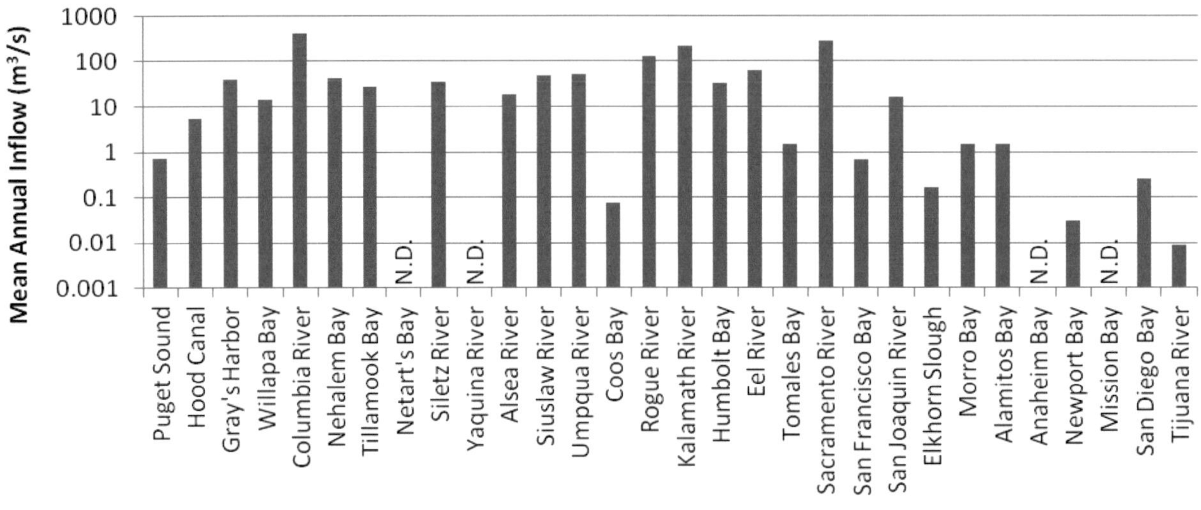

FIGURE 6.18 Mean annual inflows for U.S. west coast estuaries for 2000–2010. N.D. indicates that no data were available. SOURCE: Data from the U.S. Geological Survey, <http://waterdata.usgs.gov/nwis>.

San Joaquin rivers to the San Francisco Bay and noted that, on average, 88 percent of the annual suspended-sediment load was discharged during the wet season and 43 percent was discharged during the wettest 30-day period.

Dams and human actions have a large impact on estuaries. One of the best described examples concerns the effect of hydraulic mining and dams on sediment delivery from the Sacramento River to the northern San Francisco Bay (Gilbert, 1917; Wright and Schoellhamer, 2004; Jaffe et al., 2007). The huge pulse of sediment released during mining activities gradually moved down the Sacramento River and into San Pablo Bay. Recent losses of sediment from San Pablo Bay likely indicate the progressive movement of the sediment pulse toward the ocean.

The response of coastal marshes to sea-level rise is influenced by changes in sediment dynamics, mediated by physical forcing, biotic factors, and plant growth. Dating of buried salt marsh peats suggests that salt marsh surfaces are frequently in equilibrium with local mean sea level (see Allen, 1990, and references therein), as would be expected in areas where salt marshes survive for long periods. It is well established that the surface elevation and, in many cases, the accretion rate of marshes can change to keep pace with sea-level rise. However, it is unclear whether the elevation change is stimulated by increased inundation or whether rising sea level provides space for soil accumulation to proceed in areas where it is otherwise limited.

Much attention has been paid to the issue of landward migration of tidal marshes as a result of sea-level rise. Such migration will occur only if the landward margin of the marsh is unobstructed (e.g., Kraft et al., 1992). The rate of migration is determined by the slope of the land and the rate of rise. New marshes may develop at the landward margin, depending on the level of development. The limited availability of suitable land along the California coast is described in Heberger et al. (2009). But whether marshes at a particular location survive in the long term will be determined by their ability to build elevation. Figure 6.19 illustrates how marshes can be created by sea-level rise then lost if they cannot maintain their relative elevation. Although the figure shows an idealized uniform slope, many west coast shorelines steepen abruptly landward of the marsh, which would limit the extent of marshes as they move inland in response to sea-level rise.

Projecting the sustainability of salt marshes under future climate scenarios is complex because it depends on the relative importance of organic matter to marsh vertical development, the factors governing organic matter accumulation during rising sea level, the importance of subsurface processes in determining surface elevation change, and the role of storm events and hydrologic changes in controlling sediment deposition, soil conditions, and plant growth. A good example of this complexity and the challenge of isolating the effects of sea-level rise from other climate-related influences is described in Kirwan et al. (2009), who found evidence of an increase in the productivity of a dominant east coast salt marsh grass with increases in temperature. Other studies have found changes in the productivity of some marsh plants with increased atmospheric CO_2 levels (e.g., Cherry et al., 2009). This report's assessment of the response of estuaries and marshes to future sea levels is therefore only one part of the climate change story. Fully assessing the fate of west coast marshes under climate change, including sea-level rise, is further hampered by the lack of long-term data on many west coast marshes and the differences in species composition compared to other more-studied systems.

Response of Mudflats and Marshes to Future Sea-Level Rise and Storms

The regional projections presented in Chapter 5 show substantial differences in the magnitude of sea-level change along the west coast. If space is available for landward migration, the rate of sea-level change over biologically important timescales will determine the fate of tidal marshes. Most models of marsh response to sea-level rise ignore interannual variability in sea level and assume a consistent monotonic pattern of rise (Kirwan and Temmerman, 2009). The potential consequences of the monotonic rise in the sea levels projected in Chapter 5, as well as the effect of potential interannual variations or other conditions that could modulate those responses, are discussed below. As discussed in Chapter 4, however, local vertical land motions (subsidence or uplift) may be significantly larger than the regional land motions used in the projections, and thus relative sea-level change at any particular place along the coast may differ from the committee's regional projections.

Central and Southern California

Approximately 90–95 cm of sea-level rise is expected between 2000 and 2100 south of Cape Mendocino, but the value could be as high as ~167 cm and as low as 42 cm (Figure 5.9). The projected value of this study falls between two sea-level rise scenarios

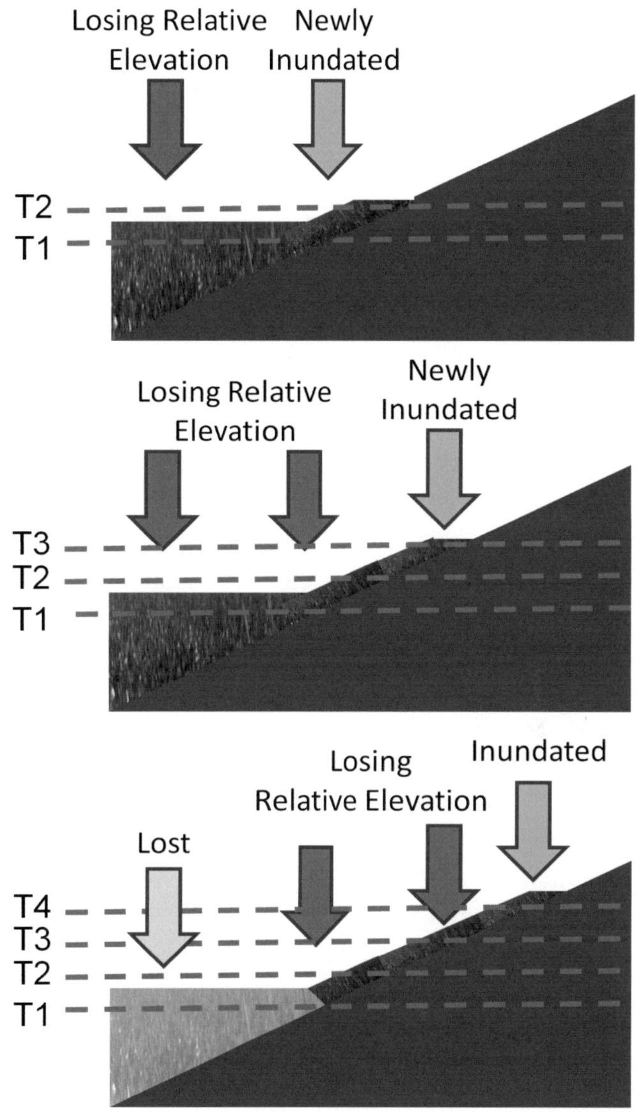

FIGURE 6.19 The change of marsh surface elevation is important to successful landward migration under sea-level rise. Depending on the difference between the rate of sea-level rise and the rate of marsh accretion, a narrow or wide band of wetlands will be present under any sea-level condition, but the area of marsh will not expand unless elevation change can keep pace with sea-level rise. (*Top*) Sea level has risen from T1 to T2. New marshes are created at the landward margin of the marsh and existing marshes lose relative elevation as the rate of sea-level rise exceeds the elevation increase. (*Middle*) Sea level rises to T3. Newly inundated marshes in T2 are now losing elevation and new marshes are created at the landward margin. (*Bottom*) Sea level continues to rise to T4. Existing marshes continue to lose elevation.

considered by Stralberg et al. (2011), who examined the fate of San Francisco Bay marshes under varying rates of organic matter accumulation and sediment supply. Their study showed that some types of marshes (e.g., those lower in the tidal frame, known as low marsh) are sustainable under even 1.65 m of sea-level rise as long as there are sufficiently high rates of suspended sediment supply. This implies that if the high estimates are realized, marshes will be sustainable by 2100 only under optimal conditions of sediment supply. Marshes respond more to rates of sea-level change over several years than they do to the absolute change in elevation. Observations suggest that marshes in San Francisco Bay can keep pace with a sea-level rise of 6 mm yr^{-1} (see Parker et al., 2011 and references therein). The committee projections for 2030 and 2050 yield rates on this order.

The supply of suspended sediment to estuarine marshes in central and southern California is driven by fluvial inputs. The few long-term studies of sediment delivery to estuaries in this area tend to show a decrease in suspended sediment in San Francisco Bay over time (Wright and Schoellhamer, 2004; Schoellhamer, 2011). Much of this decline occurred because a significant fraction of sediment that would enter the system naturally is now trapped in upstream reservoirs.

For coastal marsh accretion to occur, some of the suspended sediment carried in from rivers must be deposited (Reed, 1989). For example, at Morro Bay, high rates of sediment delivery from the adjacent watershed doubled the area of salt marsh between 1980 and 1990.[1] Sediment deposition also can be influenced by sea-level rise and storminess. Ruhl and Schoellhamer (2004) noted that wind waves can resuspend erodible bed sediment. As sea level rises, wind wave stress on bed sediment decreases, reducing the potential for sediment resuspension (Ganju and Schoellhamer, 2010). An increase of 1 m in water depth, especially in shallow subtidal areas, could have substantial effects on sediment resuspension. Larger storm events, which produce larger waves, would be required to mobilize sediments and make them available for marsh accretion.

The depth and duration of flooding control the opportunity for sediment deposition on the marsh surface. If storm events elevate water levels at times of high sediment supply, the opportunity for sediment deposition increases. Under normal tidal inundation, times of flooding may not coincide with periods of high sediment availability. Further, periodic marsh flooding during storms can allow sediment to be deposited without subjecting marsh plants to prolonged inundation stress. Zedler (2010) found that storms were important for delivering sediment and increasing the elevation of marshes in the Tijuana Estuary in southern California, although the lack of subsequent tidal flooding may lead to high soil salinities and changes in species composition in high marsh areas. In some bar-built estuaries, especially those subject to natural closure (Jacobs et al., 2010), sea-level rise and storms may alter the configuration of the estuary and either increase or decrease sediment retention (e.g., Schwarz and Orme, 2005).

For the sea-level changes projected by the committee for central and southern California, a series of storms combined with some increase in tidal inundation could allow such marshes to persist to 2100, even under the highest sea levels projected. If storm events increase both sediment resuspension and marsh flooding, then rather than causing problems for coastal marshes, they may be essential to their survival.

Northern California, Oregon, and Washington

North of Cape Mendocino, the committee projects that sea level will rise 61–65 cm by 2100, with lower rates to the north (Figure 5.9). The high end projections are ~143 cm by 2100. In the southern part of this stretch of shoreline, isolated areas of marsh exist at the mouth of several estuaries, such as Humboldt Bay and Lake Earl in California and the Rogue River in Oregon. Most of these estuaries are relatively narrow without extensive intertidal flats for storing sediment, so their ability to survive sea-level rise depends greatly on fluvial inputs of sediment. The Eel River, entering the coast south of Humboldt Bay, supplies the largest amount of sediment to the California coast (Sommerfield and Nittrouer, 1999). The Klamath River may have a higher discharge than the Eel River during a storm event, but it carries a lower sediment load because there is less erodible material in its drainage basin (Pullen and Allen, 2000).

Along this part of the coast, the supply of sediment to coastal marshes is determined by storm-induced

[1] See <http://www.mbnep.org/Library/Files/Tidings/2006/Sedimentation.pdf>.

river flooding and by management practices. For example, Willis and Griggs (2003) reported that dams on the Klamath River control 46 percent of the drainage and have reduced sand transport to the coast by 37 percent. Although sand may not be necessary for marsh survival, reduction in the supply of sand will modify the bathymetry of the estuary with potential consequences for tidal exchanges and the resuspension of sediment for transport to marshes (see discussion above regarding the potential importance of wind wave resuspension of sediment availability within estuaries).

The committee's projection of sea-level rise by 2100 is slightly lower than that used by Glick et al. (2007) to study the effects of sea-level rise on coastal habitats of the Pacific Northwest (69 cm). Using the SLAMM 5.0 model (Clough and Park, 2007), Glick et al. (2007) predicted that salt marsh would expand, partly at the expense of more inland fresh marsh areas. However, one of the drawbacks of the SLAMM 5.0 model is that it uses historic accretion rates to drive inundation and the vertical component of marsh response (Clough and Park, 2007). Rates of accretion may change with sea-level rise, and accretion is only one of several dynamic factors that determine the response of marsh elevation to sea-level change (see discussion above). For example, if historical measured rates of marsh accretion are limited by the accommodation space provided by the highest level of tidal flooding (e.g., Krone, 1987; Allen, 1990), then an increase in sea level could increase marsh accretion. Glick et al. (2007) set accretion rates at 3.6–3.75 mm yr^{-1} for coastal marshes in their study. For much of the Pacific Northwest, these rates are slightly higher than sea-level rise projected by the committee for 2030 and similar to the rise projected for 2050. If accretion rates subsequently increase in response to sea-level rise, the Glick et al. (2007) predictions for 2100 (e.g., salt marsh expands at the expense of other marsh types) will not be realized.

For 2030 and 2050, local influences, including changes in tidal hydrology and riverine sediment delivery, as well as development pressures, can be more of a threat to marsh sustainability than sea-level rise. If the highest estimates of sea-level rise are realized for this part of the coast, only marshes in areas with a high local sediment supply (e.g., at the mouth of major river estuaries) will persist in their current form.

Role of Mudflats and Marshes in Providing Protection from Future Inundation and Waves

Few controlled field studies have examined the role of coastal habitats in protecting inland areas from inundation and wave damage during sea-level rise, coastal storms, or tsunamis. Some small-scale studies (e.g., Möller et al., 1999) have detected a relationship between specific vegetative characteristics and wave attenuation, although bathymetric change appears to play a more important role. Several field studies have noted the importance of vegetation morphology or architecture in attenuating both tsunami waves (Tanaka et al., 2007) and wind-waves (Mazda et al., 1997). Field observations, measurements of wave forces, and modeling of fluid dynamics associated with the 2004 south Asian tsunami suggest that tree vegetation may shield coastlines from tsunami damage by reducing wave amplitude and energy (e.g., Danielsen et al., 2005). However, it is difficult to separate the effect of the vegetation from other aspects of coastal topography (Dahdouh-Guebas and Koedam, 2006; Feagin, 2008). The question of whether vegetation structure reduces coastal damage directly through wave attenuation or indirectly through alteration of the landscape has not been settled.

Modeling studies of hurricane storm surge and surge attenuation suggest that decreases in marsh elevation, which increases the water depth, and increases in bottom friction generally reduce storm-surge levels (e.g., Loder et al., 2009). Reductions in marsh continuity increase coastal surges. Wamsley et al. (2009) found that the extent to which wetlands attenuate surge depends on the storm and landscape characteristics.

The effect of vegetation on bottom friction or roughness can be approximated from detailed measurements of plant morphology and assumptions about stem density and flexure (see Feagin et al., 2011, for a detailed review). However, isolating this effect from the larger coastal configuration with which the storm waves or tidal flows are interacting requires numerical experiments. The depth of flooding and its interaction with plant stems and leaves is yet another nonlinear relationship as field studies of wave attenuation in seagrass beds have demonstrated (e.g., Koch et al., 2009). All of these studies point to the difficulty of generalizing the role of coastal habitats in ameliorating the effects of future storms or tsunamis on the west coast.

The morphodynamic interactions among topography and bathymetry, vegetation, sediment deposition, and turbulent flows are difficult to predict, increasing uncertainties about the extent to which coastal habitats will mitigate the effects of future sea-level rise and storms. A means for reliably determining wave damping by vegetation for engineering studies has not been developed (Augustin et al., 2009). Models that reliably predict coastal morphology (independent of the role of vegetation) over decades and under episodic storm forcing are not widely available. For these reasons, significant tolerance for future coastal habitats, vegetation, and coastal morphology configurations will have to be built into coastal protection systems.

OPPORTUNITIES FOR MARSH RESTORATION AND THE EFFECT OF MARSHES ON STORM WAVE ATTENUATION

As shown above, the response of marshes to future sea-level rise and storminess along the west coast of the United States depends on local conditions. Marsh restoration is also site specific. Consequently, the committee chose two areas where data on prior restoration are available—the California Bay Delta and the Puget Sound—to explore the potential for marsh restoration given future sea-level rise and the effect of marshes on storm and wave attenuation.

Case Study on the California Bay-Delta

California's Bay Delta estuary is one of the largest estuaries in the United States. The estuary consists of a series of interconnected bays and channels connecting San Francisco Bay to the Sacramento-San Joaquin River Delta. Salinity increases from the delta to the Golden Gate at the mouth of San Francisco Bay. At times of high river flood, fresh conditions can penetrate into the bay.

The estuary has been modified extensively by anthropogenic activities over the past 150 years (The Bay Institute, 1998; Goals Project, 1999; Brown, 2003). Approximately 80 percent of the tidal wetlands in San Francisco Bay and 95 percent of the tidal wetlands in the Sacramento-San Joaquin Delta have been lost (The Bay Institute, 1998). In the south bay, more than 90 percent of the historic tidal marsh area has been converted to salt ponds, agricultural areas, and urban developments (Foxgrover et al., 2004; Figure 6.20). Many of these areas are protected by an aging collection of levees.

The extensive loss of tidal marsh habitat has prompted calls for marsh restoration in the San Francisco Bay Delta (e.g., Goals Project, 1999; CALFED, 2000; Steere and Schaefer, 2001). Given the large investment required to restore thousands of acres of tidal marsh, it is important to understand the likely role of restored marshes in attenuating storms and waves and whether they will persist under future sea-level rise.

Potential for Marsh Restoration

One of the first steps in marsh restoration is to return the land surface to elevations that can be colonized by marsh vegetation. Many land surfaces within the delta are currently on the order of 3–5 m below water levels. Data from interferometric synthetic aperture radar show that the delta-interior regions are subsiding 3–5 mm yr^{-1} and that local regions in the delta are subsiding up to 2 cm yr^{-1} (Brooks et al., 2012). In areas where subsidence exceeds sediment accumulation, it may be necessary to fill low-lying areas to enable colonization. Sedimentation rates are low in much of the delta because fine sediments are slow to settle and waves keep them in suspension (Simenstad et al., 2000). In some shallow areas with nearly 100 years of sedimentation (e.g., Sherman Lake and Big Break), sediment accumulation has not yet been sufficient to allow vegetation to become reestablished.

Where sediment accumulation exceeds subsidence, vegetation colonization may proceed naturally. For example, high vertical accumulation rates of 3 cm yr^{-1} for 1955–1963 and 4.2 cm yr^{-1} for 1963–1983 were inferred from ^{137}Cs measurements of marsh cores at Alviso in the south bay (Patrick and DeLaune, 1990). Orr et al. (2003) found accretion rates for restored marshes in San Pablo Bay of 18–70 mm yr^{-1} for low marsh and 9–10 mm yr^{-1} for high marsh. At these rates, marsh restoration could progress under all except the committee's high projections of 2100 sea-level rise. However, high rates of past accretion may not

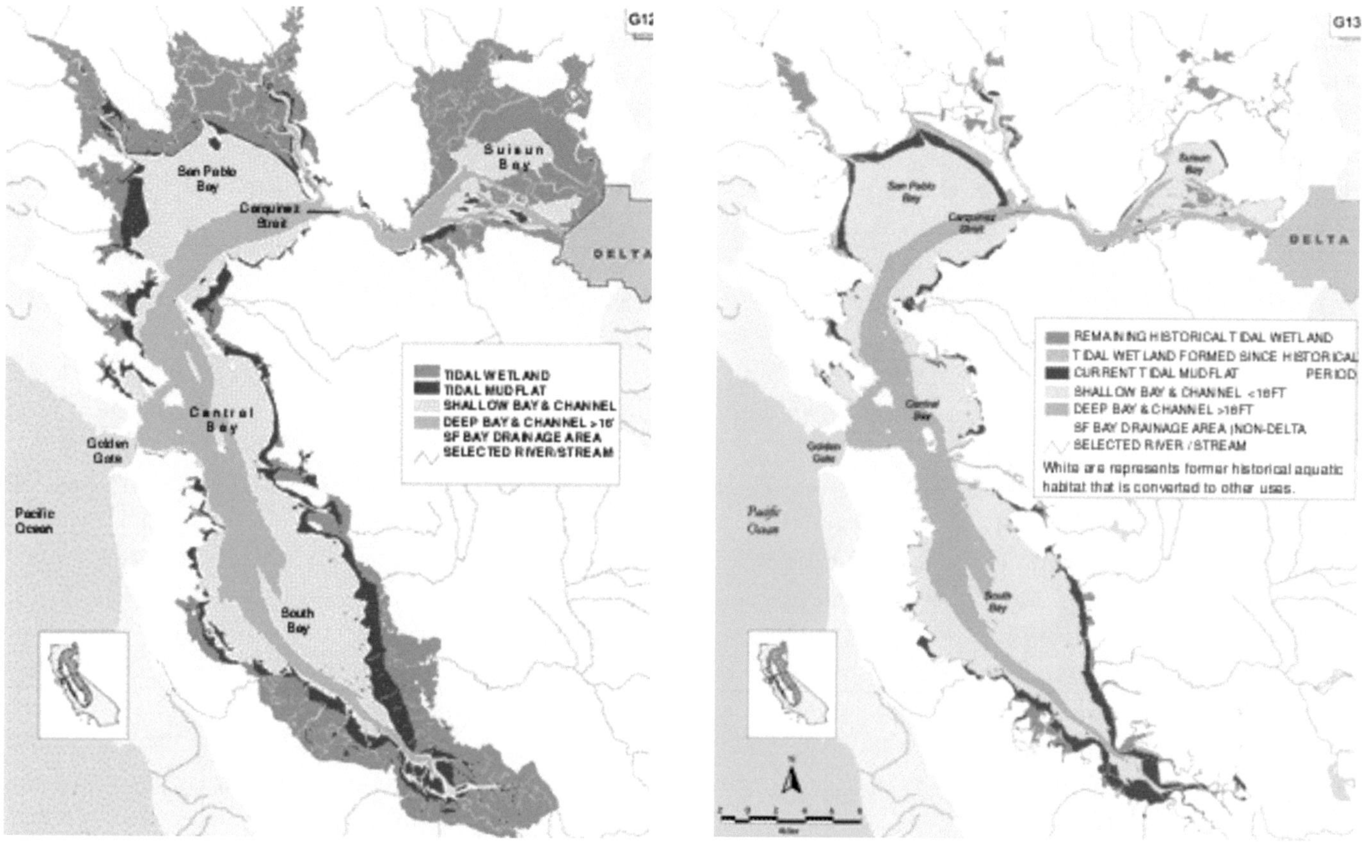

FIGURE 6.20 Extent of tidal wetlands in San Francisco Bay in the mid 19th century (*left*) compared with the extent c. 1997 (*right*). SOURCE: Courtesy of The Bay Institute.

continue in the future. Schoellhamer (2011) found a 36 percent decrease in suspended solids concentration in San Francisco Bay from water years 1991–1998 to 1999–2007. He attributed this decrease to the depletion of a large erodible sediment pool (Jaffe et al., 1998; Foxgrover et al., 2004) within the estuary. The availability of an erodible sediment pool prior to the late 1990s may have enabled higher accretion rates in restored marshes in the past than would be possible in the future. Transport of sediment from adjacent intertidal and subtidal flats into relatively quiescent restored areas where it cannot be readily suspended would promote accretion in restored marshes at the expense of the erodible sediment pool. The elimination of the sediment pool would lead to less sediment being available for development and maintenance of restored marshes around San Francisco Bay.

The committee's projected sea-level rise for the San Francisco Bay Delta is 93 cm by 2100. The studies described above illustrate that with adequate migration space and sediment supply, marshes in some areas may be able to survive future sea-level rise. However, if the highest projections for the Bay Delta are realized (1.6 m by 2100), marsh restoration will be realistic only in areas with exceptionally high and sustained sediment supply.

Effect of Restored Marshes on Wave and Storm Attenuation

The Golden Gate carries storm surges from the open coastal Pacific into San Francisco Bay and the delta (Bromirski and Flick, 2008), and the bordering low-lying lands are vulnerable to the increased water levels (Knowles, 2010). Most measurements of the effect of marsh vegetation on wave attenuation in the bay delta have focused on small waves, such as boat wakes (e.g., Bauer et al., 2002). For example, Ellis et al. (2002) measured the effect of brush bundles in attenuating waves from boat wakes and found up to a 60 percent reduction in wave energy impacting a delta levee when the bundles were in place, depending on the tides. In a study of small waves in a shallow lake, Lövstedt and Larson (2010) found an average decrease in wave height of 4–5 percent per meter within the first 5–14 m of beds of *Phragmites australis*. If these results are applicable to tules (*Schaenoplectus* spp.), which are similar in height, extensive tule restoration could result in substantial attenuation of waves (produced by wind or vessels) within the delta.

The transition from tules to *Salicornia virginica* dominated marshes in the bay is accompanied by a major change in plant morphology. *Salicornia virginica* resembles *Atriplex portulacoides* above ground, which has a lower stem density, height, and diameter than the two *Spartina* spp. (Feagin et al., 2011). This suggests that *Salicornia virginica* marshes in the bay may play less of a role on attenuating storm set-up and waves than the reed-like architecture of tule marshes in the delta.

Modeling of the propagation of long waves into the south bay (Letter and Sturm, 2010) suggests that small areas of marsh can ameliorate the effects of storm events on water level. Letter and Sturm (2010) predicted changes in water level at specific locations during simulated storm events, based on the roughness and extent of vegetation cover and other parameters. They found that water levels are lower at the edge of salt ponds fronted by some marsh than they are at the edge of mudflats. For storm tides during the January 1983 El Niño event, which set records for high sea level (see "Changes in Ocean Circulation" in Chapter 4), water elevations on levees not fronted by a small area of marsh were higher than those with marsh between the levee and the intertidal flats. The extent of marshes in the south bay is limited, so whether reductions in water levels in small areas can be extrapolated to larger landscapes will require more detailed modeling of potential future landscape configurations.

Case Study on the Puget Sound

Puget Sound includes more than 8,000 square kilometers of marine waters and nearshore environment, and 4,020 kilometers of shoreline. About 4 million people live in the Puget Sound watershed, and the population is expected to reach 5 million by 2020 and 8 million by 2040 (Puget Sound Regional Council, 2004). Commercial fish and shellfish harvesting in Puget Sound is an important industry for the state.

Tidal marshes and eelgrass beds are among the most important coastal habitats in Puget Sound. Extensive tidal marshes occur at the mouths of rivers that empty into Puget Sound. Eelgrass is found from the intertidal zone to the shallow subtidal zone in central and north Puget Sound. Loss of these habitats has been dramatic. Nearly three-quarters of the original salt marshes and essentially all river delta marshes in urbanized areas of the sound have been destroyed (Gelfenbaum et al., 2006). Eelgrass habitat is almost completely gone in Westcott Bay and several other small embayments (Mumford et al., 2003; Wyllie-Echeverria et al., 2003).

The nearshore environments of Puget Sound are maintained by a complex interplay of biological, geological, and hydrological processes that interact across the terrestrial-marine interface. Many of these processes have been significantly affected by human activities (Bortelson et al., 1980). For example, dikes have altered nearshore sedimentation patterns and eliminated the tidal influence that forms salt-marshes, and dams have reduced the magnitude and frequency of floods, limiting the sediment supply to river deltas. More than 33 percent of shoreline in the Puget Sound region has been modified (Puget Sound Action Team, 2002).

The dramatic nature of these changes and the need to accommodate future population growth without further environmental degradation has led to concerted efforts to improve coastal management and restore ecosystems (e.g., Puget Sound Partnership; Puget Sound Nearshore Ecosystem Restoration Program). Such efforts must factor in the effects of future sea-level rise, which is complicated by the strong gradients in vertical land motion in the area (Figure 6.21). Whether vertical land movements enhance or counteract the effects of regional sea-level rise has important implications for existing coastal habitats, the viability of future restoration, and the potential of these habitats to help mitigate the effects of future storms.

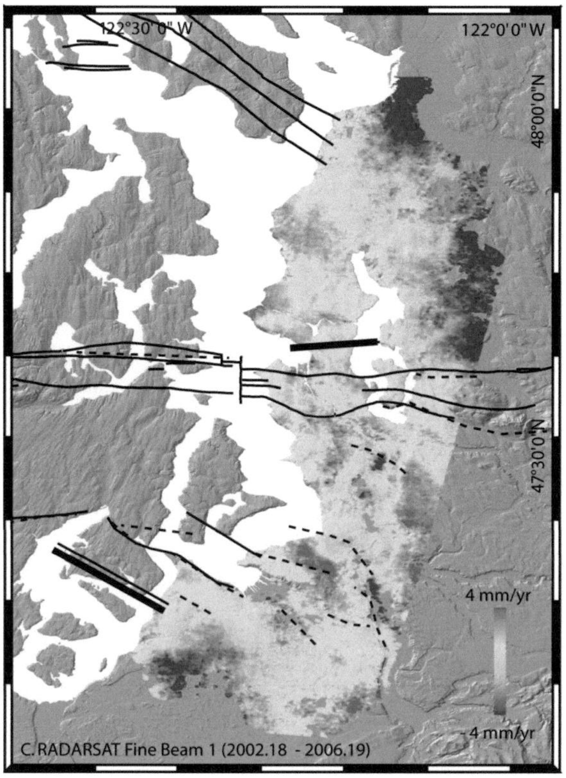

FIGURE 6.21 Vertical land movements in the Puget Sound area based on interferometric synthetic aperture radar from 2002 to 2006. Surface movements in the radar line of sight range from -4 mm yr^{-1} (subsidence, blue) to + 4 mm yr^{-1} (uplift, red). Black lines are fault locations, and dashed lines are geophysical anomalies. SOURCE: Finnegan et al. (2008).

Opportunities for Restoration

Efforts to restore tidal marshes have focused on the deltas of the major rivers draining into the sound, where many of the marshes have been diked for agriculture. A recent assessment of restoration needs in the sound (Schlenger et al., 2011) noted that delta shorelines have been so altered in the Duwamish, Puyallup, and Deschutes areas that they are now classified as artificial shoreforms. Restoring the tidal hydrology and riverine freshwater and sediment input are key elements of a delta restoration strategy. Tidal hydrology and sediment input affect many delta processes, including distributary channel migration, tidal channel formation and maintenance, sediment retention, and exchange of aquatic organisms.

Clancey et al. (2009) identified berm or dike removal or modification as the most efficient method of rapidly restoring tidal flow processes. This action could be complemented by modifying channels and making minor topographic changes such as filling ditches and removing road fill. In some areas, the tidal floodplain has been extensively filled and restoration may require resculpting of the land surface to ensure appropriate flooding and drainage of river and tidal waters.

Areas where tidal action was recently restored through these measures include portions of the Nisqually Delta and the Skokomish River. In October 2009, after a century of isolation from tidal flow, a dike was removed to inundate 308 ha of the Nisqually National Wildlife Refuge (e.g., Figure 1.13). The Nisqually Indian Tribe restored an additional 57 ha of wetlands, making the Nisqually Delta the largest tidal marsh restoration project in the Pacific Northwest. Studies show more than 3 cm of sedimentation in the first year of restoration.[2] A smaller scale restoration was carried out on the Skokomish River in September 2007, when tides were reintroduced to a 108-acre site for the first time in 75 years. For such tidal reintroduction projects to be successful, sedimentation (both mineral and organic accumulation) must both raise elevations to a level where marsh flora and fauna can flourish and maintain those elevations over time as sea-level rise increases relative water levels. Within Puget Sound, variations in vertical land motion (Figure 6.21) either increase or decrease the amount of elevation change required.

The supply of river sediment also is important for maintaining elevation of existing marsh. Dams or road crossings within a delta's watershed may indicate that river systems may not provide enough sediment to sustain the elevation of restored habitats. Rates of sediment delivery from the Puget Sound watershed vary over time and place, depending on runoff patterns and land use changes. For example, the Skagit River carries more than 2 million tons of sediment per year, and streams draining the Olympic Peninsula (excluding the Skokomish) carry generally less than 15,000 tons per year (Figure 6.22). The spatial patterns of sediment delivery, combined with general trends in vertical land motion, can be used to identify areas where restored coastal marshes would most likely survive future sea-level rise. In general, areas with high fluvial sediment supply and low subsidence or marginal uplift

[2] See <http://nisquallydeltarestoration.org/science_geomorphology_sedimentation.php>.

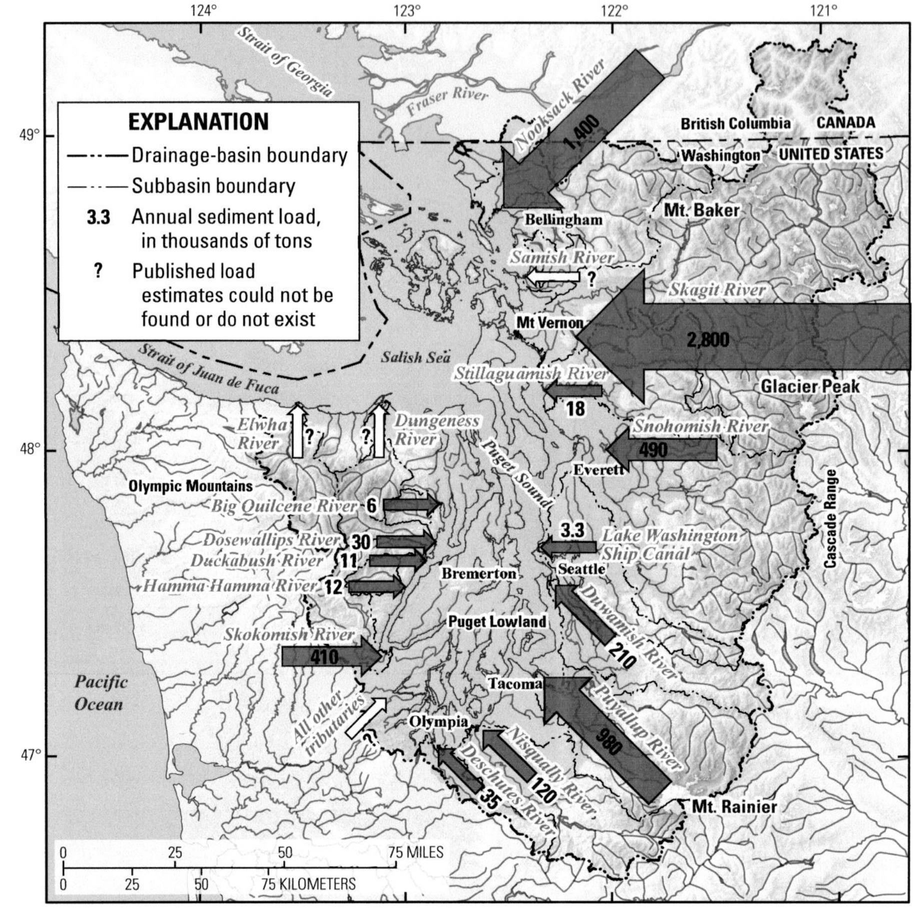

FIGURE 6.22 Annual sediment load of major rivers draining into Puget Sound measured at or near the river mouth. The size of the arrow is scaled to the annual sediment load. SOURCE: Czuba et al. (2011).

(e.g., north and western regions of the sound) are the most promising locations for sustainable coastal marsh restoration, at least under the committee's projected sea-level rise for 2030 and 2050. Under the highest sea-level projections for 2100, a high sediment supply and uplift may not be enough for restoration to succeed, and additional steps will have to be taken (e.g., filling previously subsided areas).

Linking restoration plans in these areas with land use and watershed management plans would improve the sustainability of coastal habitats. Land use plans could include, for example, conservation easements or limits on construction to accommodate the lateral migration of coastal marshes as sea level rises. Watershed management plans could include changes in dam operations to increase the amount of sediment that reaches Puget Sound deltas.

Efforts to restore eelgrass in some areas of the sound have had only limited success (Thom, 1990; Carlisle, 2004; Mumford, 2007). Stamey (2004) found an overall success rate of 13–80 percent, concluding that eelgrass transplantation cannot yet be used reliably for mitigation in Puget Sound. Eelgrass restoration costs are high, between $100,000 and $1 million per acre (Fonseca et al., 1998). However, if appropriate substrate and water quality conditions can be established and maintained, the effects of sea-level rise on eelgrass is likely minimal.

Potential for Wave Attenuation

Eelgrass beds play an important role in nearshore ecosystems. The plant blades slow water currents and dampen waves, thereby trapping sediments, detritus, and larvae. Lacy and Wyllie-Echeverria (2011) studied the influence of eelgrass (*Zostera marina*) on near-bed currents, turbulence, and drag in the San Juan archipelago of Puget Sound. *Zostera marina* grows at water depths less than 5 m relative to mean lower low water along 43 percent of Puget Sound's shoreline (Berry et al., 2003). Lacy and Wyllie-Echeverria (2011) measured velocity profiles up to 1.5 m above the sea floor over a spring-neap tidal cycle, including measurements above and within the canopy. They found that eelgrass attenuated currents by a minimum of 40 percent, and by more than 70 percent at the most densely vegetated site, with attenuation decreasing with increasing current speed. Even sparse canopies influenced near-bed flow and significantly attenuated currents.

Most Puget Sound shorelines are sheltered, and waves are generated by local winds with little or no energy component from ocean swell. The topographic confines of Puget Sound limit the height of waves (Finlayson, 2006). Large waves (greater than 0.4 m significant wave height) occur only during infrequent wind storms. Consequently, the effect of eelgrass beds, and to some extent coastal marsh vegetation, on wave attenuation can be substantial.

CONCLUSIONS

Sea-level rise and storms along the west coast of the United States have caused significant coastal retreat. Cliff and bluff retreat, caused mainly by wave erosion and terrestrial processes (e.g., landslides, slumps, rockfalls, runoff), ranges from a few centimeters to tens of centimeters or more annually, with weaker rocks and areas of lower topography retreating more than resistant bedrock cliffs and headlands. Cliff retreat is not reversible. Although coastal armoring can buy time, today's seawalls and revetments will eventually be overwhelmed by sea-level rise and increasing wave heights.

Sand dunes and beaches, which consist primarily of unconsolidated sand, provide little resistance to severe wave attack, especially at times of elevated sea level. Consequently, beaches and barrier spits may grow and shrink several meters or more per year. Because beaches are nearly flat, a small rise in sea level can cause a large retreat of a beach. Where beaches and barrier spits are prevented from migrating by coastal armor or structures, they will eventually be inundated by future sea-level rise.

Rising sea levels and increasing wave heights will exacerbate coastal erosion and shoreline retreat in all geomorphic environments along the west coast. Projections of future cliff and bluff retreat are limited by sparse data in Oregon and Washington and by a high degree of geomorphic variability along the coast. Projections using only historic rates of cliff erosion predict 10–30 meters or more of retreat along the west coast by 2100. An increase in the rate of sea-level rise combined with larger waves could significantly increase these rates. Future retreat of beaches will depend on the rate of sea-level rise and, to a lesser extent, the amount of sediment input and loss.

Some of the coastal damage expected from sea-level rise and storminess may be mitigated in some areas by coastal mudflats and marshes. Mudflats and marshes protect inland areas from inundation and wave damage, but the specific effect depends on local conditions. Some studies have found that certain plants, such as eelgrass, slow water currents. Other studies have found that marsh vegetation with high roughness, stem height, and density—along with coastal topography and bathymetry—reduces wave height and energy. However, this relationship has not been specifically demonstrated for many of the species populating west coast marshes.

West coast tidal marshes can survive sea-level rise by building elevation to keep pace with rising water levels, which requires an adequate supply of sediment and/or organic matter accumulation. They may migrate inland if the area is unobstructed, but unless they maintain elevation under sea-level rise, the area of marsh will be limited by the slope of the land surface and the tidal range. Storms are an important agent for delivering sediment and increasing the elevation of marshes. For the sea-level changes projected by the committee for 2030 and 2050 in central and southern California, frequent storms that increase tidal inundation and promote sediment deposition could allow marshes to survive. In northern California and southern Oregon, fluvial inputs of sediment, which depend on storms and water management practices, also are important for sediment deposition. Entrapment of sediment behind dams makes marshes less able to survive sea-level rise in this area. Coastal areas in Oregon and Washington are projected to have lower rates of sea-level rise, in part because the land is rising. In some areas, the rising land surface will help coastal marshes maintain their elevation as sea level rises, making sea-level rise a less important threat in this area than other parts of the coast. Should the highest sea-level projections for 2100 be realized, marsh survival will be possible only in areas with high local sediment supply.

A detailed assessment of the response of west coast marshes to sea-level rise is hampered by the lack of long-term and/or comparable data and by the variety of geological (e.g., vertical land motion, sediment supply), hydrological (e.g., floods, storms, dams), and biological (e.g., accumulation of organic matter) factors that govern marsh survival, all of which combine to cause significant spatial variability along the coast. In general, most marshes with natural sediment delivery and unimpaired hydrology will survive the sea levels projected by the committee for 2030 and 2050. For 2100, marshes will need room to migrate, a high sediment supply, and uplift or low subsidence to survive projected sea-level rise.

References

Ablain, M., A. Cazenave, S. Guinehut, and G. Valladeau, 2009, A new assessment of global mean sea level from altimeters highlights a reduction of global slope from 2005 to 2008 in agreement with in-situ measurements, *Ocean Science Discussions*, **6**, 31-56.

Alexander, M.A., D.J. Vimont, P. Chang, and J.D. Scott, 2010, The impact of extratropical atmospheric variability on ENSO: Testing the seasonal footprinting mechanism using coupled model experiments, *Journal of Climate*, **23**, 2885-2901.

Allan, J.C., and P.D. Komar, 2006, Climate controls on U.S. west coast erosion processes, *Journal of Coastal Research*, **22**, 511-529.

Allen, J.R.L., 1990, Salt-marsh growth and stratification: A numerical model with special reference to the Severn Estuary, southwest Britain, *Marine Geology*, **95**, 77-96.

Allen, J.R.L., 2000, Holocene coastal lowlands in NW Europe: Autocompaction and the uncertain ground, *Geological Society Special Publication*, **175**, 239-252.

Alley, R.B., P.U. Clark, P. Huybrechts, and I. Joughin, 2005, Ice-sheet and sea-level changes, *Science*, **310**, 456-460.

AMAP (Arctic Monitoring and Assessment Programme), 2011, *Snow, Water, Ice and Permafrost in the Arctic (SWIPA): Climate Change and the Cryosphere*, Arctic Monitoring and Assessment Programme, Oslo, Norway, 537 pp.

Amundson, J.M., M. Fahnestock, M. Truffer, J. Brown, M.P. Lüthi, and R.J. Motyka, 2010, Ice mélange dynamics and implications for terminus stability, Jakobshavn Isbræ, Greenland, *Journal of Geophysical Research*, **115**, F01005, doi:1029/2009JF001405.

Anderson, R.S., 1990, Evolution of the northern Santa Cruz Mountains by advection of crust past a San Andreas Fault bend, *Science*, **249**, 397-401.

Antonov, J.I., S. Levitus, and T.P. Boyer, 2002, Steric sea level variations during 1957-1994: Importance of salinity, *Journal of Geophysical Research*, **107**, 8013, doi:10.1029/2001JC000964.

Antonov, J.I., S. Levitus, and T.P. Boyer, 2005, Steric variability of the world ocean, 1955-2003, *Geophysical Research Letters*, **32**, L12602, doi:10.1029/2005GL023112.

Argus, D.F., and R.G. Gordon, 2001, Present tectonic motion across the Coast Ranges and San Andreas Fault System in central California, *Geological Society of America Bulletin*, **113**, 1580-1592.

Argus, D.F., and W.R. Peltier, 2010, Constraining models of postglacial rebound using space geodesy: A detailed assessment of model ICE-5G (VM2) and its relatives, *Geophysical Journal International*, **181**, 697-723.

Argus, D.F., M.B. Heflin, A. Donnellan, F.H. Webb, D. Dong, K.J. Hurst, D.C. Jefferson, G.A. Lyzenga, M.M. Watkins, and J.F. Zumberge, 1999, Shortening and thickening of metropolitan Los Angeles measured and inferred by using geodesy, *Geology*, **27**, 703-706.

Argus, D.F., M.B. Heflin, G. Peltzer, F. Crampe, and F.H. Webb, 2005, Interseismic strain accumulation and anthropogenic motion in metropolitan Los Angeles, *Journal of Geophysical Research*, **110**, B04401, doi:10.1029/2003JB002934.

Armstrong, G., and R.E. Flick, 1989, Storm damage assessment for the January 1988 storm along the southern California shoreline, *Shore & Beach*, **57**, 18-23.

Atwater, B.F., 1982, Geologic maps of the Sacramento-San Joaquin Delta, California, U.S. Geological Survey, MF-1401, Menlo Park, California.

Atwater, B.F., 1987, Evidence for great Holocene earthquakes along the outer coast of Washington state, *Science*, **236**, 942-944.

Atwater, B.F., and E. Hemphill-Haley, 1997, *Recurrence Intervals for Great Earthquakes of the Past 3500 Years at Northeastern Willapa Bay, Washington*, U.S. Geological Survey Professional Paper 1576, U.S. Government Printing Office, Washington, D.C., 108 pp.

Atwater, B.F., C.W. Hedel, and E.J. Helley, 1977, *Late Quaternary Depositional History, Holocene Sea-Level Changes, and Vertical Crust Movement, Southern San Francisco Bay, California*, U.S. Government Printing Office, Washington, D.C., 15 pp.

Atwater, B.F., S.G. Conard, J.N. Dowden, C.W. Hedel, R.L. MacDonald, and W. Savage, 1979, History, landforms, and vegetation of the estuary's tidal marshes, in *San Francisco Bay: The Urbanized Estuary*, T.J. Conomos, A.E. Leviton, and M. Berson, eds., American Association for the Advancement of Science, San Francisco, California, pp. 347-385.

Atwater, B.F., M. Satoko, S. Kenji, T. Yoshinobu, U. Kazue, and D.K. Yamaguchi, 2005, *The Orphan Tsunami of 1700*, U.S. Geological Survey Professional Paper 1707, Reston, Virginia, 133 pp.

Augustin, L.N., J.L. Irish, and P. Lynett, 2009, Laboratory and numerical studies of wave damping by emergent and near-emergent wetland vegetation, *Coastal Engineering*, **56**, 332-340.

Bahr, D.B., M.F. Meier, and S.D. Peckham, 1997, The physical basis of glacier volume-area scaling, *Journal of Geophysical Research*, **102**, 20,355-20,362.

Bamber, J.L., R.L. Layberry, and S.P. Gogenini, 2001, A new ice thickness and bed data set for the Greenland Ice Sheet 1: Measurement, data reduction, and errors, *Journal of Geophysical Research*, **106**, 33,773-33,780.

Barnard, P.L., J. Allan, J.E. Hansen, G.M. Kaminsky, P. Ruggiero, and A. Doria, 2011, The impact of the 2009–10 El Niño Modoki on U.S. west coast beaches, *Geophysical Research Letters*, **38**, L13604, doi:10.1029/2011GL047707.

Barnhart, R.A., M.J. Boyd, and J.E. Pequegnat, 1992, *Ecology of Humboldt Bay, California: An Estuarine Profile*, U.S. Fish and Wildlife Service, Biological Report 1, Washington, D.C., 121 pp.

Bartholdy, J., 2001, Storm surge effects on a back-barrier tidal flat of the Danish Wadden Sea, *Geo-Marine Letters*, **20**, 133-141.

Baschek, B., and J. Imai, 2011, Rogue wave observations off the U.S. west coast, *Oceanography*, **24**, 158-165.

Bauer, B.O., M.S. Lorang, and D.J. Sherman, 2002, Estimating boat-wake-induced levee erosion using sediment suspension measurements, *Journal of Waterway, Port, Coastal, and Ocean Engineering*, **128**, 152-163.

Bawden, G.W., W.R. Thatcher, R.S. Stein, K.W. Hudnut, and G. Peltzer, 2001, Tectonic contraction across Los Angeles after removal of groundwater pumping effects, *Nature*, **412**, 812-815.

The Bay Institute, 1998, *From the Sierra to the Sea: The Ecological History of the San Francisco Bay-Delta Watershed*, Novato, CA, 272 pp.

Bearman, J.A., C.T. Friedrichs, B.E. Jaffe, and A.C. Foxgrover, 2010, Spatial trends in tidal flat shape and associated environmental parameters in south San Francisco Bay, *Journal of Coastal Research*, **26**, 342-349.

Bennett, R.A., J.L. Davis, and B.P. Wernicke, 1999, Present-day pattern of Cordilleran deformation in the western United States, *Geology*, **27**, 371-374.

Bentley, C.R., 1987, Antarctic ice streams: A review, *Journal of Geophysical Research*, **92**, 8843-8858.

Benumof, B.T., and G.B. Griggs, 1999, The dependence of seacliff erosion rates on cliff material properties and physical processes, San Diego, California, *Shore & Beach*, **67**, 29-41.

Benumof, B.T., C.D. Storlazzi, R.J. Seymour, and G.B. Griggs, 2000, The relationship between incident wave energy and seacliff erosion rates: San Diego County, California, *Journal of Coastal Research*, **16**, 1162-1178.

Berry, H.D., A.T. Sewell, S. Wyllie-Echeverria, B.R. Reeves, T.F. Mumford, Jr., J.R. Skalski, R.C. Zimmerman, and J. Archer, 2003, *Puget Sound Submerged Vegetation-Monitoring Project: 2000-2003 Monitoring Report*, Washington State Department of Natural Resources, Olympia, Washington, 60 pp. plus appendixes.

Biancamaria, S., A. Cazenave, N.M. Mognard, W. Llovel, and F. Frappart, 2011, Satellite-based high latitude snow volume trend, variability and contribution to sea level over 1989/2006, *Global and Planetary Change*, **75**, 99-107.

Bindoff, N.L., J. Willebrand, V. Artale, A. Cazenave, J. Gregory, S. Gulev, K. Hanawa, C. Le Quéré, S. Levitus, Y. Nojiri, C.K. Shum, L.D. Talley, and A. Unnikrishnan, 2007, Observations: Oceanic climate change and sea level, in *Climate Change 2007: The Physical Science Basis*, Contribution of Working Group I to the Fourth Assessment Report of the Intergovernmental Panel on Climate Change, S. Solomon, D. Qin, M. Manning, Z. Chen, M. Marquis, K.B. Averyt, M. Tignor and H.L. Miller, eds., Cambridge University Press, Cambridge, pp. 385-432.

Bindschadler, R.A., 1985, Contribution of the Greenland Ice Cap to changing sea level, present and future, in *Glaciers, Ice Sheets, and Sea Level: Effects of a CO_2-Induced Climatic Change*, National Academy Press, Washington, D.C., pp 258-266.

Blaszczyk, M., J. Jania, and J. Hagen, 2009, Tidewater glaciers of Svalbard: Recent changes and estimates of calving fluxes, *Polish Polar Research*, **30**, 85-142.

Bond, N.A., J.E. Overland, M. Spillane, and P. Stabeno, 2003, Recent shifts in the state of the North Pacific, *Geophysical Research Letters*, **30**, 2183, doi:10.1029/2003GL018597.

Bortelson, G.C., M.J. Chrzastowski, and A.K. Helgerson, 1980, *Historical Changes of Shoreline and Wetland at Eleven Major Deltas in the Puget Sound Region, Washington*, Atlas HA-617, U.S. Geological Survey, Denver, Colorado, 11 sheets.

Brain, M.J., A.J. Long, D.N. Petley, B.P. Horton, and R.J. Allison, 2011, Compression behaviour of minerogenic low energy intertidal sediments, *Sedimentary Geology*, **233**, 28-41.

Breaker, L.C., and A. Ruzmaikin, 2010, The 154-year record of sea level at San Francisco: Extracting the long-term trend, recent changes, and other tidbits, *Climate Dynamics*, **36**, 545-559.

Bromirski, P.D., and R.E. Flick, 2008, Storm surge in the San Francisco Bay/Delta and nearby coastal locations, *Shore & Beach*, **76**, 29-37.

Bromirski, P., R.E. Flick, and D.R. Cayan, 2003, Storminess variability along the California coast: 1858-2000, *Journal of Climate*, **16**, 982-993.

Bromirski, P.D., D.R. Cayan, and R.E. Flick, 2005, Wave spectral energy variability in the northeast Pacific, *Journal of Geophysical Research*, **110**, C03005, doi:10.1029/2004/2004JC002398.

Bromirski, P.D., A.J. Miller, R.E. Flick, and G. Auad, 2011, Dynamical suppression of sea level rise along the Pacific coast of North America: Indications for imminent acceleration, *Journal of Geophysical Research*, **116**, C07005, doi:10.1029/2010JC006759.

Brooks, B.A., M.A. Merrifield, J. Foster, C.L. Werner, F. Gomez, M. Bevis, and S. Gill, 2007, Space geodetic determination of spatial variability in relative sea level change, Los Angeles Basin, *Geophysical Research Letters*, **34**, L01611, doi:10.1029/2006GL028171.

Brooks, B.A., G. Bawden, D. Manjunath, C. Werner, N. Knowles, J. Foster, J. Dudas, and D. Cayan, 2012, Contemporaneous subsidence and levee overtopping potential, Sacramento-San Joaquin Delta, California, *San Francisco Estuary and Watershed Science*, **10**(1), 18 pp.

Brown, L.R., 2003, Will tidal wetland restoration enhance populations of native fishes? *San Francisco Estuary and Watershed Science*, **1**(1), 42 pp.

Burgette, R.J., R.J. Weldon, and D.A. Schmidt, 2009, Interseismic uplift rates for western Oregon and along-strike variation in locking on the Cascadia subduction zone, *Journal of Geophysical Research*, **114**, B01408, doi:10.1029/2008JB005679.

Bürgmann, R., G. Hilley, A. Ferretti, and F.Novali, 2006, Resolving vertical tectonics in the San Francisco Bay area from GPS and permanent scatterer InSAR analysis, *Geology*, **34**, 221-224.

Burkett, V., C. Groat, and D. Reed, 2007, Hurricanes not the key to a sustainable coast, *Science*, **315**, 1366-1367.

Byrd, K.B., and M. Kelly, 2006, Salt marsh vegetation response to edaphic and topographic changes from upland sedimentation in a Pacific estuary, *Wetlands*, **26**, 813-829.

Cabanes, C., A. Cazenave, and C. Le Provost, 2001, Sea level rise during past 40 years determined from satellite and in situ observations, *Science*, **294**, 840-842.

Cahoon, D.R., 2006, A review of major storm impacts on coastal wetland elevation, *Estuaries and Coasts*, **29**, 889-898.

Cahoon, D.R., J.C. Lynch, P. Hensel, R. Boumans, B.C. Perez, B. Segura, and J.W. Day, Jr., 2002, A device for high precision measurement of wetland sediment elevation: I. Recent improvements to the sedimentation-erosion table, *Journal of Sedimentary Research*, **72**, 730-733.

Caires, S., A. Sterl, G. Komen, and V. Swail, 2004, The web-based KNMI/ERA-40 global wave climatology atlas, *Bulletin of the World Meteorological Organization*, **53**, 142-146.

CALFED (CALFED Bay-Delta Program), 2000, *Ecosystem Restoration Program Plan: Final Programmatic Environmental Impact Statement/Environmental Impact Report*, Sacramento, California, 218 pp.

Carlisle, E.L., 2004, *Seagrass Research and Management in Washington State*, Master's Thesis, University of Washington, Seattle, Washington, 176 pp.

Carton, J.A., and B.S. Giese, 2008, A reanalysis of ocean climate using Simple Ocean Data Assimilation (SODA), *Monthly Weather Review*, **136**, 2999-3017.

Carton, J.A., B.S. Giese, and S.A. Grodsky, 2005, Sea level rise and the warming of the oceans in the Simple Ocean Data Assimilation (SODA) ocean reanalysis, *Journal of Geophysical Research*, **110**, C09006, doi:10.1029/2004JC002817.

Cayan, D.R., P.D. Bromirski, K. Hayhoe, M. Tyree, M.D. Dettinger, and R.E. Flick, 2008, Climate change projections of sea level extremes along the California coast, *Climatic Change*, **87**, S57-S73.

Cayan, D., M. Tyree, M. Dettinger, H. Hidalgo, T. Das, E. Maurer, P. Bromirski, N. Graham, and R. Flick, 2009, *Climate Change Scenarios and Sea Level Rise Estimates for the California 2009 Climate Change Scenarios Assessment*, California Climate Change Center, CEC-500-2009-014-F, Sacramento, CA, 50 pp.

Cazenave, A., and W. Llovel, 2010, Contemporary sea level rise, *Annual Review of Marine Science*, **2**, 145-173.

Cazenave, A., K. Dominh, S. Guinehut, E. Berthier, W. Llovel, G. Ramillien, M. Ablain, and G. Larnicol, 2009, Sea level budget over 2003-2008: A reevaluation from GRACE space gravimetry, satellite altimetry and Argo, *Global and Planetary Change*, **65**, 83-88.

Chambers, D., J. Wahr, and R. Nerem, 2004, Preliminary observations of global ocean mass variations with GRACE, *Geophysical Research Letters*, **31**, L13310, doi:10.1029/2004GL020461.

Chao, B.F., Y.H. Wu, and Y.S. Li, 2008, Impact of artificial reservoir water impoundment on global sea level, *Science*, **320**, 212-214.

Chelton, D.B., and R.E. Davis, 1982, Monthly mean sea-level variability along the west coast of North America, *Journal of Physical Oceanography*, **12**, 757-784.

Chen, J.L., C.R. Wilson, D. Blankenship, and B.D. Tapley, 2009, Accelerated Antarctic ice loss from satellite gravity measurements, *Nature Geoscience*, **2**, 859-862.

Chenoweth, M., and C. Landsea, 2004, The San Diego hurricane of 2 October 1858, *Bulletin of the American Meteorological Society*, **85**, 1689-1697.

Cherry, J.A., K.L. McKee, and J.B. Grace, 2009, Elevated CO_2 enhances biological contributions to elevation change in coastal wetlands by offsetting stressors associated with sea-level rise, *Journal of Ecology*, **97**, 67-77.

Church, J.A., 2001, How fast are sea levels rising? *Science*, **294**, 802-803.

Church, J.A., and N.J. White, 2006, A 20th century acceleration in global sea-level rise, *Geophysical Research Letters*, **33**, L01602, doi:10.1029/2005GL024826.

Church, J.A., and N.J. White, 2011, Sea-level rise from the late 19th to the early 21st century, *Surveys in Geophysics*, **32**, 585-602.

Church, J.A., D. Roemmich, C.M. Domingues, J.K. Willis, N.J. White, J.E. Gilson, D. Stammer, A. Kohl, D.P. Chambers, F.W. Landerer, J. Marotzke, J.M. Gregory, T. Suzuki, A. Cazenave, and P.-Y. Le Traon, 2010, Ocean temperature and salinity contributions to global and regional sea-level change, in *Understanding Sea-Level Rise and Variability*, J.A. Church, P.L. Woodworth, T. Aarup, and W.S. Wilson, eds., Wiley-Blackwell, UK, pp. 143-176.

Church, J.A., N.J. White, L.F. Konikow, C.M. Domingues, J.G. Cogley, E. Rignot, J.M. Gregory, M.R. van den Broeke, A.J. Monaghan, and I. Velicogna, 2011, Revisiting the Earth's sea-level and energy budgets from 1961 to 2008, *Geophysical Research Letters*, **38**, L18601, doi:10.1029/2011GL048794.

Clancey, M., I. Logan, J. Lowe, J. Johannessen, A. MacLennan, F.B. Van Cleve, J. Dillon, B. Lyons, R. Carman, P. Cereghino, B. Barnard, C. Tanner, D. Myers, R. Clark, J. White, C. Simenstad, M. Gilmer, and N. Chin, 2009, *Management Measures for Protecting and Restoring the Puget Sound Nearshore*, Puget Sound Nearshore Partnership Report 2009-01, Washington Department of Fish and Wildlife, Olympia, Washington, 307 pp.

Clark, J.A., W.E. Farrell, and W.R. Peltier, 1978, Global changes in postglacial sea level: A numerical calculation, *Quaternary Research*, **9**, 265-287.

Clark, P.U., A.S. Dyke, J.D. Shakun, A.E. Carlson, J. Clark, B. Wohlfarth, J.X. Mitrovica, S.W. Hostetler, and A.M. McCabe, 2009, The Last Glacial Maximum, *Science*, **325**, 710-714.

Cloern, J.E., N. Knowles, L.R. Brown, D. Cayan, M.D. Dettinger, T.L. Morgan, D.H. Schoellhamer, M.T. Stacey, M. van der Wegen, R.W. Wagner, and A.D. Jassby, 2011, Projected evolution of California's San Francisco Bay-Delta-River System in a century of climate change, *PLoS ONE*, **6**, e24465, doi:10.1371/journal.pone.0024465.

Clough, J.S., and R.A. Park, 2007, *Technical Documentation for SLAMM 5.0*, Warren Pinnacle Consulting, Inc., Warren, VT.

CO-CAT (Sea-Level Rise Task Force of the Coastal and Ocean Working Group of the California Climate Action Team), 2010, *State of California Sea-Level Rise Interim Guidance Document*, October 2010, 18 pp.

Cogley, J.G., 2009, A more complete version of the World Glacier Inventory, *Annals of Glaciology*, **50**, 32-38.

Cogley, J.G., 2012, The future of the world's glaciers, in *Future Climates of the World*, 2nd edition, A. Henderson-Sellers and K. McGuffie, eds., Elsevier, Waltham, MA, pp. 197-222.

Cogley, J.G., J.S. Kargel, G. Kaser, and C.J. van der Veen, 2010, Tracking the source of glacier misinformation, *Science*, **337**, 522.

Collins, J.N., and R.M. Grossinger, 2004, *Synthesis of Scientific Knowledge Concerning Estuarine Landscapes and Related Habitats of the South Bay Ecosystem*, Technical Report of the South Bay Salt Pond Restoration Project, San Francisco Estuary Institute, Oakland, California, 91 pp.

Cooke, M., and S.T. Marshall, 2006, Fault slip rates from three-dimensional models of the Los Angeles metropolitan area, California, *Geophysical Research Letters*, **33**, L21313, doi:10.1029/2006GL027850.

Cooper, W.S., 1958, *Coastal Sand Dunes of Oregon and Washington*, Geological Society of America Memoir 72, Washington, D.C., 169 pp.

Cooper, W.S., 1967, *Coastal Sand Dunes of California*, Geological Society of America Memoir 104, Boulder, Colorado, 131 pp.

Corell, R., 2005, Arctic climate impact assessment, *Bulletin of the American Meteorological Society*, **86**, 860-861.

Crowell, M., H. Leikin, and M.K. Buckley, 1999, Evaluation of coastal erosion hazards study: An overview, *Journal of Coastal Research*, Special Issue 28, 2-9.

Cummins, P.F., and H.J. Freeland, 2007, Variability of the North Pacific current and its bifurcation, *Progress in Oceanography*, **75**, 253-265.

Czuba, J.A., C.S. Magirl, C.R. Czuba, E.E. Grossman, C.A. Curran, A.S. Gendaszek, and R.S. Dinicola, 2011, Sediment load from major rivers into Puget Sound and its adjacent waters, USGS Fact Sheet 2011-3083, available at <http://pubs.usgs.gov/fs/2011/3083/>.

Dahdouh-Guebas, F., and N. Koedam, 2006, Coastal vegetation and the Asian tsunami, *Science*, **311**, 37-38.

Danielsen, F., M.K. Sørensen, M.F. Olwig, V. Selvam, F. Parish, N. Burgess, T. Hiraishi, V. Karunagaran, M. Rasmussen, L. Hansen, A. Quarto, and N. Suryadiputra, 2005, The Asian tsunami: A protective role for coastal vegetation, *Science*, **310**, 643.

Dare, J., 2005, Coastal erosion and armor database for California, California Coastal Commission, CD-ROM, San Francisco, CA.

Darienzo, M.E., and C.D. Peterson, 1995, Magnitude and frequency of subduction-zone earthquakes along the northern Oregon coast in the past 3,000 years, *Oregon Geology*, **57**, 3-12.

Dean, R.G., and R.A. Dalrymple, 1991, *Water Wave Mechanics for Engineers and Scientists*, World Scientific Press, Singapore, 353 pp.

Di Lorenzo, E., N. Schneider, K.M. Cobb, P.J.S. Franks, K. Chhak, A.J. Miller, J.C. McWilliams, S.J. Bograd, H. Arango, E. Curchitser, T.M. Powell, and P. Riviere, 2008, North Pacific Gyre Oscillation links ocean climate and ecosystem change, *Geophysical Research Letters*, **35**, L08607, doi:10.1029/2007GL032838.

Di Lorenzo, E., K.M. Cobb, J.C. Furtado, N. Schneider, B.T. Anderson, A. Bracco, M.A. Alexander, and D.J. Vimont, 2010, Central Pacific El Niño and decadal climate change in the North Pacific Ocean, *Nature Geoscience*, **3**, 762-765.

Domingues, C.M., J.A. Church, N.J. White, P.J. Gleckler, S.E. Wijffels, P.M. Barker, and J.R. Dunn, 2008, Improved estimates of upper-ocean warming and multi-decadal sea-level rise, *Nature*, **453**, 1090-1093.

Domurat, G.W., and A.T. Shak, 1989, The storm of 1988: Damage to coastal structures, *Shore & Beach*, **57**, 24-27.

Donaldson, E.C., G.V. Chilingarian, and H.H. Rieke, 1995, Stresses in sediments, in *Subsidence Due to Fluid Withdrawal: Developments in Petroleum Science*, G.V. Chilingarian, E.C. Donaldson, and T.F. Yen, eds., Elsevier Science, Amsterdam, pp. 165-190.

Dong, D., T.A. Herring, and R.W. King, 1998, Estimating regional deformation from a combination of space and terrestrial geodetic data, *Journal of Geodesy*, **72**, 200-214.

Donnellan, A., B. Hager, R. King, and T. Herring, 1993, Geodetic measurement of deformation in the Ventura Basin region, southern California, *Journal of Geophysical Research*, **98**, 21,727-21,739.

Donnelly, J.P., P. Cleary, P. Newby, and R. Ettinger, 2004, Coupling instrumental and geological records of sea-level change: Evidence from southern New England of an increase in the rate of sea-level rise in the late 19th century, *Geophysical Research Letters*, **31**, L05203, doi:10.1029/2003GL018933.

Douglas, B.C., 1991, Global sea level rise, *Journal of Geophysical Research*, **96**, 6981-6992.

Douglas, B.C., 1992, Global sea level acceleration, *Journal of Geophysical Research*, **97**, 12,699-12,706.

Dyurgerov, M., 2002, *Glacier Mass Balance and Regime: Data of Measurements and Analysis*, Occasional paper 55, Institute of Arctic and Alpine Research, University of Colorado, Boulder, 275 pp.

Dyurgerov, M.B., 2010, *Reanalysis of Glacier Changes: From the IGY to the IPY, 1960–2008*, Data of Glaciological Studies, Publication 108, Moscow, 116 pp.

Dyurgerov, M., and M.F. Meier, 2005, *Glaciers and the Changing Earth System: A 2004 Snapshot*, Occasional Paper 58, Institute of Arctic and Alpine Research, University of Colorado, Boulder, 117 pp.

Eagleson, P.S., 1994, The evolution of modern hydrology (from watershed to continent in 30 years), *Advances in Water Resources*, **17**, 3-18.

Edelman, T., 1972, Dune erosion during storm conditions, in *Proceedings of the 13th Conference on Coastal Engineering*, American Society of Civil Engineers, pp. 1305-1312.

Ekman, M., 1988, The world's longest continued series of sea level observations, *Pure and Applied Geophysics*, **127**, 73-77.

Ellis, J.T., D.J. Sherman, B.O. Bauer, and J. Hart, 2002, Assessing the impact of an organic restoration structure on boat wake energy, *Journal of Coastal Research*, **36**, 256-265.

Engelhart, S.E., W.R. Peltier, and B.P. Horton, 2011, Holocene relative sea-level changes and glacial isostatic adjustment of the U.S. Atlantic coast, *Geology*, **39**, 751-754.

Famiglietti, J.S., L. Murdoch, V. Lakshmi, and R.P. Hooper, 2009, Towards a framework for community modeling in hydrologic science: Blueprint for a community hydrologic modeling platform, in *Proceedings of the 2nd Workshop on a Community Hydrologic Modeling Platform*, Memphis, Tennessee, March 31-April 1, 2009, available at <http://www.cuahsi.org/docs/dois/CUAHSI-TR9.pdf>.

Farrell, W.E., and J.A. Clark, 1976, On postglacial sea level, *Geophysical Journal of the Royal Astronomical Society*, **46**, 647-667.

Feagin, R.A., 2008, Vegetation's role in coastal protection, *Science*, **320**, 176-177.

Feagin, R.A., J.L. Irish, I. Möller, A.M. Williams, R.J. Colón-Rivera, and M.E. Mousavi, 2011, Short communication: Engineering properties of wetland plants with application to wave attenuation, *Coastal Engineering*, **58**, 251-255.

Feist, B.E., and C. Simenstad, 2000, Expansion rates and recruitment frequency of exotic smooth cordgrass, *Spartina alterniflora* (Loisel), colonizing unvegetated littoral flats in Willapa Bay, *Estuaries and Coasts*, **23**, 267-274.

Finlayson, D., 2006, *The Geomorphology of Puget Sound Beaches*, Puget Sound Nearshore Partnership Report 2006-02, Washington Sea Grant Program, University of Washington, Seattle, Washington, 45 pp.

Finnegan, N.J., M.E. Pritchard, R.B. Lohman, and P.R. Lundgren, 2008, Constraints on surface deformation in the Seattle, WA, urban corridor from satellite radar interferometry time-series analysis, *Geophysical Journal International*, **174**, 29-41.

Flechtner, F., 2007, *AOD1B Product Description Document for Product Releases 01 to 04, GRACE 327-750*, CSR publication GR-GFZ-AOD-0001, Rev. 3.1, University of Texas, Austin, 43 pp.

Fleming, K., P. Johnston, D. Zwartz, Y. Yokoyama, K. Lambeck, and J. Chappell, 1998, Refining the eustatic sea-level curve since the Last Glacial Maximum using far- and intermediate-field sites, *Earth and Planetary Science Letters*, **163**, 327-342.

Flick, R.E., 1998, Comparison of tides, storm surges, and mean sea level during the El Niño winters of 1982–83 and 1997–98, *Shore & Beach*, **66**, 7-17.

Flick, R.E., 2000, Time-of-day of peak tides in a mixed-tide regime, *Shore & Beach*, **68**, 15-17.

Flick, R.E., J.F. Murray, and L.C. Ewing, 2003, Trends in United States tidal datum statistics and tide range, *Journal of Waterway, Port, Coastal, and Ocean Engineering*, **129**, 155-164.

Flück, P., R.D. Hyndman, and K. Wang, 1997, Three-dimensional dislocation model for great earthquakes of the Cascadia Subduction Zone, *Journal of Geophysical Research*, **102**, 20,539-20,550.

Fonseca, M.S., W.J. Kenworthy, and G.W. Thayer, 1998, *Guidelines for the Conservation and Restoration of Seagrasses in the United States and Adjacent Waters*, NOAA Coastal Ocean Program Decision Analysis Series, No. 12, NOAA Coastal Ocean Offices, Silver Spring, Maryland, 222 pp.

Foxgrover, A.C., S.A. Higgins, M.K. Ingraca, B.E. Jaffe, and R.E. Smith, 2004, *Deposition, Erosion, and Bathymetric Change in South San Francisco Bay: 1858-1983*, U.S. Geological Survey, Open-File Report 2004-1192, Reston, Virginia, 25 pp.

French, J.R., and D.J. Reed, 2001, Physical contexts for saltmarsh conservation, in *Habitat Conservation: Managing the Physical Environment*, A. Warren and J.R. French, eds., John Wiley & Son, Chichester, pp. 179-228.

Fu, L.-L., and B. Qiu, 2002, Low-frequency variability of the North Pacific Ocean: The roles of boundary-and wind-driven baroclinic Rossby waves, *Journal of Geophysical Research*, **107**, 3220, doi:10.1029/2001JC001131.

Galloway, D.L., D.R. Jones, and S.E. Ingebritsen, eds., 2001, *Land Subsidence in the United States*, U.S. Geological Survey Circular 1182, Denver, Colorado, 175 pp.

Ganju, N.K., and D.H. Schoellhamer, 2010, Decadal-timescale estuarine geomorphic change under future scenarios of climate and sediment supply, *Estuaries and Coasts*, **33**, 15-29.

Gardner, A.S., G. Moholdt, B. Wouters, G.J. Wolken, D.O. Burgess, M.J. Sharp, J.G. Cogley, C. Braun, and C. Labine, 2011, Sharply increased mass loss from glaciers and ice caps in the Canadian Arctic Archipelago, *Nature*, **473**, 357-360.

Gehrels, W.R., 2010, Sea-level changes since the Last Glacial Maximum: An appraisal of the IPCC Fourth Assessment Report, *Journal of Quaternary Science*, **25**, 26-38.

Gehrels, W.R., G.A. Milne, J.R. Kirby, R.T. Patterson, and D.F. Belknap, 2004, Late Holocene sea-level changes and isostatic crustal movements in Atlantic Canada, *Quaternary International*, **120**, 79-89.

Gehrels, W.R., W.A. Marshall, M.J. Gehrels, G. Larsen, J.R. Kirby, J. Eiriksson, J. Heinemeier, and T. Shimmield, 2006, Rapid sea-level rise in the North Atlantic Ocean since the first half of the 19th century, *The Holocene*, **16**, 948-964.

Gehrels, W.R., B.W. Hayward, R.M. Newnham, and K.E. Southall, 2008, A 20th century sea-level acceleration in New Zealand, *Geophysical Research Letters*, **35**, L02717, doi:10.1029/2007GL032632.

Gelfenbaum, G., T. Mumford, J. Brennan, H. Case, M. Dethier, K. Fresh, F. Goetz, M. van Heeswijk, T.M., Leschine, M. Logsdon, D. Myers, J. Newton, H. Shipman, C.A. Simenstad, C. Tanner, and D. Woodson, 2006, *Coastal Habitats in Puget Sound: A Research Plan in Support of the Puget Sound Nearshore Partnership*, Puget Sound Nearshore Partnership Report 2006-1, U.S. Geological Survey, Seattle, Washington, 44 pp.

Gemmrich, J., B. Thomas, and R. Bouchard, 2011, Observational changes and trends in northeast Pacific wave records, *Geophysical Research Letters*, **38**, L22601, doi:10.1029/2011GL049518.

Gilbert, G.K., 1917, *Hydraulic Mining in the Sierra Nevada*, U.S. Geological Survey Professional Paper 105, U.S. Government Printing Office, Washington, D.C., 154 pp.

Gille, S.T., 2008, Decadal-scale temperature trends in the southern hemisphere ocean, *Journal of Climate*, **21**, 4749-4765.

Glick, P., J. Clough, and B. Nunley, 2007, *Sea-level Rise and Coastal Habitats in the Pacific Northwest: An Analysis for Puget Sound, Southwestern Washington, and Northwestern Oregon*, National Wildlife Federation, Seattle, Washington, 94 pp.

Goals Project, 1999, *Baylands Ecosystem Habitat Goals: A Report of Habitat Recommendations Prepared by the San Francisco Bay Area Wetlands Ecosystem Goals Project*, Environmental Protection Agency, San Francisco, California, and San Francisco Bay Regional Water Quality Control Board, Oakland, California, 295 pp.

Goldfinger, C., C.H. Nelson, and J.F. Johnson, 2003, Holocene earthquake records from the Cascadia Subduction Zone and northern San Andreas Fault based on precise dating of offshore turbidites, *Annual Review of Earth and Planetary Sciences*, **31**, 555-577.

Goldfinger, C., K. Grijalva, R. Bürgmann, A. Morey, J.E. Johnson, C.H. Nelson, J. Gutiérrez-Pastor, A. Ericsson, E. Karabanov, J.D. Chaytor, J. Patton, and E. Gràcia, 2008, Late Holocene rupture of the northern San Andreas Fault and possible stress linkage to the Cascadia Subduction Zone, *Bulletin of the Seismological Society of America*, **98**, 861-889.

Gornitz, V., 2001, Impoundment, groundwater mining, and other hydrologic transformations: Impacts on sea level rise, in *Sea Level Rise, History and Consequences*, B.C. Douglas, M.S. Kearnery, and S.P. Leatherman, eds., Academic Press, San Diego, pp. 97-119.

Gornitz, V., and S. Lebedeff, 1987, Global sea-level changes during the past century, in *Sea-Level Fluctuation and Coastal Evolution*, D. Nummedal, O.H. Pilkey, and J.D. Howard, eds., SEPM Special Publication 41, Society for Sedimentary Geology, Tulsa, Oklahoma, pp. 3-16.

Gouretski, V., and K.P. Koltermann, 2007, How much is the ocean really warming? *Geophysical Research Letters*, **34**, L01610, doi:10.1029/2006GL027834.

Gouretski, V., and F. Reseghetti, 2010, On depth and temperature biases in bathythermograph data: Development of a new correction scheme based on analysis of a global database, *Deep Sea Research* (Part I, Oceanographic Research Papers), **57**, 812-833.

Graham, N.E., and H.F. Diaz, 2001, Evidence for intensification of North Pacific winter cyclones since 1948, *Bulletin of the American Meteorological Society*, **82**, 1869-1893.

Griggs, G.B., 1994, California's coastal hazards, *Journal of Coastal Research*, **12**, 1-15.

Griggs, G.B., 1999, The protection of California's coast: Past, present and future, *Shore & Beach*, **67**, 18-28.

Griggs, G.B., 2005, The impacts of coastal armoring, *Shore & Beach*, **73**, 13-22.

Griggs, G.B., 2010, *Introduction to California's Beaches and Coast*, University of California Press, Berkeley, CA, 311 pp.

Griggs, G.B., and K.B. Patsch, 2004, California's coastal cliffs and bluffs, in *Formation, Evolution, and Stability of Coastal Cliffs, Status and Trends*, M.A. Hampton and G.B. Griggs, eds., U.S. Geological Survey Professional Paper 1693, Denver, CO, pp. 53-64.

Griggs, G.B., and A.S. Trenhaile, 1994, Coastal cliffs and platforms, in *Coastal Evolution: Late Quaternary Shoreline Morphodynamics*, R.W.G. Carter and C.D. Woodroffe, eds., Press Syndicate of the University of Cambridge, Cambridge, pp. 425-450.

Griggs, G.B., K.B. Patsch, and L.E. Savoy, 2005, *Living with the Changing California Coast*, University of California Press, Berkeley, CA, 540 pp.

Grinsted, A., J.C. Moore, and S. Jevrejeva, 2009, Reconstructing sea level from paleo and projected temperatures 200 to 2100 AD, *Climate Dynamics*, **34**, 461-472.

Gruber, S., 2011, Derivation and analysis of a high-resolution estimate of global permafrost zonation, *The Cryosphere Discussions*, **5**, 1547-1582.

Gulev, S.K., and V. Grigorieva, 2006, Variability of the winter wind waves and swell in the North Atlantic and North Pacific as revealed by the voluntary observing ship data, *Journal of Climate*, **19**, 5667-5685.

Hager, B.H., G.A. Lyzenga, A. Donnellan, and D. Dong, 1999, Reconciling rapid strain accumulation with deep seismogenic fault planes in the Ventura Basin, California, *Journal of Geophysical Research*, **104**, 25,207-25,220.

Hansen, J.E., 2007, Scientific reticence and sea level rise, *Environmental Research Letters*, **2**, 024002, doi:10.1088/1748-9326/2/2/024002.

Hapke, C.J., and D. Reid, 2007, *National Assessment of Shoreline Change, Part 4: Historical Coastal Cliff Retreat along the California Coast*, U.S. Geological Survey Open File Report 2007-1133, 51 pp.

Hapke, C.J., D. Reid, B.M. Richmond, P. Ruggiero, and J. List, 2006, *National Assessment of Shoreline Change Part 3: Historical Shoreline Change and Associated Coastal Land Loss along Sandy Shorelines of the California Coast*, U.S. Geological Survey Open File Report 2006-1219, 72 pp.

Hawkes, A.D., B.P. Horton, and A.R. Nelson, 2011, Coastal subsidence in Oregon, USA, during the giant Cascadia earthquake of AD 1700, *Quaternary Science Reviews*, **30**, 364-376.

Hawkins, E., and R.T. Sutton, 2009, The potential to narrow uncertainty in regional climate predictions, *Bulletin of the American Meteorological Society*, **90**, 1095-1107.

Hayhoe, K., D. Cayan, C.B. Field, P.C. Frumhoff, E.P. Maurer, N.L. Miller, S.C. Moser, S.H. Schneider, K. Nicholas Cahill, E.E. Cleland, L. Dale, R. Drapek, R.M. Hanemann, L.S. Kalkstein, J. Lenihan, C.K. Lunch, R.P. Neilson, S.C. Sheridan, and H. Verville, 2004, Emissions pathways, climate change, and impacts on California, *Proceedings of the National Academy of Sciences*, **101**, 12,422-12,427.

He, J., K. Wang, H. Dragert, and M.M. Miller, 2003, Spherical viscoelastic finite element model for Cascadia interseismic deformation, *Eos Transactions of the American Geophysical Union*, **84**, Fall Meeting Supplement, Abstract S42A-0141.

Heberger, M., H. Cooley, P. Herrera, P.H. Gleick, and E. Moore, 2009, *The Impacts of Sea-Level Rise on the California Coast*, California Climate Change Center, CEC-500-2009-024-F, Sacramento, California, 101 pp.

Heberger, M., H. Cooley, P. Herrera, P.H. Gleick, and E. Moore, 2011, Potential impacts of increased coastal flooding in California due to sea-level rise, *Climatic Change*, **109**, 229-249.

Hegerl, G.C., F.W. Zwiers, P. Braconnot, N.P. Gillett, Y. Luo, J.A. Marengo Orsini, N. Nicholls, J.E. Penner, and P.A. Stott, 2007, Understanding and attributing climate change, in *Climate Change 2007: The Physical Science Basis*, Contribution of Working Group I to the Fourth Assessment Report of the Intergovernmental Panel on Climate Change, S. Solomon, D. Qin, M. Manning, Z. Chen, M. Marquis, K.B. Averyt, M. Tignor and H.L. Miller, eds., Cambridge University Press, Cambridge, pp. 663-745.

Holgate, S.J., 2007, On the decadal rates of sea level change during the twentieth century, *Geophysical Research Letters*, **34**, L01602, doi:10.1029/2006GL028492.

Holgate, S.J., and P.L. Woodworth, 2004, Evidence for enhanced coastal sea level rise during the 1990s, *Geophysical Research Letters*, **31**, L07305, doi:10.1029/2004GL019626.

Holland, D.M., R.H. Thomas, B. De Young, M.H. Ribergaard, and B. Lyberth, 2008, Acceleration of Jakboshavn Isbræ triggered by warm subsurface ocean waters, *Nature Geoscience*, **1**, 659-664.

Horton, B.P., and R.J. Edwards, 2006, *Quantifying Holocene Sea Level Change Using Intertidal Foraminifera: Lessons from the British Isles*, Cushman Foundation for Foraminiferal Research, Special Publication 40, 97 pp.

Horton, B.P., and Y. Sawai, 2010, Diatoms as indicators of coastal evolution, in *The Diatoms: Applications for the Environmental and Earth Sciences*, 2nd edition, J.P. Smol and E.F. Stoermer, eds., Cambridge University Press, Cambridge, pp. 357-372.

Horton, B.P., and I. Shennan, 2009, Compaction of Holocene strata and the implications for relative sea-level change, *Geology*, **37**, 1083-1086.

Hosack, G., B.R. Dumbauld, J.L. Ruesink, and D. Amstrong, 2006, Habitat associations of estuarine species: Comparisons of intertidal mudflat, seagrass (*Zostera marina*), and oyster (*Crassostrea gigas*) habitats, *Estuaries and Coasts*, **29**, 1150-1160.

Houston, J.R., and R.G., Dean, 2011, Sea-level acceleration based on U.S. tide gauges and extensions of previous global-gauge analyses, *Journal of Coastal Research*, **27**, 409-417.

Howat, I.M., I. Joughin, and T.A. Scambos, 2007, Rapid changes in ice discharge from Greenland outlet glaciers, *Science*, **315**, 1559-1561.

Huntington, T.G., 2008, Can we dismiss the effect of changes in land-based water storage on sea-level rise? *Hydrological Processes*, **22**, 717-723.

Hurd, W.E., 1939, North Pacific Ocean, September 1939, *Monthly Weather Review*, **67**, 357-358.

Huybrechts, P., 2002, Sea-level changes at the LGM from ice-dynamic reconstructions of the Greenland and Antarctic ice sheets during the glacial cycles, *Quaternary Science Reviews*, **21**, 203-231.

Hyndman, R.D., and K. Wang, 1993, Thermal constraints on the zone of major thrust earthquake failure: The Cascadia Subduction Zone, *Journal of Geophysical Research*, **98**, 2039-2060.

Hyndman, R.D., and K. Wang, 1995, The rupture zone of Cascadia great earthquakes from current deformation and the thermal regime, *Journal of Geophysical Research*, **100**, 22,133-22,154.

Ingleby, B., and M. Huddleston, 2007, Quality control of ocean temperature and salinity profiles—Historical and real-time data, *Journal of Marine Systems*, **65**, 158-175.

IPCC (Intergovernmental Panel on Climate Change), 2000, *Emission Scenarios*, N. Nakicenovic and R. Swart, eds., Cambridge University Press, Cambridge, UK, 570 pp.

IPCC, 2007, *Climate Change 2007: The Physical Science Basis*, Contribution of Working Group I to the Fourth Assessment Report of the IPCC, S. Solomon, D. Qin, M. Manning, Z. Chen, M. Marquis, K.B. Averyt, M. Tignor and H.L. Miller, eds., Cambridge University Press, Cambridge, 996 pp.

Ishii, M., and M. Kimoto, 2009, Revaluation of historical ocean heat content variations with time-varying XBT and MBT depth bias corrections, *Journal of Oceanography*, **65**, 287-299.

Ishii, M., M. Kimoto, K. Sakamoto, and S.I. Iwasaki, 2006, Steric sea level changes estimated from historical ocean subsurface temperature and salinity analyses, *Journal of Oceanography*, **62**, 155-170.

Jacob, T., J. Wahr, W.T. Pfeffer, and S. Swenson, 2012, Recent contributions of glaciers and ice caps to sea level rise, *Nature*, **482**, 514-518.

Jacobs, D.K., E.D. Stein, and T. Longcore, 2010, *Classification of California Estuaries Based on Natural Closure Patterns: Templates for Restoration and Management*, Southern California Coastal Water Research Project, Technical Report 619, 50 pp., available at <ftp://ftp.sccwrp.org/pub/download/DOCUMENTS/TechnicalReports/619_EstuarineClassificationRestorationDesign.pdf>.

Jaffe, B.E., R.E. Smith, and L. Torresan, 1998, *Sedimentation and Bathymetric Change in San Pablo Bay, 1856-1983*, U.S. Geological Survey Open-File Report 98-759, Menlo Park, California, 1 pp.

Jaffe, B.E., R.E. Smith, and A.C. Foxgrover, 2007, Anthropogenic influence on sedimentation and intertidal mudflat change in San Pablo Bay, California: 1856-1983, *Estuarine, Coastal and Shelf Science*, **73**, 175-187.

James, T.S., J.J. Clague, K. Wang, and I. Hutchinson, 2000, Postglacial rebound at the northern Cascadia Subduction Zone, *Quaternary Science Reviews*, **19**, 1527-1541.

Jelgersma, S., 1961, *Holocene Sea Level Changes in the Netherlands*, Van Aelst, Maastricht, 101 pp.

Jevrejeva, S., A. Grinsted, J.C. Moore, and S. Holgate, 2006, Nonlinear trends and multiyear cycles in sea level records, *Journal of Geophysical Research*, **111**, C09012, doi:10.1029/2005JC003 229.

Jevrejeva, S., J.C. Moore, A. Grinsted, and P.L. Woodworth, 2008, Recent global sea level acceleration started over 200 years ago? *Geophysical Research Letters*, **35**, L08715, doi:10.1029/2008GL033611.

Johnson, D.R., T.P. Boyer, H.E. Garcia, R.A. Locarnini, A.V. Mishonov, M.T. Pitcher, O.K. Baranova, J.I. Antonov, and I.V. Smolyar, 2006, *World Ocean Database 2005 Documentation*, S. Levitus, ed., NODC Internal Report 18, U.S. Government Printing Office, Washington, D.C., 163 pp.

Joughin, I., W. Abdalati, and M. Fahnestock, 2004, Large fluctuations in speed on Greenland's Jakobshavn Isbrae glacier, *Nature*, **432**, 608-610.

Kalnay, E., M. Kanamitsu, R. Kistler, W. Collins, D. Deaven, L. Gandin, M. Iredell, S. Saha, G. White, J. Woollen, Y. Zhu, M. Chelliah, W. Ebisuzaki, W. Higgins, J. Janowiak, K.C. Mo, C. Ropelewski, J. Wang, A. Leetmaa, R. Reynolds, R. Jenne, and D. Joseph, 1996, The NCEP/NCAR 40-year reanalysis project, *Bulletin of the American Meteorological Society*, **77**, 437-470.

Kaminsky, G.M., R.C. Daniels, R. Huxford, D. McCandless, and P. Ruggiero, 1999, Mapping erosion hazard areas in Pacific County, Washington, *Journal of Coastal Research*, Special Issue 28, 158-170.

Kaser, G., J. Cogley, M. Dyurgerov, M. Meier, and A. Ohmura, 2006, Mass balance of glaciers and ice caps: Consensus estimates for 1961-2004, *Geophysical Research Letters*, **33**, L19501, doi:10.1029/2006GL027511.

Katsman, C.A., and G.J. van Oldenborgh, 2011, Tracing the upper ocean's "missing heat," *Geophysical Research Letters*, **38**, L14610, doi:10.1029/2011GL048417.

Kaye, C.A., and E.S. Barghoorn, 1964, Quaternary sea-level change and crustal rise at Boston, Massachusetts, with notes on the autocompaction of peat, *Geological Society of America Bulletin*, **75**, 63-80.

Kelsey, H.M., R.C. Witter, and E. Hemphill-Haley, 2002, Plate-boundary earthquakes and tsunamis of the past 5500 years, Sixes River Estuary, southern Oregon, *Geological Society of America Bulletin*, **114**, 298-314.

Kemp, A.C., B.P. Horton, S.J. Culver, D.R. Corbett, O. van de Plassche, W.R. Gehrels, B.C. Douglas, and A.C. Parnell, 2009, Timing and magnitude of recent accelerated sea-level rise (North Carolina, United States), *Geology*, **37**, 1035-1038.

Kemp, A., B.P. Horton, J.P. Donnelly, M.E. Mann, M. Vermeer, and S. Rahmstorf, 2011, Climate related sea-level variations over the past two millennia, *Proceedings of the National Academy of Sciences*, **108**, 11,017-11,022.

Kerr, R.A., 2007, Pushing the scary side of global warming, News focus, *Science*, **316**, 1412-1415.

Kirwan, M.L., and A.B. Murray, 2007, A coupled geomorphic and ecological model of tidal marsh evolution, *Proceedings of the National Academy of Sciences*, **104**, 6118-6122.

Kirwan, M., and S. Temmerman, 2009, Coastal marsh response to historical and future sea-level acceleration, *Quaternary Science Reviews*, **28**, 1801-1808.

Kirwan, M., G. Guntenspergen, and J Morris, 2009, Latitudinal trends in *Spartina alterniflora* productivity and the response of coastal marshes to global change, *Global Change Biology*, **15**, 1982-1989.

Knowles, N., 2010, Potential inundation due to rising sea levels in the San Francisco Bay region, *San Francisco Estuary and Watershed Science*, **8**(1), 19 pp.

Knutti, R., G. Abramowitz, M. Collins, V. Eyring, P.J. Gleckler, B. Hewitson, and L. Mearns, 2010, *Good Practices Guidance Paper on Assessing and Combining Multi Model Climate Projections*, IPCC Expert Meeting, Boulder, January 25-27, 2010, 13 pp., available at <http://www.ipcc.ch/pdf/supporting-material/IPCC_EM_MME_GoodPracticeGuidancePaper.pdf>.

Koch, E., B. Silliman, E. Barbier, E. Granek, J. Primavera, N. Muthiga, D. Reed, G. Perillo, C. Kappel, D. Bael, C. Kennedy, S. Hacker, S. Polasky, E. Wolanski, S. Aswani, L. Cramer, and D. Stoms, 2009, Non-linearity in ecosystem services: Temporal, spatial and biological variability in coastal protection, *Frontiers in Ecology and the Environment*, **7**, 29-37.

Kohl, A., and D. Stammer, 2008, Decadal sea level changes in the 50-year GECCO ocean synthesis, *Journal of Climate*, **21**, 1876-1890.

Kollet, S.J., and R.M. Maxwell, 2008, Capturing the influence of groundwater dynamics on land surface processes using an integrated, distributed watershed model, *Water Resources Research*, **44**, W02402, doi:10.1029/ 2007WR006004.

Komar, P.D., 1997, *The Pacific Northwest Coast – Living with the Shores of Oregon and Washington*, Duke University Press, Durham, 195 pp.

Komar, P.D., and S.M. Shih, 1991, Sea-cliff erosion along the Oregon Coast, in *Proceedings of Coastal Sediments '91*, American Society of Civil Engineers, pp. 1558-1970.

Komar, P.D., J.C. Allan, and P. Ruggiero, 2011, Sea level variations along the U.S. Pacific northwest coast: Tectonic and climate controls, *Journal of Coastal Research*, **27**, 808-823.

Konikow, L.F., 2011, Contribution of global groundwater depletion since 1900 to sea-level rise, *Geophysical Research Letters*, **38**, L17401, doi:10.1029/2011GL048604.

Kopp, R.E., F.J. Simons, J.X. Mitrovica, A.C. Maloof, and M. Oppenheimer, 2009, Probabilistic assessment of sea level during the last interglacial stage, *Nature*, **462**, 863-867.

Kouketsu, S., T. Doi, T. Kawano, S. Masuda, N. Sugiura, Y. Sasaki, T. Toyoda, H. Igarashi, Y. Kawai, K. Katsumata, H. Uchida, M. Fukasawa, and T. Awaji, 2011, Deep ocean heat content changes estimated from observation and reanalysis product and their influence on sea level change, *Journal of Geophysical Research*, **116**, C03012, doi:10.1029/2010JC006464.

Kraft, J.C., H.-I. Yi, and M. Khalequzzaman, 1992, Geological and human factors in the decline of the tidal salt marsh lithosome: The Delaware Estuary and Atlantic coastal zone, *Sedimentary Geology*, **80**, 233-246.

Krone, R.B., 1987, A method for simulating historic marsh elevations, in *Coastal Sediments '87: Proceedings of the Specialty Conference on Quantitative Approaches to Coastal Sediment Processes*, May 12-14, 1987; New Orleans, Louisiana, pp. 316-323.

Kuhn, G.G., and F.P. Shepard, 1984, *Seacliffs, Beaches, and Coastal Valleys of San Diego County*, University of California Press, Berkeley, CA, 193 pp.

Lacy, J.R., and S. Wyllie-Echeverria, 2011, The influence of current speed and vegetation density on flow structure in two macrotidal eelgrass canopies, *Limnology and Oceanography: Fluids and Environments*, **1**, 38-55.

Lagerloef, G.S.E., 1995, Interdecadal variations in the Alaska Gyre, *Journal of Physical Oceanography*, **25**, 2242-2258.

Lambeck, K., M. Anzidei, F. Antonioli, A. Benini, and A. Esposito, 2004, Sea level in Roman time in the Central Mediterranean and implications for recent change, *Earth and Planetary Science Letters*, **224**, 563-575.

Lanari, R., P. Lundgren, M. Manzo, and F. Casu, 2004, Satellite radar interferometry time series analysis of surface deformation for Los Angeles, California, *Geophysical Research Letters*, **31**, L23613, doi:10.1029/2004GL021294.

Larsen, C.F., R.J. Motyka, A.A. Arendt, K.A. Echelmeyer, and P.E. Geissler, 2007, Glacier changes in southeast Alaska and northwest British Columbia and contribution to sea level rise, *Journal of Geophysical Research*, **112**, F01007, doi:10.1029/2006JF000586.

Le Brocq, A.M., A.J. Payne, and A. Vieli, 2010, An improved Antarctic dataset for high resolution numerical ice sheet models (ALBMAP v1), *Earth System Science Data*, **2**, 247-260.

Leclercq, P., J. Oerlemans, and J.G. Cogley, 2011, Estimating the glacier contribution to sea-level rise for the period 1800-2005, *Surveys in Geophysics*, **32**, 519-535.

Lee, L., C.J. Pinckney, and C. Bemis, 1976, Sea cliff base erosion, in *National Water Resources Ocean Engineering Conference*, April 5-8, 1976, San Diego, California, American Society of Civil Engineers, preprint 2708, 13 pp.

Lemke, P., J. Ren, R. Alley, I. Allison, J. Carrasco, G. Flato, Y. Fujii, G. Kaser, P. Mote, R. Thomas, and T.J. Zhang, 2007, Observations: Changes in snow, ice and frozen ground, in *Climate Change 2007: The Physical Science Basis*, Contribution of Working Group I to the Fourth Assessment Report of the Intergovernmental Panel on Climate Change, S. Solomon, D. Qin, M. Manning, Z. Chen, M. Marquis, K.B. Averyt, M. Tignor and H.L. Miller, eds., Cambridge University Press, Cambridge, pp. 337-383.

Leonard, L.J., R.D. Hyndman, and S. Mazzotti, 2004, Coseismic subsidence in the 1700 great Cascadia earthquake coastal estimates versus elastic dislocation models, *Geological Society of America Bulletin*, **116**, 655-670.

Leonard, L.J., C.A. Currie, S. Mazzotti, and R.D. Hyndman, 2010, Rupture area and displacement of past Cascadia great earthquakes from coastal coseismic subsidence, *Geological Society of America Bulletin*, **122**, 2079-2096.

Lettenmaier, D.P., and P.C.D. Milly, 2009, Land waters and sea level, *Nature Geoscience*, **2**, 452-454.

Letter, J.V., and A.K. Sturm, 2010, *Long Wave Modeling for South San Francisco Bay Shoreline Study: Without Project Conditions*, U.S. Army Corps of Engineers, ERDC/CHL Draft May 2010, 138 pp.

Leuliette, E.W., and L. Miller, 2009, Closing the sea level rise budget with altimetry, Argo, and GRACE, *Geophysical Research Letters*, **36**, L04608, doi:10.1029/2008GL036010.

Leuliette, E.W., and J.K. Willis, 2011, Balancing the sea-level budget, *Oceanography*, **24**, 122-129.

Levitus, S., J.I. Antonov, T.P. Boyer, R.A. Locarnini, H.E. Garcia, and A.V. Mishonov, 2009, Global ocean heat content 1955–2008 in light of recently revealed instrumentation problems, *Geophysical Research Letters*, **36**, L07608, doi:10.1029/2008GL037155.

Loder, N.M., J.L. Irish, M.A. Cialone, and T.V. Wamsley, 2009, Sensitivity of hurricane surge to morphological parameters of coastal wetlands, *Estuarine, Coastal and Shelf Science*, **84**, 625-636.

Loeb, N.G., J.M. Lyman, G.C. Johnson, R.P. Allan, D.R. Doelling, T. Wong, B.J. Soden, and G.L. Stephens, 2012, Observed changes in top-of-the-atmosphere radiation and upper-ocean heating consistent within uncertainty, *Nature Geoscience*, doi:10.1038/ngeo1375.

Lombard, A., D. Garcia, G. Ramillien, A. Cazenave, R. Biancale, J.M. Lemoine, F. Flechtner, R. Schmidt, and M. Ishii, 2007, Estimation of steric sea level variations from combined GRACE and Jason-1 data, *Earth and Planetary Science Letters*, **254**, 194-202.

Long, A.J., and I. Shennan, 1998, Models of rapid relative sea-level change in Washington and Oregon, USA, *The Holocene*, **8**, 129-142.

Lövstedt, C.B., and M. Larson, 2010, Wave damping in reed: Field measurements and mathematical modeling, *Journal of Hydraulic Engineering*, **136**, 222-233.

Lowe, J.A., P.L. Woodworth, T. Knutson, R.E. McDonald, K.I. McInnes, K. Woth, H. von Storch, J. Wolf, V. Swail, N.B. Berier, S. Gulev, K.J. Horsburgh, A.S. Unnikrishnan, J.R. Hunter, and R. Weisse, 2010, Past and future changes in extreme sea levels and waves, in *Understanding Sea-Level Rise and Variability*, J.A. Church, P.L. Woodworth, T. Aarup, and W.S. Wilson, eds., Wiley-Blackwell, UK, pp. 326-375.

Luthcke, S.B., A.A. Arendt, D.D. Rowlands, J.J. McCarthy, and C.F. Larsen, 2008, Recent glacier mass changes in the Gulf of Alaska region from GRACE mascon solutions, *Journal of Glaciology*, **54**, 767-777.

Lyman, J.M., S.A. Good, V.V. Gouretski, M. Ishii, G.C. Johnson, M.D. Palmer, D.M. Smith, and J.K. Willis, 2010, Robust warming of the global upper ocean, *Nature*, **465**, 334-337.

MA (Millennium Ecosystem Assessment), 2005, *Ecosystems and Human Well-being: Current State and Trends*, Volume 1, R. Hassan, R. Scholes, and N. Ash, eds., Island Press, Washington, D.C., 917 pp.

Mantua, N.J., and S.R. Hare, 2002, The Pacific Decadal Oscillation, *Journal of Oceanography*, **58**, 35-44.

Marshall, S.J., and K.M. Cuffey, 2000, Peregrinations of the Greenland Ice Sheet divide in the Last Glacial Cycle: Implications for central Greenland ice cores, *Earth and Planetary Science Letters*, **179**, 73-90.

Matsuo, K., and K. Heki, 2010, Time-variable ice loss in Asian high mountains from satellite gravimetry, *Earth and Planetary Science Letters*, **290**, 30-36.

Mayuga, M.N., and D.R. Allen, 1969, Subsidence in the Wilmington oil field, Long Beach, California, USA, in *Land Subsidence*, L.J. Tison, ed., International Association of Hydrological Sciences Publication 88, pp. 66-79, available at <http://iahs.info/>.

Mazda, Y., E. Wolanski, B. King, A. Sase, D. Ohtsuka, and M. Magi, 1997, Drag force due to vegetation in mangrove swamps, *Mangroves and Salt Marshes*, **1**, 193-199.

Mazzotti, S., A. Lambert, N. Courtier, L. Nykolaishen, and H. Dragert, 2007, Crustal uplift and sea level rise in northern Cascadia from GPS, absolute gravity, and tide gauge data, *Geophysical Research Letters*, **34**, L15306, doi:10.1029/2007GL030283.

Mazzotti, S., C. Jones, and R.E. Thomson, 2008, Relative and absolute sea level rise in western Canada and northwestern United States from a combined tide gauge-GPS analysis, *Journal of Geophysical Research*, **113**, C11019, doi:10.1029/2008JC004835.

McHarg, I., 1969, *Design with Nature*, Natural History Press, Garden City, N.J., 208 pp.

McKee, L.J., N.K. Ganju, and D.H. Schoellhamer, 2006, Estimates of suspended sediment entering San Francisco Bay from the Sacramento and San Joaquin Delta, San Francisco Bay, California, *Journal of Hydrology*, **323**, 335-352.

Meehl, G.A., W.D. Collins, P. Friedlingstein, A.T. Gaye, J.M. Gregory, A. Kitoh, R. Knutti, J.M. Murphy, A. Noda, S.C.B. Raper, I.G. Watterson, A.J. Weaver, and Z.-C. Zhao, 2007, Global climate projections, in *Climate Change 2007: The Physical Science Basis*, Contribution of Working Group I to the Fourth Assessment Report of the Intergovernmental Panel on Climate Change, S. Solomon, D. Qin, M. Manning, Z. Chen, M. Marquis, K.B. Averyt, M. Tignor and H.L. Miller, eds., Cambridge University Press, Cambridge, pp. 747-845.

Meehl, G.A., J.M. Arbalster, J.T. Fasullo, A. Hu, and K.E. Trenberth, 2011, Model-based evidence of deep-ocean heat uptake during surface-temperature hiatus periods, *Nature Climate Change*, **1**, 360-364.

Meier, M.F., and A. Post, 1987, Fast tidewater glaciers, *Journal of Geophysical Research*, **92**, 9051-9058.

Meier, M., M.B. Dyurgerov, U.K. Rick, S. O'Neel, W.T. Pfeffer, R.S. Anderson, S.P. Anderson, and A.F. Glazovsky, 2007, Glaciers dominate eustatic sea-level rise in the 21st century, *Science*, **317**, 1064-1067.

Menéndez, M., F.J. Méndez, I.J. Losada, and N.E. Graham, 2008, Variability of extreme wave heights in the northeast Pacific Ocean based on buoy measurements, *Geophysical Research Letters*, **35**, L22607, doi:10.1029/2008GL035394.

Mercer, J.H., 1978, West Antarctic Ice Sheet and CO_2 greenhouse effect: A threat of disaster, *Nature*, **272**, 321-325.

Merrifield, M.A., 2011, A shift in western tropical Pacific sea level trends during the 1990s, *Journal of Climate*, **24**, 4126-4138.

Merrifield, M.A., S.T. Merrifield, and G.T. Mitchum, 2009, An anomalous recent acceleration of global sea level rise, *Journal of Climate*, **22**, 5772-5781.

Miller, L., and B.C. Douglas, 2007, Gyre-scale atmospheric pressure variations and their relation to 19th and 20th century sea level rise, *Geophysical Research Letters*, **34**, L16602, doi:10.1029/2007GL030862.

Milly, P.C.D., A. Cazenave, and M.C. Gennero, 2003, Contribution of climate-driven change in continental water storage to recent sea-level rise, *Proceedings of the National Academy of Sciences*, **100**, 13,158-13,161.

Milly, P.C.D., A. Cazenave, J.S. Famiglietti, V. Gornitz, K. Laval, D.P. Lettenmaier, D.L. Sahagian, J.M. Wahr, and C.R. Wilson, 2010, Terrestrial water-storage contributions to sea-level rise and variability, in *Understanding Sea-Level Rise and Variability*, J.A. Church, P.L. Woodworth, T. Aarup, and W.S. Wilson, eds., Wiley-Blackwell, UK, pp. 226-255.

Milne, G.A., W.R. Gehrels, C.W. Hughes, and M.E. Tamisiea, 2009, Identifying the causes of sea-level change, *Nature Geoscience*, **2**, 471-478.

Mitchum, G., R. Nerem, M. Merrifield, and W. Gehrels, 2010, Modern sea level estimates, in *Understanding Sea-Level Rise and Variability*, J.A. Church, P.L. Woodworth, T. Aarup, and W.S. Wilson, eds., Wiley-Blackwell, UK, pp. 122-428.

Mitrovica, J.X., M.E. Tamisiea, J.L. Davis, and G.A. Milne, 2001, Polar ice mass variations and the geometry of global sea level change, *Nature*, **409**, 1026-1029.

Mitrovica, J.X., N. Gomez, E. Morrow, C. Hay, K. Latychev, and M.E. Tamisiea, 2011, On the robustness of predictions of sea level fingerprints, *Geophysical Journal International*, **187**, 729-742.

Mitsch, W.J., and J.G. Gosselink, 2000, *Wetlands*, 3rd edition, John Wiley & Sons, New York, 920 pp.

Möller, I., T. Spencer, J.R. French, D.J. Leggett, and M. Dixon, 1999, Wave transformation over salt marshes: A field and numerical modelling study from North Norfolk, England, *Estuarine, Coastal and Shelf Science*, **49**, 411-426.

Monaghan, A.J., D.H. Bromwich, and S.-H. Wang, 2006, Recent trends in Antarctic snow accumulation from Polar MM5 simulations, *Philosophical Transactions of the Royal Society, A*, **364**, 1683-1708.

Moon, T., I. Joughin, B. Smith, and I. Howat, 2012, 21st-century evolution of Greenland outlet glacier velocities, *Science*, **336**, 576-578.

Moore, J.C., A. Grinsted, and S. Jevrejeva, 2005, The new tools for analyzing the time series relationships and trends, *Eos Transactions of the American Geophysical Union*, **86**, 24.

Moore, L.J., B.T. Benumof, and G.B. Griggs, 1999, Coastal erosion hazards in Santa Cruz and San Diego counties, California, *Journal of Coastal Research*, Special Issue 28, 121-139.

Moss, R.H., J.A. Edmonds, K. Hibbard, M. Manning, S.K. Rose, D. Van, T. Carter, S. Emori, M. Kainuma, T. Kram, G. Meehl, J. Mitchell, N. Nakicenovic, K. Riahi, S.J. Smith, R. Stouffer, A.M. Thomson, J. Weyant, and T. Wilbanks, 2010, The next generation of scenarios for climate change research and assessment, *Nature*, **463**, 747-756.

Mote, P.W., A. Petersen, S. Reeder, H. Shipman, and L.C. Whitely Binder, 2008, *Sea Level Rise in the Coastal Waters of Washington State*, Climate Impacts Group, University of Washington, Seattle, and the Washington Department of Ecology, 11 pp.

Motyka, R.J., M. Truffer, M. Fahnestock, J. Mortensen, S. Rysgaard, and I. Howat, 2011, Submarine melting of the 1985 Jakobshavn Isbrae floating tongue and the triggering of the current retreat, *Journal of Geophysical Research*, **116**, F01007, doi:10.1029/2009JF001632.

Mumford, T.F., 2007, *Kelp and Eelgrass in Puget Sound*, Puget Sound Nearshore Partnership Report 2007-05, U.S. Army Corps of Engineers, Seattle, Washington, 27 pp.

Mumford, T.F., Jr., S. Wyllie-Echeverria, R.T. Thom, and D.E. Penttila, 2003, Loss of eelgrass (*Zostera marina L.*) in small embayments, Puget Sound, Washington: Similarity to worldwide early signs of collapse, in *Estuaries on the Edge: Convergence of Ocean, Land and Culture*, Proceedings of the Estuarine Research Foundation Conference, September 14-18, 2003, Seattle, Washington.

Munk, W., 2003, Ocean freshening, sea level rising, *Science*, **300**, 2041-2043.

Nagel, N.B., 2001, Compaction and subsidence issues within the petroleum industry: From Wilmington to Ekofisk and beyond, *Physics and Chemistry of the Earth, Part A: Solid Earth and Geodesy*, **26**, 3-14.

Nakada, M., and H. Inoue, 2005, Rates and causes of recent global sea-level rise inferred from long tide gauge records, *Quaternary Science Reviews*, **24**, 1217-1222.

Namson, J., and T. Davis, 1991, Balanced cross sections of the Western Transverse Ranges and southern Coast Ranges, California, Point Conception to the San Andreas Fault, Davis-Namson Consulting Geologists, California, available at <http://www.davisnamson.com/downloads/index.htm>.

Nelson, A.R., 2007, Tectonics and relative sea-level change, in *Encyclopedia of Quaternary Science*, S. Elias, ed., Elsevier, pp. 3072-3087.

Nelson, A.R., I. Shennan, and A.J. Long, 1996, Identifying coseismic subsidence in tidal-wetland stratigraphic sequences at the Cascadia Subduction Zone of western North America, *Journal of Geophysical Research*, **101**, 6115-6135.

Nerem, R.S., D.P. Chambers, C. Choe, and G.T. Mitchum, 2010, Estimating mean sea level change from the TOPEX and Jason altimeter missions, *Marine Geodesy*, **33**, 435-446.

Newman, M., G.P. Compo, and M.A. Alexander, 2003, ENSO-forced variability of the Pacific Decadal Oscillation, *Journal of Climate*, **16**, 3853-3857.

Nick, F.M., A. Vieli, I.M. Howat, and I. Joughin, 2009, Large-scale changes in Greenland outlet glacier dynamics triggered at the terminus, *Nature Geoscience*, **2**, 110-114.

Nummedal, D., O.H. Pilkey, and J.D. Howard, eds., 1987, *Sea-Level Fluctuations and Coastal Evolution*, Society of Economic Paleontologists and Mineralogists, Special Publication 41, Tulsa, OK, 267 pp.

Oerlemans, J., 1989, A projection of future sea level, *Climatic Change*, **15**, 151-174.

Ohmura, A., 2004, Cryosphere during the twentieth century, in *The State of the Planet: Frontiers and Challenges in Geophysics*, R.S.J. Sparks and C.J. Hawkesworth, eds., Geophysical Monograph Series, 150, American Geophysical Union, Washington, D.C., pp. 239-257.

Okada, Y., 1985, Surface deformation due to shear and tensile faults in a half-space, *Bulletin of the Seismological Society of America*, **75**, 1135-1154.

Oleson, K.W., and 27 others, 2010, *Technical Description of Version 4.0 of the Community Land Model (CLM)*, NCAR Technical Note, NCAR/TN-478+STR, National Center for Atmospheric Research, Boulder, Colorado, 257 pp.

O'Reilly, W.C., and R.T. Guza, 1991, A comparison of spectral refraction and refraction-diffraction wave propagation models, *Journal of Waterway, Port, Coastal, and Ocean Engineering*, **117**, 199-215.

Orme, A.R., 1991, Mass movement and seacliff retreat along the southern California coast, *Bulletin of the Southern California Academy of Sciences*, **90**, 58-79.

Orme, A.R., 1992, Late Quaternary deposits near Point Sal, south-central California: A time frame for coastal-dune emplacement, in *Quaternary Coasts of the United States: Marine and Lacustrine Systems*, C.H. Fletcher and J.F. Wehmiller, eds., Society of Sedimentary Geology, Special Publication 48, Tulsa, Oklahoma, pp. 309-315.

Orme, A.R., 1998, Late Quaternary tectonism along the Pacific coast of the Californias: A contrast in style, in *Coastal Tectonics*, I.S. Stewart and C Vita-Finzi, eds., Geological Society of London, Special Publication 146, London, pp. 179-197.

Orme, A.R., G.B. Griggs, D.L. Revell, J.G. Zoulas, C.C. Grandy, and H. Koo, 2011, Beach changes along the southern California coast during the 20th century: A comparison of natural and human forcing factors, *Shore & Beach*, **79**, 38-50.

Orr, M., S. Crooks, and P. Williams, 2003, Will restored tidal marshes be sustainable? *San Francisco Estuary and Watershed Science*, **1**(1), 33 pp.

Pardaens, A.K., J.M. Gregory, and J.A. Lowe, 2010, A model study of factors influencing projections of sea level over the twenty-first century, *Climate Dynamics*, **36**, 2015-2033.

Parker, V.T., J.C. Callaway, L.M. Schile, M.C. Vasey, and E.R. Herbert, 2011, Climate change and San Francisco Bay-Delta tidal wetlands, *San Francisco Estuary and Watershed Science*, **9**(3), 15 pp.

Patrick, W.H., Jr., and R.D. DeLaune, 1990, Subsidence, accretion, and sea level rise in south San Francisco Bay marshes, *Limnology and Oceanography*, **35**, 1389-1395.

Paulson, A., S.J. Zhong, and J. Wahr, 2007, Inference of mantle viscosity from GRACE and relative sea level data, *Geophysical Journal International*, **171**, 497-508.

Peltier, W.R., 1976, Glacial isostatic adjustment: The inverse problem, *Geophysical Journal of the Royal Astronomical Society*, **46**, 669-706.

Peltier, W., 1998, Postglacial variations in the level of the sea: Implications for climate dynamics and solid-earth geophysics, *Reviews of Geophysics*, **36**, 603-689.

Peltier, W., 2001, Global glacial isostatic adjustment and modern instrumental records of relative sea level history, in *Sea Level Rise: History and Consequences*, B. Douglas, M. Kearney, and S. Leatherman, eds., International Geophysics Series, **75**, Academic Press, pp. 65-95.

Peltier, W., 2002a, Global glacial isostatic adjustment: Palaeo-geodetic and space-geodetic tests of the ICE-4G (VM2) model, *Journal of Quaternary Science*, **17**, 491-510.

Peltier, W.R., 2002b, On eustatic sea level history: Last Glacial Maximum to Holocene, *Quaternary Science Reviews*, **21**, 377-396.

Peltier, W., 2004, Global glacial isostasy and the surface of the ice-age Earth: The ICE-5G (VM2) model and GRACE, *Annual Review of Earth and Planetary Sciences*, **32**, 111-149.

Peltier, W.R., 2010, Closing the budget of global sea level rise: The GRACE correction for GIA over the oceans, in *Workshop Report of the Intergovernmental Panel on Climate Change Workshop on Sea Level Rise and Ice Sheet Instabilities*, T.F. Stocker, D. Qin, G.-K. Plattner, M. Tignor, S. Allen, and P.M. Midgley, eds., University of Bern, IPCC Working Group Technical Support Unit, Bern, Switzerland, pp. 157-159.

Peltier, W.R., and R. Drummond, 2008, Rheological stratification of the lithosphere: A direct inference based upon the geodetically observed pattern of the glacial isostatic adjustment of the North American continent, *Geophysical Research Letters*, **35**, L16314, doi:10.1029/2008GL034586.

Peltier, W.R., and R.G. Fairbanks, 2006, Global glacial ice volume and Last Glacial Maximum duration from an extended Barbados sea level record, *Quaternary Science Reviews*, **25**, 3322-3337.

Peltier, W.R., and A.M. Tushingham, 1989, Global sea level rise and the greenhouse effect: Might they be connected? *Science*, **244**, 806-810.

Peltier, W.R., I. Shennan, R. Drummond, and B.P. Horton, 2002, On the post-glacial isostatic adjustment of the British Isles and the shallow visco-elastic structure of the Earth, *Geophysical Journal International*, **148**, 443-475.

Perillo, G.M.E., 2008, Tidal courses: Classification, origin and functionality, in *Coastal Wetlands: An Integrated Ecological Approach*, G.M.E. Perillo, E. Wolanski, D. Cahoon, and M. Brinson, eds., Elsevier, Amsterdam, pp. 185-209.

Pestrong, R., 1965, *The Development of Drainage Patterns on Tidal Marshes*, Stanford University Publications, Geological Sciences, **10**, Stanford, California, 87 pp.

Pfeffer, W.T., 2007, A simple mechanism for irreversible tidewater glacier retreat, *Journal of Geophysical Research*, **112**, F03S25, doi:10.1029/2006JF000590.

Pfeffer, W.T., J. Harper, and S. O'Neel, 2008, Kinematic constraints on glacier contributions to 21st-century sea-level rise, *Science*, **321**, 1340-1343.

Pierce, D.W., T.P. Barnett, K.M. Achutarao, P.J. Gleckler, J.M. Gregory, and W.M. Washington, 2006, Anthropogenic warming of the oceans: Observations and model results, *Journal of Climate*, **19**, 1873-1900.

Pilkey, O., and T.W. Davis, 1987, An analysis of coastal recession models: North Carolina coast, in *Sea-Level Fluctuation and Coastal Evolution*, Society of Economic Paleontologists and Mineralogists, Special Publication 41, Tulsa, OK, pp. 59-70.

Pollard, D., and R.M. DeConto, 2009, Modelling West Antarctic Ice Sheet growth and collapse through the past five million years, *Nature*, **458**, 329-332.

Price, S.F., A.J. Payne, I.M. Howat, and B.E. Smith, 2011, Committed sea-level rise for the next century from Greenland Ice Sheet dynamics during the past decade, *Proceedings of the National Academy of Sciences*, **108**, 8978-8983.

Priest, G.R., 1999, Coastal shoreline change, northern and southern Lincoln County, Oregon, *Journal of Coastal Research*, Special Issue 28, 140-157.

Puget Sound Action Team, 2002, *2002 Puget Sound Update: Eighth Report of the Puget Sound Ambient Monitoring Program*, Olympia, Washington, 144 pp.

Puget Sound Regional Council, 2004, *Population Change and Net Migration, Puget Sound Trends No. D7*, available at <http://www.psrc.org/datapubs/pubs/trends/d7jul04.pdf>.

Pullen, J.D., and J.S. Allen, 2000, Modeling studies of the coastal circulation off northern California: Shelf response to a major Eel River flood event, *Continental Shelf Research*, **20**, 2213-2238.

Purkey, S.G., and G.C. Johnson, 2010, Warming of global abyssal and deep southern ocean waters between the 1990s and 2000s: Contributions to global heat and sea level rise budgets, *Journal of Climate*, **23**, 6336-6351.

PWA (Phillip Williams & Associates, Ltd.), 2009, *California Coastal Erosion Response to Sea Level Rise - Analysis and Mapping*, San Francisco, CA, 27 pp. plus appendixes.

Quincey, D.J., and A. Luckman, 2009, Progress in satellite remote sensing of ice sheets, *Progress in Physical Geography*, **33**, 547-567.

Radic, V., and R. Hock, 2010, Regional and global volumes of glaciers derived from statistical upscaling of glacier inventory data, *Journal of Geophysical Research*, **115**, F01010, doi:10.1029/2009JF001373.

Rahmstorf, S., 2007, A semi-empirical approach to projecting future sea-level rise, *Science*, **315**, 368-370.

Rahmstorf, S., 2010, A new view on sea level rise, *Nature Climate Change*, **4**, 44-45.

Rahmstorf, S., and M. Vermeer, 2011, Discussion of "Houston, J.R. and R.G. Dean, 2011, Sea-level acceleration based on U.S. tide gauges and extensions of previous global-gauge analyses," *Journal of Coastal Research*, **27**, 784-787.

Ramillien, G., S. Bohours, A. Lombard, A. Cazenave, F. Flechner, and R. Schmidt, 2008, Land water storage contribution to sea level form GRACE geoid data over 2003-2006, *Global and Planetary Change*, **60**, 381-392.

Rasmussen, L.A., H. Conway, R.M. Krimmel, and R. Hock, 2011, Surface mass balance, thinning and iceberg production, Columbia Glacier, Alaska, 1948-2007, *Journal of Glaciology*, **57**, 431-440.

Reed, D.J., 1989, Patterns of sediment deposition to subsiding coastal salt marshes, Terrebonne Bay, Louisiana: The role of winter storms, *Estuaries*, **12**, 222-227.

Reed, D.J., 1995, The response of coastal marshes to sea-level rise: Survival or submergence? *Earth Surface Processes and Landforms*, **20**, 39-48.

Reed, D.J., A. Commagere, and M. Hester, 2009, Marsh elevation response to hurricanes Katrina and Rita and the effect of altered nutrient regimes, *Journal of Coastal Research*, Special Issue 54, 166-173.

Revell, D.L., R. Battalio, B. Spear, P. Ruggiero, and J. Vandever, 2011, A methodology for predicting future coastal hazards due to sea-level rise on the California Coast, *Climatic Change*, **109**, S251-S276.

Ridley, J., J.M. Gregory, P. Huybrechts, and J. Lowe, 2010, Thresholds for irreversible decline of the Greenland Ice Sheet, *Climate Dynamics*, **35**, 1049-1057.

Rignot, E., and S. Jacobs, 2002, Rapid bottom melting widespread near Antarctic Ice Sheet grounding lines, *Science*, **296**, 2020-2023.

Rignot, E., and P. Kanagaratnam, 2006, Changes in the velocity structure of the Greenland Ice Sheet, *Science*, **311**, 986-990.

Rignot, E., J.E. Box, E. Burgess, and E. Hanna, 2008, Mass balance of the Greenland Ice Sheet from 1958 to 2007, *Geophysical Research Letters*, **35**, L20502, doi:10.1029/2008GL035417.

Rignot, E., I. Velicogna, M.R. van den Broeke, A. Monaghan, and J. Lenaerts, 2011a, Acceleration of the contribution of the Greenland and Antarctic ice sheets to sea level rise, *Geophysical Research Letters*, **38**, L05503, doi:10.1029/2011GL046583.

Rignot, E., J. Mouginot, and B. Scheuchl, 2011b, Ice flow of the Antarctic Ice Sheet, *Science*, **333**, 1427-1430.

Ruggiero, P., P.D. Komar, and J.C. Allan, 2010, Increasing wave heights and extreme-value projections: The wave climate of the U.S. Pacific Northwest, *Coastal Engineering*, **57**, 539-552.

Ruhl, C., and D. Schoellhamer, 2004, Spatial and temporal variability of suspended-sediment concentrations in a shallow estuarine environment, *San Francisco Estuary and Watershed Science*, **2**(2), 10 pp.

Salathé, E.P., Jr., 2006, Influences of a shift in North Pacific storm tracks on western North American precipitation under global warming, *Geophysical Research Letters*, **33**, L19820, doi:10.1029/2006GL026882.

Sallenger, A.H., W. Krabill, J. Brock, R. Swift, S. Manizade, and H. Stockdon, 2002, Sea-cliff erosion as a function of beach changes and extreme wave runup during the 1997-1998 El Niño, *Marine Geology*, **187**, 279-297.

Sasgen, I., V. Klemann, and Z. Martinec, 2012, Towards the joint inversion of GRACE gravity fields for present-day ice-mass changes and glacial-isostatic adjustment in North America and Greenland, *Journal of Geodynamics*, in press.

Satake, K., and B.F. Atwater, 2007, Long-term perspectives on giant earthquakes and tsunamis at subduction zones, *Annual Review of Earth and Planetary Sciences*, **35**, 349-274.

Satake, K., K. Shimazaki, Y. Tsuji, and K. Ueda, 1996, Time and size of a giant earthquake in Cascadia inferred from Japanese tsunami record of January 1700, *Nature*, **379**, 246-249.

Satake, K., K.L. Wang, and B.F. Atwater, 2003, Fault slip and seismic moment of the 1700 Cascadia earthquake inferred from Japanese tsunami descriptions, *Journal of Geophysical Research*, **108**, 2535, doi:10.1029/2003JB002521.

Savage, J.C., 1983, A dislocation model of strain accumulation and release at a subduction zone, *Journal of Geophysical Research*, **88**, 4984-4996.

Savage, J.C., M. Lisowski, and W.H. Prescott, 1981, Geodetic strain measurements in Washington, *Journal of Geophysical Research*, **86**, 4929-4940.

Scambos, T.A., J. Bohlander, C. Shuman, and P. Skvarca, 2004, Glacier acceleration and thinning after ice shelf collapse in the Larsen B embayment, Antarctica, *Geophysical Research Letters*, **31**, doi:10.1029/2004GL020670.

Schlenger, P., A. MacLennan, E. Iverson, K. Fresh, C. Tanner, B. Lyons, S. Todd, R. Carman, D. Myers, S. Campbell, and A. Wick, 2011, *Strategic Needs Assessment: Analysis of Nearshore Ecosystem Process Degradation in Puget Sound*, Prepared for the Puget Sound Nearshore Ecosystem Restoration Project, Technical Report 2011-02, Olympia, WA, 430 pp.

Schmidt, D.A., and R. Bürgmann, 2003, Time-dependent land uplift and subsidence in the Santa Clara Valley, California, from a large interferometric synthetic aperture radar data set, *Journal of Geophysical Research*, **108**, 2416, doi:10.1029/2002JB002267.

Schneider, N., and B.D. Cornuelle, 2005, The forcing of the Pacific Decadal Oscillation, *Journal of Climate*, **18**, 4355-4373.

Schoellhamer, D.H., 2011, Sudden clearing of estuarine waters upon crossing the threshold from transport to supply regulation of sediment transport as an erodible sediment pool is depleted: San Francisco Bay, 1999, *Estuaries and Coasts*, **34**, 885-899.

Schoof, C., 2007, Ice sheet grounding line dynamics: Steady states, stability, and hysteresis, *Journal of Geophysical Research*, **112**, F03S28, doi:10.1029/2006JF000664.

Schrama, E.J.O., and B. Wouters, 2011, Revisiting Greenland Ice Sheet mass loss observed by GRACE, *Journal of Geophysical Research*, **116**, B02407, doi:10.1029/2009JB006847.

Schwarz, K.M., and A.R. Orme, 2005, Opening and closure of a seasonal river mouth: The Malibu Estuary-barrier-lagoon System, California, *Zeitschrift für Geomorphologie, Supplementbände*, **141**, 91-109.

Seager, R., N.H. Naik, M.F. Ting, M.A. Cane, N. Harnik, and Y. Kushnir, 2010, Adjustment of the atmospheric circulation to tropical Pacific SST anomalies: Variability of transient eddy propagation in the Pacific-North America sector, *Quarterly Journal of the Royal Meteorological Society*, **136**, 277-296.

Seliskar, D.M., and J.L. Gallagher, 1983, *The Ecology of Tidal Marshes of the Pacific Northwest Coast: A Community Profile*, U.S. Fish and Wildlife Service, FWS/OBS-82/32, Washington, D.C., 65 pp.

Sella, G.F., S. Stein, T.H. Dixon, M. Craymer, T.S. James, S. Mazzotti, and R.K. Dokka, 2007, Observation of glacial isostatic adjustment in "stable" North America with GPS, *Geophysical Research Letters*, **34**, L02306, doi:10.1029/2006GL027081.

Seymour, R.J., 1998, Effect of El Niño on the west coast wave climate, *Shore & Beach*, **66**, 3-6.

Seymour, R.J., 2011, Evidence for changes to the northeast Pacific wave climate, *Journal of Coastal Research*, **27**, 194-201.

Seymour, R.J., R.R. Strange III, D.R. Cayan, and R.A. Nathan, 1984, Influence of El Niños on California's wave climate, in *Proceedings of the 19th International Conference on Coastal Engineering*, Houston, Texas, September 3-7, 1984, B.L. Edge, ed., American Society of Civil Engineers, pp. 577-592.

Shaw, J., and P. Shearer, 1999, An elusive blind-thrust fault beneath metropolitan Los Angeles, *Science*, **283**, 1516-1518.

Shaw, J., and J. Suppe, 1994, Active faulting and growth folding in the eastern Santa Barbara Channel, California, *Geological Society of America Bulletin*, **106**, 607-626.

Shaw, J., and J. Suppe, 1996, Earthquake hazards of active blind-thrust faults under the central Los Angeles Basin, California, *Journal of Geophysical Research*, **101**, 8623-8642.

Shaw, J., P. Gareau, and R.C. Courtney, 2002, Palaeogeography of Atlantic Canada 13-0 kyr. *Quaternary Science Reviews*, **21**, 1861-1878.

Shennan, I., and B.P. Horton, 2002, Relative sea-level changes and crustal movements of the UK, *Journal of Quaternary Science*, **16**, 511-526.

Shepard, F.P., 1963, *Submarine Geology*, Harper & Row, New York, 557 pp.

Sherman, D.J., K.M. Barron, and J.T. Ellis, 2002, Retention of beach sand by dams and debris basins in southern California, *Journal of Coastal Research*, Special Issue 36, 662-674.

Shih, S., and P.D. Komar, 1994, Sediments, beach morphology and sea cliff erosion within an Oregon coast littoral cell, *Journal of Coastal Research*, **10**, 144-157.

Shiklomanov, I.A., ed., 1997, *Comprehensive Assessment of the Freshwater Resources of the World: Assessment of Water Resources and Water Availability in the World*, World Meteorological Association, Geneva, 88 pp.

Shipman, H., M. Dethier, G. Gelfenbaum, K.L. Fresh, and R.S. Dinicola, 2010, *Puget Sound Shorelines and the Impacts of Armoring—Proceedings of a State of the Science Workshop, May 2009*, Scientific Investigations Report 2010-5254, U.S. Geological Survey, Denver, CO, 278 pp.

Shum, C., and C. Kuo, 2011, Observation and geophysical causes of present-day sea level rise, in *Climate Change and Food Security in South Asia*, R. Lal, M. Sivakumar, S.M.A. Faiz, A.H.M. Mustafizur-Rahman, and K.R. Islam, eds., Springer, pp 85-104.

Shum, C., C. Kuo, and J. Guo, 2008, Role of Antarctic ice mass balances in present-day sea level change, *Polar Science*, **2**, 149-161.

Simenstad, C., J. Toft, H. Higgins, J. Cordell, M. Orr, P. Williams, L. Grimaldo, Z. Hymanson, and D. Reed, 2000, *Sacramento-San Joaquin Delta breached levee wetland study (BREACH), Preliminary Report*, Wetland Ecosystem Team, University of Washington, Seattle, Washington, 46 pp.

Slagel, M.J., and G.B. Griggs, 2008, Cumulative losses of sand to the major littoral cells of California by impoundment behind coastal dams, *Journal of Coastal Research*, **252**, 50-61.

Slangen, A.B.A., and R.S.W. van de Wal, 2011, An assessment of uncertainties in using volume-area modelling for computing the twenty-first century glacier contribution to sea-level change, *The Cryosphere Discussions*, **5**, 1655-1695.

Smithers, S.G., and C.D. Woodroffe, 2001, Coral microatolls and 20th century sea level in the eastern Indian Ocean, *Earth and Planetary Science Letters*, **191**, 173-184.

Sommerfield, C.K., and C.A. Nittrouer, 1999, Modern accumulation rates and a sediment budget for the Eel Shelf: A flood-dominated depositional environment, *Marine Geology*, **154**, 227-241.

Song, Y.T., and F. Colberg, 2011, Deep ocean warming assessed from altimeters, Gravity Recovery and Climate Experiment, in situ measurements, and a non-Boussinesq ocean general circulation model, *Journal of Geophysical Research*, **116**, C02020, doi:10.1029/2010JC006601.

Sørensen, L.S., S.B. Simonsen, K. Nielsen, P. Lucas-Picher, G. Spada, G. Adalgeirsdottir, R. Forsberg, and C.S. Hvidberg, 2011, Mass balance of the Greenland Ice Sheet (2003–2008) from ICESat data – The impact of interpolation, sampling and firn density, *The Cryosphere*, **5**, 173-186.

Stamey, M.T., 2004, *An Analysis of Eelgrass Transplantation Performance in Puget Sound, WA, 1990-2000*, Master's Thesis, University of Washington, Seattle, Washington, 120 pp.

Steere, J.T., and N. Schaefer, 2001, *Restoring the Estuary: Implementation Strategy of the San Francisco Bay Joint Venture*, San Francisco Bay Joint Venture, Oakland, California, 111 pp.

Steers, J.A., 1948, *The Coastline of England & Wales*, Cambridge University Press, New York, 644 pp.

Storlazzi, C.D., and G.B. Griggs, 1998, The 1997-98 El Niño and erosion processes along the central coast of California, *Shore & Beach*, **66**, 12-17.

Storlazzi, C.D., and G.B. Griggs, 2000, The influence of El Niño-Southern Oscillation (ENSO) events on the evolution of central California's shoreline, *Geological Society of America Bulletin*, **112**, 236-249.

Storlazzi, C.D., and D.K. Wingfield, 2005, *Spatial and Temporal Variations in Oceanographic and Meteorologic Forcing Along the Central California Coast, 1980-2002*, Scientific Investigations Report 2005-5085, U.S. Geological Survey, Denver, CO, 39 pp.

Stralberg, D., M. Brennan, J.C. Callaway, J.K. Wood, L.M. Schile, D. Jongsomjit, M. Kelly, V.T. Parker, and S. Crooks, 2011, Evaluating tidal marsh sustainability in the face of sea-level rise: A hybrid modeling approach applied to San Francisco Bay, *PLoS ONE*, **6**, e27388, doi:10.1371/journal.pone.0027388.

Straneo, F., G.S. Hamilton, D.A. Sutherland, L.A. Stearns, F. Davidson, M.O. Hammill, G.B. Stenson, and A. Rosing-Asvid, 2010, Rapid circulation of warm subtropical waters in a major glacial fjord in East Greenland, *Nature Geoscience*, **3**, 182-186.

Sturges, W., and B.C. Douglas, 2011, Wind effects on estimates of sea level rise, *Journal of Geophysical Research*, **116**, C06008, doi:10.1029/2010JC006492.

Sun, H., D. Grandstaff, and R. Shagam, 1999, Land subsidence due to groundwater withdrawal: Potential damage of subsidence and sea level rise in southern New Jersey, USA, *Environmental Geology*, **34**, 290-296.

Sunamura, T., 1992, *Geomorphology of Rocky Coasts*, John Wiley and Sons, Chichester, 302 pp.

Swenson, S., D. Chambers, and J. Wahr, 2008, Estimating geocenter variations from a combination of GRACE and ocean model output, *Journal of Geophysical Research*, **113**, B08410, doi:10.1029/2007JB005338.

Tamisiea, M., 2011, Ongoing glacial isostatic contributions to observations of sea level change, *Geophysical Journal International*, **186**, 1036-1044.

Tamisiea, M.E., J.X. Mitrovica, J.L. Davis, and G.A. Milne, 2003, Long wavelength sea level and solid surface perturbations driven by polar mass variations: Fingerprinting Greenland and Antarctic ice sheet flux, *Space Science Reviews*, **108**, 82-93.

Tamisiea, M.E., E.W. Leuliette, J.L. Davis, and J.X. Mitrovica, 2005, Constraining hydrological and cryospheric mass flux in southeastern Alaska using space-based gravity measurements, *Geophysical Research Letters*, **32**, L20501, doi:10.1029:2005GL023961.

Tanaka, N., Y. Sasaki, M.I.M. Mowjood, K.B.S.N. Jinadasa, and S. Homchuen, 2007, Coastal vegetation structures and their functions in tsunami protection: Experience of the recent Indian Ocean tsunami, *Landscape and Ecological Engineering*, **3**, 33-45.

Tapley, B.D., S. Bettadpur, J. Ries, P. Thompson, and M. Watkins, 2004, GRACE measurements of mass variability in the Earth system, *Science*, **305**, 503-505.

Tarasov, L., and W.R. Peltier, 2002, Greenland glacial history and local geodynamic consequences, *Geophysical Journal International*, **150**, 198-229.

Tebaldi, C., B.H. Strauss, and C.E. Zervas, 2012, Modelling sea level rise impacts on storm surges along US coasts, *Environmental Research Letters*, **7**, doi:10.1088/1748-9326/7/01432.

Thieler, E.R., and E.S. Hammar-Klose, 2000, *National Assessment of Coastal Vulnerability to Sea-Level Rise: Preliminary Results for the U.S. Pacific Coast*, U.S. Geological Survey Open-File Report 00-178, available at <http://pubs.usgs.gov/of/2000/of00-178/>.

Thom, R.M., 1990, A review of eelgrass (*Zostera marina L.*) transplanting projects in the Pacific Northwest, *Northwest Environmental Journal*, **6**, 121-137.

Thom, R.M., A.B. Borde, N.R. Evans, C.W. May, G.E. Johnson, and J.A. Ward, 2004, *A Conceptual Model for the Lower Columbia River Estuary*, Prepared by Pacific Northwest National Laboratory for the U.S. Army Corps of Engineers, Portland District, available from the Lower Columbia Estuary Partnership, Portland, Oregon.

Thomas, R.H., and C.R. Bentley, 1978, A model for Holocene retreat of the West Antarctic Ice Sheet, *Quaternary Research*, **10**, 150-170.

Thomas, R., T. Sanderson, and K. Rose, 1979, Effect of climatic warming on the West Antarctic Ice Sheet, *Nature*, **277**, 355-358.

Timmermann, A., S. McGregor, and F.-F. Jin, 2010, Wind effects on past and future regional sea level trends in the southern Indo-Pacific, *Journal of Climate*, **23**, 4429-4437.

Tiwari, V.M., J. Wahr, and S. Swenson, 2009, Dwindling groundwater resources in northern India, from satellite gravity observations, *Geophysical Research Letters*, **36**, L18401, doi:10.1029/2009GL039401.

Tokinaga, H., and S.-P. Xie, 2011, Wave- and anemometer-based sea surface wind (WASWind) for climate change analysis, *Journal of Climate*, **24**, 267-285.

Toniazzo, T., J. Gregory, and P. Huybrechts, 2004, Climatic impact of a Greenland deglaciation and its possible irreversibility, *Journal of Climate*, **17**, 21-33.

Törnqvist, T.E., D.J. Wallace, J.E.A. Storms, J. Wallinga, R.L. Dam, M. Blaauw, M.S. Derksen, C.J.W. Klerks, C. Meijneken, and E.M.A. Snijders, 2008, Mississippi Delta subsidence primarily caused by compaction of Holocene strata, *Nature Geoscience*, **1**, 173-176.

Tregoning, P., G. Ramillien, H. McQueen, and D. Zwartz, 2009, Glacial isostatic adjustment and nonstationary signals observed by GRACE, *Journal of Geophysical Research*, **114**, B06406, doi:10.1029/2008JB006161.

Trenberth, K.E., P.D. Jones, P. Ambenje, R. Bojariu, D. Easterling, A. Klein Tank, D. Parker, F. Rahimzadeh, J.A. Renwick, M. Rusticucci, B. Soden, and P. Zhai, 2007, Observations: Surface and atmospheric climate change, in *Climate Change 2007: The Physical Science Basis*, Contribution of Working Group I to the Fourth Assessment Report of the Intergovernmental Panel on Climate Change, S. Solomon, D. Qin, M. Manning, Z. Chen, M. Marquis, K.B. Averyt, M. Tignor, and H.L. Miller, eds., Cambridge University Press, Cambridge, pp. 235-336.

Trupin, A.S., and J.M. Wahr, 1992, Spectroscopic analysis of global tide gauge sea level data, *Geophysical Journal International*, **108**, 1-15.

van den Broeke, M., J. Bamber, J. Ettema, E. Rignot, E. Schrama, W.J. van de Berg, E. van Meijgaard, I. Velicogna, and B. Wouters, 2009, Partitioning recent Greenland mass loss, *Science*, **326**, 984-986.

van den Broeke, M., C. Bus, J. Ettema, and P. Smeets, 2010, Temperature thresholds for degree-day modelling of Greenland Ice Sheet melt rates, *Geophysical Research Letters*, **37**, L18501, doi:10.1029/2010GL044123.

van der Veen, C.J., and ISMASS (Ice Sheet Mass Balance and Sea Level) Working Group, 2010, *Ice Sheet Mass Balance and Sea Level: A Science Plan*, Scientific Committee on Antarctic Research, Report 38, Cambridge, United Kingdom, 35 pp.

van der Wal, W., A. Braun, P. Wu, and M. Sideris, 2009, Prediction of decadal slope changes in Canada by glacial isostatic adjustment modeling, *Canadian Journal of Earth Sciences*, **46**, 587-595.

Vaughan, D., 2006, Recent trends in melting conditions on the Antarctic Peninsula and their implications for ice-sheet mass balance and sea level, *Arctic, Antarctic, and Alpine Research*, **38**, 147-152.

Velicogna, I., 2009, Increasing rates of ice mass loss from the Greenland and Antarctic ice sheets revealed by GRACE, *Geophysical Research Letters*, **36**, L19503, doi:10.1029/2009GL040222.

Velicogna, I., and J. Wahr, 2006a, Measurements of time variable gravity shows mass loss in Antarctica, *Science*, **311**, 1754-1756.

Velicogna, I., and J. Wahr, 2006b, Significant acceleration of Greenland ice mass loss in spring 2004, *Nature*, **443**, 329-331.

Vermeer, M., and S. Rahmstorf, 2009, Global sea level linked to global temperature, *Proceedings of the National Academy of Sciences*, **106**, 21,527-21,532.

Vimont, D.J., M. Alexander, and A. Fontaine, 2009, Midlatitude excitation of tropical variability in the Pacific: The role of thermodynamic coupling and seasonality, *Journal of Climate*, **22**, 518-534.

Wada, Y., L.P.H. van Beek, C.M. van Kempen, J. Reckman, S. Vasak, and M.F.P. Bierkens, 2010, Global depletion of groundwater resources, *Geophysical Research Letters*, **37**, L20402, doi:10.1029/2010GL044571.

Wada, Y., L.P.H. van Beek, and M.F.P. Bierkens, 2012a, Nonsustainable groundwater sustaining irrigation: A global assessment, *Water Resources Research*, **48**, W00L06, doi:10.1029/2011WR010562.

Wada, Y., L.P.H. van Beek, F.C. Sperna Weiland, B.F. Chao, Y.-H. Wu, and M.F.P. Bierkens, 2012b, Past and future contribution of global groundwater depletion to sea-level rise, *Geophysical Research Letters*, **39**, L09402, doi:10.1029/2012GL051230.

Wahr, J., S. Swenson, V. Zlotnicki, and I. Velicogna, 2004, Time-variable gravity from GRACE: First results, *Geophysical Research Letters*, **31**, L11501, doi:10.1029/2004GL019779.

Wahr, J., S. Swenson, and I. Velicogna, 2006, Accuracy of GRACE mass estimates, *Geophysical Research Letters*, **33**, L06401, doi:10.1029/2005GL025305.

Wamsley, T.V., M.A. Cialone, J.M. Smith, J.H. Atkinson, and J.D. Rosati, 2009, The potential of wetlands in reducing storm surge, *Ocean Engineering*, **37**, 59-68.

Wang, H., and P. Wu, 2006, Effects of lateral variations in lithospheric thickness and mantle viscosity on glacially induced surface motion on a spherical, self-gravitating Maxwell Earth, *Earth and Planetary Science Letters*, **244**, 576-589.

Wang, K., 2007, Elastic and viscoelastic models of subduction earthquake cycles, in *The Seismogenic Zone of Subduction Thrust Faults*, T.H. Dixon and J.C. Moore, eds., Columbia University Press, New York, pp. 540-575.

Wang, K., R. Wells, S. Mazzotti, R.D. Hyndman, and T. Sagiya, 2003, A revised dislocation model of interseismic deformation of the Cascadia Subduction Zone, *Journal of Geophysical Research*, **108**, 2016-2029.

Wang, X.L., and V.R. Swail, 2001, Changes of extreme wave heights in Northern Hemisphere oceans and related atmospheric circulation regimes, *Journal of Climate*, **14**, 2204-2221.

Ward, K.M., J.C. Callaway, and J.B. Zedler, 2003, Episodic colonization of an intertidal mudflat by native cordgrass (*Spartina foliosa*) at Tijuana Estuary, *Estuaries*, **26**, 116-130.

Wentz, F.J., and L. Ricciardulli, 2011, Comment on "Global trends in wind speed and wave height," *Science*, **334**, 905.

Wentz, F.J., L. Ricciardulli, K. Hilburn, and C. Mears, 2007, How much more rain will global warming bring? *Science*, **317**, 233-235.

Widdows, J., and M. Brinsley, 2002, Impact of biotic and abiotic processes on sediment dynamics and the consequences to the structure and functioning of the intertidal zone, *Journal of Sea Research*, **48**, 143-156.

Widdows, J., A. Blauw, C.H.R. Heip, P.M.J. Herman, C.H. Lucas, J.J. Middelburg, S. Schmidt, M.D. Brinsley, F. Twisk, and H. Verbeek, 2004, Role of physical and biological processes in sediment dynamics of a tidal flat in Westerschelde Estuary, SW Netherlands, *Marine Ecology Progress Series*, **274**, 41-56.

Wijffels, S.E., J. Willis, C.M. Domingues, P. Barker, N.J. White, A. Gronell, K. Ridgway, and J.A. Church, 2008, Changing expendable bathythermograph fall rates and their impact on estimates of thermosteric sea level rise, *Journal of Climate*, **21**, 5657-5672.

Williams, H.F.L., 2010, Storm surge deposition by Hurricane Ike on the McFaddin National Wildlife Refuge, Texas: Implications for paleotempestology studies, *Journal of Foraminiferal Research*, **40**, 210-219.

Williams, P.B., and M.K. Orr, 2002, Physical evolution of restored breached levee salt marshes in the San Francisco Bay Estuary, *Restoration Ecology*, **10**, 527-542.

Willis, C.M., and G.B. Griggs, 2003, Reductions in fluvial sediment discharge by California coastal dams and implications for beach sustainability, *Journal of Geology*, **111**, 167-182.

Willis, J.K., J.M. Lyman, G.C. Johnson, and J. Gilson, 2009, In situ data biases and recent ocean heat content variability, *Journal of Atmospheric and Oceanic Technology*, **26**, 846-852.

Willis, J.K., D.P. Chambers, C.-Y. Kuo, and C.K. Shum, 2010, Global sea level rise: Recent progress and challenges for the decade to come, *Oceanography*, **23**, 26-35.

Wills, C.J., R.J. Weldon II, and W.A. Bryant, 2006, Fault Section Database 2.0, Developed for the Working Group on California Earthquake Probabilities, available at <http://gravity.usc.edu/WGCEP/modelComponents/deformModels/index.html>.

Wingham, D.J., A. Shepherd, A. Muir, and G.J. Marshall, 2006, Mass balance of the Antarctic Ice Sheet, *Philosophical Transactions of the Royal Society A*, **364**, 1627-1635.

Witter, R.C., H.M. Kelsey, and E. Hemphill-Haley, 2003, Great Cascadia earthquakes and tsunamis of the past 6700 years, Coquille River Estuary, southern coastal Oregon, *Geological Society of America Bulletin*, **115**, 1289-1306.

Wolters, M., A. Gabutt, and J.P. Bakker, 2005, Plant colonization after managed realignment: The relative importance of diaspore dispersal, *Journal of Applied Ecology*, **42**, 770-777.

Wood, E.F., and 21 others, 2011, Hyperresolution global land surface modeling: Meeting a grand challenge for monitoring Earth's terrestrial water, *Water Resources Research*, **47**, W05301, doi:10.1029/2010WR010090.

Woodworth, P.L., 1999, High waters at Liverpool since 1768: The UK's longest sea level record, *Geophysical Research Letters*, **26**, 1589-1592.

Woodworth, P.L., and D.L. Blackman, 2004, Evidence for systematic changes in extreme high waters since the mid-1970's, *Journal of Climate*, **17**, 1190-1197.

Woodworth, P.L., N.J. White, S. Jevrejeva, S.J. Holgate, J.A. Church, and W.R. Gehrels, 2009, Evidence for the accelerations of sea level on multi-decade and century timescales, *International Journal of Climatology*, **29**, 777-789.

Wöppelmann, G., B.M. Miguez, M.N. Bouin, and Z. Altamimi, 2007, Geocentric sea-level trend estimates from GPS analyses at relevant tide gauges world-wide, *Global Planetary Change*, **57**, 396-406.

Wöppelmann, G., N. Pouvreau, A. Coulomb, B. Simon, and P. Woodworth, 2008, Tide gauge datum continuity at Brest since 1711: France's longest sea-level record, *Geophysical Research Letters*, **35**, L22605, doi:10.1029/2008GL035783.

Wright, S.A., and D.H. Schoellhamer, 2004, Trends in the sediment yield of the Sacramento River, California, 1957–2001, *San Francisco Estuary and Watershed Science*, **2**(2), 14 pp.

Wu, P., 2006, Sensitivity of relative sea levels and crustal velocities in Laurentide to radial and lateral viscosity variations in the mantle, *Geophysical Journal International*, **165**, 401-413.

Wu, X., M.B. Heflin, H. Schotman, B.L.A. Vermeersen, D. Dong, R.S. Gross, E.R. Ivins, A.W. Moore, and S.E. Owen, 2010, Simultaneous estimation of global present-day water transport and glacial isostatic adjustment, *Nature Geoscience*, **3**, 642-646.

Wyllie-Echeverria, S., T. Mumford, J. Gaydos, and S. Buffum, 2003, *Zostera marina* declines in San Juan County, Washington, in *Westcott Bay Taskforce Mini-Workshop*, July 26, 2003, San Juan Island, Washington, 18 pp., available at <http://www.sanjuans.org/pdf_document/eelgrass-decline-report.pdf>.

Yamaguchi, D.K., B.F. Atwater, D.E. Bunker, B.E. Benson, and M.S. Reid, 1997, Tree-ring dating the 1700 Cascadia earthquake, *Nature*, **389**, 922-923.

Yeats, R., 1993, Converging more slowly, *Nature*, **366**, 300-301.

Yeats, R.S., and G.J. Huftile, 1995, The Oak Ridge Fault System and the 1994 Northridge earthquake, *Nature*, **373**, 418-420.

Young, I.R., S. Zieger, and A.V. Babanin, 2011a, Global trends in wind speed and wave height, *Science*, **332**, 451-455.

Young, I., S. Zieger, and A.V. Babanin, 2011b, Response to comment on "Global trends in wind speed and wave height," *Science*, **334**, 905.

Yuill, B., D. Lavoie, and D.J. Reed, 2009, Understanding subsidence processes in coastal Louisiana, *Journal of Coastal Research*, **54**, 23-36.

Zedler, J., 2010, How frequent storms affect wetland vegetation: A preview of climate-change impacts, *Frontiers in Ecology and the Environment*, **8**, 540-547.

Zemp, M., M. Hoelzle, and W. Haeberli, 2009, Six decades of glacier mass-balance observations: A review of the worldwide monitoring network, *Annals of Glaciology*, **50**, 101-111.

Zervas, C., 2009, *Sea Level Variations of the United States 1854–2006*, NOAA Technical Report NOS CO-OPS 053, Silver Spring, MD, 187 pp.

Zetler, B.D., and R.E. Flick, 1985, Predicted extreme high tides for California, 1983-2000, *Journal of Waterway, Port, Coastal, and Ocean Engineering*, **111**, 758-765.

Zhang, T., R.G. Barry, K. Knowles, F. Ling, and R.L. Armstrong, 2003, Distribution of seasonally and perennially frozen ground in the Northern Hemisphere, in *Proceedings of the 8th International Conference on Permafrost*, July 21-25, 2003, Zurich, Switzerland, M. Phillips, S.M. Springman, and L.U. Arenson, eds., A.A. Balkema, Lisse, The Netherlands, pp. 1289-1294.

Zwally, H.J., L. Jun, A.C. Brenner, M. Beckley, H.G. Cornejo, J. Dimarzio, M.B. Giovinetto, T.A. Neumann, J. Robbins, J.L. Saba, D. Yi, and W. Wang, 2011, Greenland Ice Sheet mass balance: Distribution of increased mass loss with climate warming; 2003-07 versus 1992-2002, *Journal of Glaciology*, **57**, 88-102.

Appendix A

Vertical Land Motion and Sea-Level Data Along the West Coast of the United States

As summarized in Chapter 4 ("Analysis of West Coast Tide Gage Records"), the committee determined rates of historical sea-level change along the California, Oregon, and Washington coasts using 12 tide gages. The rates were then corrected for vertical land motion and atmospheric pressure effects to compare sea-level rise along the west coast of the United States with the global mean sea-level rise. Details of these analyses are given below.

VERTICAL LAND MOTION FROM CONTINUOUS GPS

Tide gages are generally corrected for vertical land motions using glacial isostatic adjustment (GIA) models. However, GIA models capture only a small component of the total vertical land motion in coastal areas that are tectonically active or undergoing subsidence or uplift associated with sediment compaction and/or fluid withdrawal or recharge. The Global Positioning System (GPS) began to be used to adjust tide gage data for vertical land motion around 1997 (Ashkenazi et al., 1993; Nerem et al., 1997; Zerbini, 1997). Continuous GPS (CGPS) records along the U.S. west coast have been available since the early 1990s, although significantly more stations have been installed since 2003 as part of the National Science Foundation's Plate Boundary Observatory.[1] Continuous GPS solutions allow vertical land motions to be estimated from shorter temporal records and with more confidence than episodic sampling of the land motion signal. The committee used CGPS data to estimate and remove the vertical land motion component from the sea levels recorded by west coast tide gages. The locations of the tide gages and CGPS stations analyzed in this report are shown in Figure A.1.

Several published GPS solutions are available for estimating west coast vertical land motions, each of which uses a different number of stations, timespan, and/or processing software (e.g., Donnellan et al., 1993a,b; Dong et al., 1998; Argus et al., 1999, 2005; Bennett et al., 1999; Argus and Gordon, 2001; Spinler et al., 2010). The committee used the Scripps Orbit and Permanent Array Center (SOPAC) velocity model,[2] which is a routinely updated, publicly available solution with the longest time span for each station as well as the greatest GPS station density for the U.S. west coast. The SOPAC processing details are described in Nikolaidis (2002). Briefly, GAMIT and GLOBK software (Dong et al., 1998) are used to calculate daily site positions, which are input to the velocity estimation model. Using the entire time series for a specific site, the model accounts for offsets, linear velocity, annual and semi-annual fluctuations (for stations with at least 2 years and 1 year of data, respectively), and post-seismic relaxation. Noise analysis (Williams, 2003) using white noise plus flicker noise covariances provides realistic uncertainty estimates.

A vertical land motion value was assigned to each tide gage site by taking the closest CGPS station with a velocity estimate within 15 km from the tide gage. The

[1] See <http://pboweb.unavco.org/>.

[2] See <http://sopac.ucsd.edu/processing/refinedModelDoc.html>.

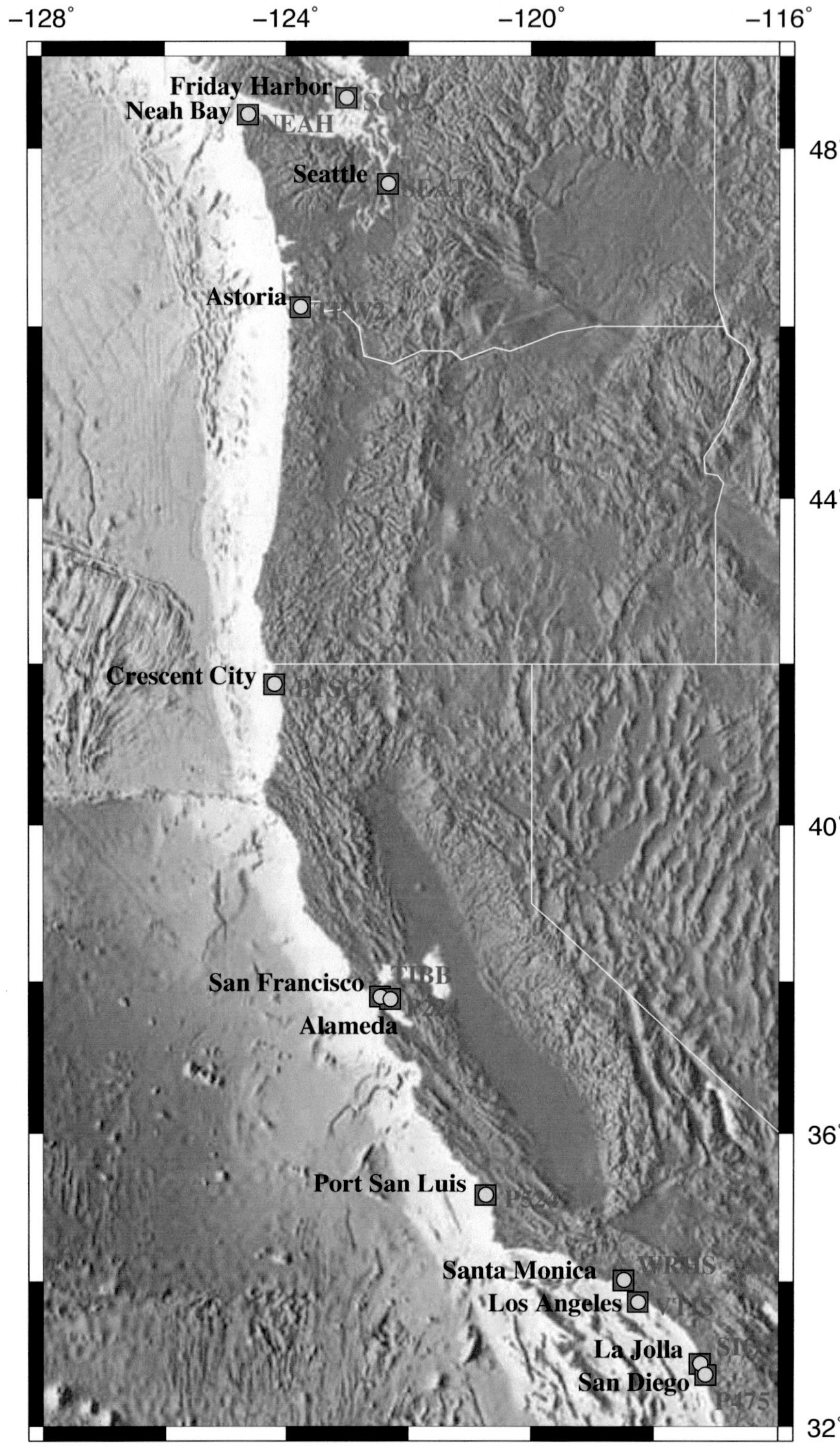

FIGURE A.1 Map showing names and locations of the 12 tide gages and CGPS stations analyzed in this report.

15-km value is somewhat arbitrary, but it is similar to the distance threshold typically used in previous joint GPS and tide gage analyses (e.g., Mazzotti et al., 2007; Wöppelmann et al., 2007). The committee's estimated rates of vertical land motions are given in Table A.1. Positive values of vertical land motion mean that the land is rising. Table A.1 also reports the standard deviation of vertical land motion within a 15-km radius of each tide gage station. This value was used in the uncertainty estimate described below.

To test the importance of the CGPS solution on the calculated rate of vertical land motion, the committee compared the SOPAC-derived rates with rates published in the literature. Although most published reports for coastal areas south of the Mendocino Triple Junction present only horizontal rates (e.g., Spinler et al., 2010), the SOPAC rates for Cascadia are similar to vertical rates for the same sites published in Mazzotti et al. (2008). Mazzotti et al. (2008) estimated rates using different GPS processing software (BERNESE) than that used by SOPAC. The correspondence in rates produced by two independent approaches suggests that the committee's results are reasonable.

Uncertainty in Vertical Land Motion

The estimated error in the vertical land motion adjustment to the tide gage records can be expressed as $e_{TG} = e_v + e_{sv} + e_{ref}$, where e_v is the error in the velocity estimation, e_{sv} is the error associated with spatial variability in vertical land motion between a CGPS station and the tide gage, and e_{ref} is the error associated with the definition of the GPS reference frame. Of these error sources, e_v is well defined (Nikolaidis, 2002; Williams et al., 2004) and is taken to be the vertical land motion error associated with the nearest CGPS station to the tide gage (and fifth and sixth columns of Table A.1).

Error in the spatial variability of vertical land motion between a CGPS station and a tide gage, e_{sv}, is locally variable and cannot be determined without more detailed studies. For example, Brooks et al. (2007) reported a variation of ~ ± 3 mm yr^{-1} for e_{sv} in the Los Angeles Basin. To estimate e_{sv}, the committee used the standard deviation of vertical land motion within a 15-km radius of each tide gage (right two columns in Table A.1). Using this value, rather than the formal error estimate associated with any given GPS station's velocity, seemed justified, given the potential for significant variability in vertical land motion at the km scale. For instance, the Santa Monica tide gage has seven CGPS stations within a 15-km radius, with vertical land motion estimates ranging from -1.5 mm yr^{-1} to 1.8 mm yr^{-1}. The nearest CGPS station, WRHS, has the minimum vertical land motion estimate (-1.5 mm yr^{-1}) and may be the most appropriate value for the correction. However, confidence in this value is diminished by the local spatial variability, which is reflected in the 15-km standard deviation value (Table A.1).

The reference frame error (e_{ref}) is a classical geodetic problem (Strang and Borre, 1997). The SOPAC

TABLE A.1 Parameters for Vertical Land Motion Correction

				Vertical Land Motion, Nearest Station		Vertical Land Motion, 15-km Radius[a]	
Tide Gage	Nearest CGPS Station	Start Date	Distance (km)	Rate (mm yr^{-1})	Error (1σ)	Rate (mm yr^{-1})	Error (1σ)
Friday Harbor	SC02	2001.860	0.70	0.90	0.70		
Neah Bay	NEAH	1996.000	7.71	3.00	0.40		
Seattle	SEAT	1996.000	6.25	0.20	0.50	-1.10	0.94
Astoria	TPW2	2000.247	1.08	0.60	0.00	1.20	0.40
Crescent City	PTSG	1999.820	5.85	2.60	0.40		
San Francisco	TIBB	1994.460	10.23	-1.40	0.50	-1.44	1.97
Alameda	P224	2005.174	12.90	-0.20	0.60	-1.58	1.20
Port San Luis	P524	2007.048	14.50	1.70	0.30		
Santa Monica	WRHS	1999.770	9.36	-1.50	0.60	-0.01	1.34
Los Angeles	VTIS	1998.938	2.54	-0.50	0.50	-0.27	2.34
La Jolla	SIO3	1993.522	0.26	2.10	0.50	0.73	1.11
San Diego	P475	2007.601	9.17	-3.00	0.20	-4.50	0.81

[a] Rates and errors are not reported for 15-km areas with only 1 CGPS station.

TABLE A.2 Tide Gage Records from PSMSL, Corrected for Atmospheric Pressure and Vertical Land Motion

				Trend (mm yr^{-1})			Confidence Limit		Trend (mm yr^{-1})
Tide Gage	Latitude	Longitude	Period	Original Record	With IB Adjustment	With IB and GPS Adjustment	Upper 95%	Lower 95%	With IB and GIA Adjustment
Friday Harbor	48.550	-123.000	1934–2008	+1.04	+1.14	+2.04	+3.44	+0.64	+1.31
Neah Bay	48.367	-124.617	1934–2008	-1.77	-1.65	+1.35	+2.22	+0.48	-1.33
Seattle	47.760	-122.333	1900–2008	+2.01	+2.10	+2.30	+3.29	+1.31	+1.67
Astoria	46.217	-123.767	1925–2008	-0.38	-0.30	+0.30	+0.61	-0.01	-1.37
Crescent City	41.750	-124.200	1933–2008	-0.73	-0.65	+1.95	+2.78	+1.12	-0.87
San Francisco[a]	37.800	-122.467	1900–2008	+1.92	+1.98	+0.58	+0.69	+0.47	+1.99
Alameda	37.767	-122.300	1939–2008	+0.70	+0.82	+0.62	+1.82	-0.58	+0.85
Port San Luis	35.167	-120.750	1945–2008	+0.68	+0.76	+2.46	+3.10	+1.82	+0.68
Santa Monica	34.017	-118.500	1933–2008	+1.41	+1.44	-0.09	+0.72	-0.90	+1.43
Los Angeles	33.717	-118.267	1923–2008	+0.84	+0.83	+0.33	+1.32	-0.66	+0.80
La Jolla	32.867	-117.250	1924–2008	+2.08	+2.07	+4.17	+5.17	+3.17	+2.02
San Diego	32.717	-117.167	1906–2008	+2.04	+2.07	-0.96	+0.03	-1.95	+2.01

NOTE: IB = inverse barometer.
[a] Although the San Francisco record starts in 1854, the IB correction starts at 1900.

velocity solution uses the International Terrestrial Reference Frame's ITRF2005 realization,[3] one of the most accurate and rigorously constrained versions of a global reference frame. ITRF2005 was developed from a combination of space geodetic observations from four independent platforms: GPS, Very Long Baseline Interferometry, Satellite Laser Ranging, and Doppler Orbit Determination and Radio-Positioning Integrated on Satellite. ITRF2005 is an improvement over previous frame realizations because it uses better antenna phase center models and its use of time series enables treatment of nonlinear and discontinuous behavior. A detailed description of the ITRF2005 realization appears in Altamimi et al. (2007).

Altamimi et al. (2007) suggested that the disagreement in origin definition between ITRF2005 and ITRF2000 (the previous frame realization) could be used as a metric of the origin accuracy. They reported a translation misfit of 0.1 mm, 0.8 mm, and 5.8 mm along the x-, y-, and z-axes, respectively, with a formal error of 0.3 mm for each component. The misfit of translation rate was reported as 0.2 mm yr^{-1}, 0.1 mm yr^{-1}, and 1.8 mm yr^{-1}, with a formal error of 0.3 mm yr^{-1} (Altamimi et al., 2007). Mazzotti et al. (2007), using ITRF2000, estimated that their vertical land motion values could contain a reference frame bias of -0.5–0.8 mm yr^{-1}. Given these two results and the difficulty of estimating e_{ref} in an absolute sense, the committee adopted a conservative value of ± 1 mm yr^{-1} for e_{ref}. The committee's error estimates for the vertical land motion correction to the tide gages are given in Table A.2.

ANALYSIS OF SEA-LEVEL TREND FROM TIDE GAGES

Sea-level records are archived at the Permanent Service for Mean Sea Level (PSMSL).[4] PSMSL reduces the data reported from each tide gage to monthly mean values and adjusts them to a common datum to produce a revised local reference dataset (Woodworth and Player, 2003). The sea-level data used in this report are all revised local reference datasets. PSMSL gives sea level in mm and, to avoid negative numbers, adds 7,000 mm to each monthly mean.

The committee examined the sea-level records from the 28 PSMSL tide gages along the west coast of the United States (Table A.2). For its analysis, the committee chose 12 tide gages that are currently operating and have records at least 60 years long. Shorter records are subject to decadal bias (Box A.1; see also "Tide Gage Measurements" in Chapter 2) and were not analyzed. Records with gaps in the data such as Santa Monica were simply treated as nonuniformly spaced in time.

[3] See <http://itrf.ensg.ign.fr/>.

[4] See <http://psmsl.org>.

BOX A.1
Effect of Record Length on Sea-Level Trend

As noted in Chapter 2, different lengths of record yield different sea-level trends because of decadal variability. The committee investigated the effect of record length on the rate of sea-level rise by fitting a linear trend over varying lengths of record from the San Francisco tide gage. The top figure shows the trends based on a starting year of 1900 and using longer and longer record lengths. The resulting rates of sea-level rise range from 1.12 mm yr^{-1} to 2.10 mm yr^{-1}. The bottom figure shows the reverse analysis, fixing the fit line at 2009 and adding progressively older data. For this case, rates of sea-level rise range from -0.05 mm yr^{-1} to 2.66 mm yr^{-1}, and the recent (1980–1990) downtrend in sea level noted in Bromirski et al. (2011) can be seen. The wide range in sea-level trends depending on the starting time and the length of the record also shows the difficulty of describing a complex time-varying signal with a simple linear relationship.

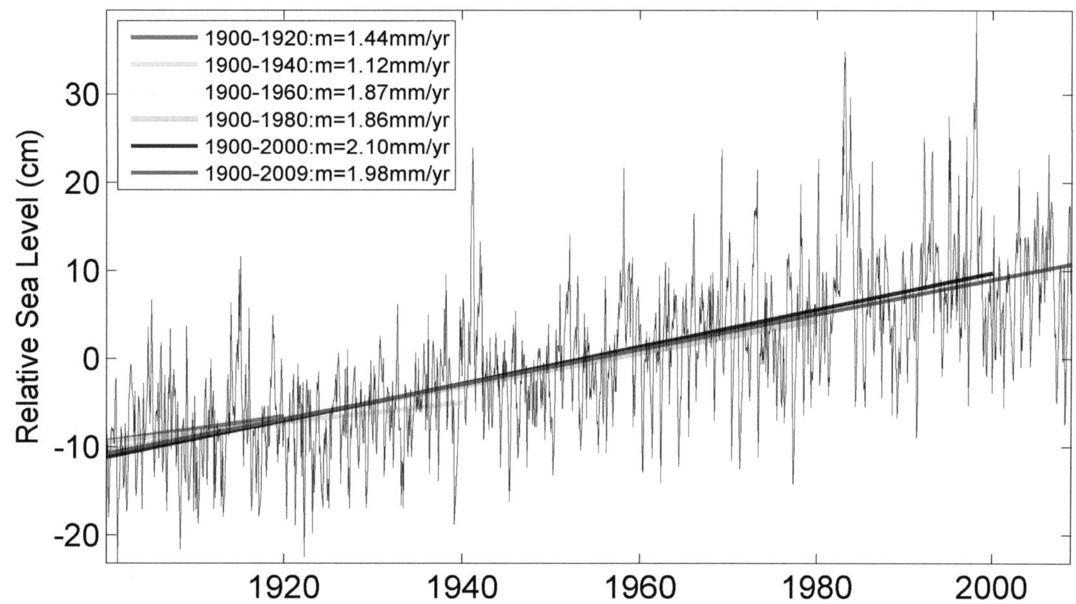

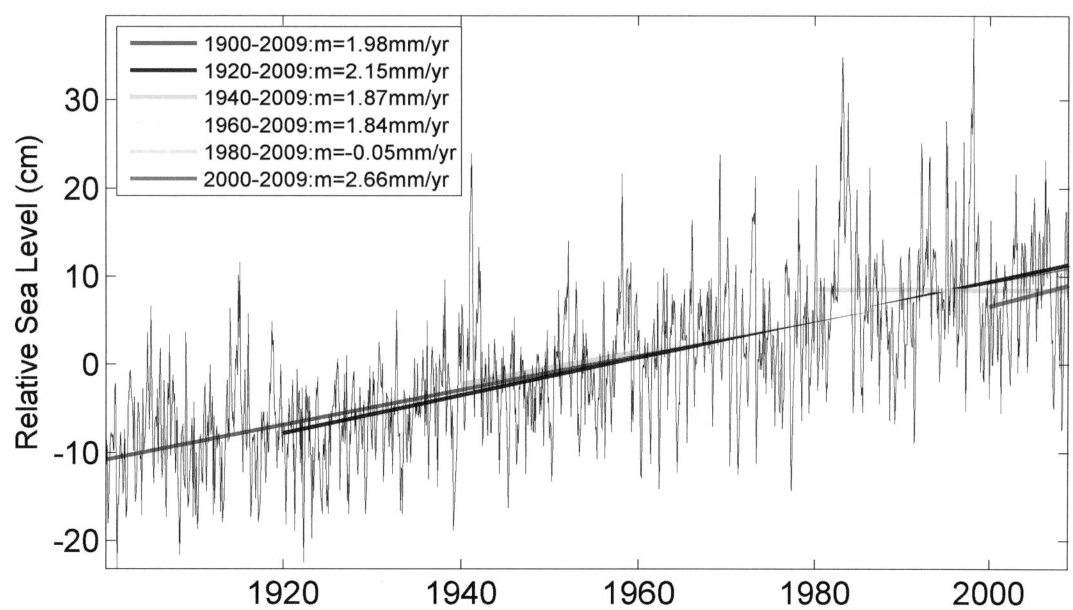

FIGURE Sea-level trends for the San Francisco tide gage data as a function of data length. (*Top*) Straight line fits from 1900. (*Bottom*) Straight line fits from 2009.

Although this approach has little or no effect on the estimated rate of sea-level rise, it slightly reduces the uncertainty (variability) when the gaps are relatively long.

The committee began the analysis of the PSMSL data by removing the 7,000 mm offset, thus reintroducing negative values. Long-term trends for the original tide gage data were obtained by fitting a straight line to the monthly average values for each record using a least squares method expressed as $y_t = mx_t + b$, where y_t (in mm) is the monthly mean tide level at time x_t, where $t = 1, 2, \ldots, N$, and N equals the total number of observations in the record. The slope m, which represents the relative sea-level rise, is given in mm yr^{-1}, and b is the y-intercept in mm. The linear sea-level trends (m) were determined using the entire record length of each tide gage.

Adjustments to the Original Tide Gage Trends

The committee adjusted the original tide gage trends to remove the effects of atmospheric pressure and vertical land motions, as described below.

Atmospheric Pressure

In many studies of sea-level rise, tide gage records are not adjusted for the barotropic response of the ocean due to variations in atmospheric pressure. However, observations suggest that an ocean response to atmospheric pressure loads is expected everywhere except the tropics and western boundary current extension regions (Wunsch and Stammer, 1997). The so-called inverse barometer (IB) adjustment is a good approximation of the barotropic response of sea level.

The committee adjusted the 12 tide gage records for the inverse barometer effect following the procedures of Ponte et al. (1991). The adjustment can be expressed as $(P(t) - P(t)_{ref})/\rho g$, where ρ is the density of sea water and g is the acceleration of gravity. The term $-1/\rho g$ is a scale factor that converts local air pressure anomalies to water level equivalent with a value of -9.948 mm mb^{-1}. $P(t)$ represents the monthly averaged sea-level pressure for a specific tide gage and month at time t, and $P(t)_{ref}$ is the mean surface pressure taken over the global ocean for the same month. This adjustment is then subtracted from the monthly averaged sea level for that month.

Sea-level pressure data were obtained from the National Oceanic and Atmospheric Administration Earth System Research Laboratory (NOAA ESRL) 20th Century Reanalysis V2 database.[5] These data are available on a 2° × 2° global grid from 1871 through 2008. The sea-level pressure for each tide gage ($P(t)$) was extracted from the nearest 2° × 2° box. The mean ocean surface pressure for each month ($P(t)_{ref}$) was obtained by averaging over all 2° × 2° boxes. The results of the IB adjustment are shown in Table A.2. The effect of the adjustment is relatively small, changing the slope by less than 10 percent in the majority of cases. The changes appear to be smallest in southern California.

Vertical Land Motion

The tide gage records were corrected for local site motion using GPS data (see "Vertical Land Motion from Continuous GPS" above). The CGPS rates of vertical land motion were simply added to the tide gage rate of relative sea-level rise (the slope m) to obtain a new local motion-adjusted record. The GPS data extend back in time only one to two decades, while some tide gage records are more than 100 years long. For correction purposes, the committee made the crucial assumption that the GPS adjustment remains constant over the entire length of the sea-level record. This assumption is more likely valid where the vertical land motion is dominated by GIA, but it is open to question where subsidence or uplift are the primary geophysical causes. In most cases, the GPS adjustments are significantly larger than the GIA adjustments, confirming the importance of tectonics and subsidence to relative sea-level rise in the area (Table A.2).

The GPS adjustment is generally larger than the IB adjustment. The magnitude of the two adjustments varies among tide gages, with relatively large changes to the original trend in a few places due primarily to the GPS adjustment. For example, these corrections changed the slopes from -1.77 mm yr^{-1} to +1.35 mm yr^{-1} at Neah Bay, from -0.73 mm yr^{-1} to +1.95 mm yr^{-1} in Crescent City, and from +2.04 mm yr^{-1} to -0.96 mm yr^{-1} at San Diego (Table A.2). The large change at San Diego underscores the importance of small-scale spatial variability in vertical motion in the region.

[5] See <www.esrl.noaa.gov/psd/data/gridded/data.20thC_ReanV2.html>.

Results

Figure A.2 shows the distributions of slopes for the original tide gage data, the data with IB adjustments, and the data with IB and GPS adjustments for the 12 tide gages. The mean value of all slopes is 0.82 mm yr^{-1} for the original sea-level records, 0.88 mm yr^{-1} with the IB adjustment, and 1.25 mm yr^{-1} with the IB and GPS adjustments. The standard deviation increases from about 1.23 mm yr^{-1} to 1.39 mm yr^{-1} when the GPS adjustment is included. This slight increase suggests that the local variability in vertical motion is often significant and uncorrelated with the vertical motion experienced by the tide gage. It also may indicate that vertical land motions over the past one or two decades are not representative of the ground motion over the lifetime of the tide gage.

To suppress the influence of possible outliers, the slopes were plotted as a function of latitude then fit with robust or weighted lines of regression (solid red line in Figure A.3). In weighted regression, values that lie far from the line of regression are given less weight than values that lie closer to the line. The weighted regression employs iteratively reweighted least squares with a selection of weighting functions depending on what form and how much weighting is desired (Holland and Welsch, 1977). The form of weighting chosen was the Welsch weight function. The weights, $w(r)$, are given by $\exp(-r^2)$, where r is related to the distance between a given value and the line of regression. For a more detailed explanation, see Huber (1981). The process for deciding what weighting to use is somewhat subjective, but tests using several different weight functions did not significantly affect the results. The calculated slopes seldom varied by more than ± 10 percent.

To examine the variation in sea-level trends along the coastline, a weighted regression was applied to all 12 gages. The linear trend in the weighted regression for the entire coast shows a significant increase in slope from south to north (Figure A.3, solid red line), whereas the original data (dashed red line) show a decrease from south to north. The change in trend reflects vertical land motion along the coast.

Uncertainty in Adjusted Sea-Level Trends

The 95 percent confidence limits in the sea-level trends were calculated using measures of uncertainty for the original records and the GPS adjustments. For data where the individual observations are statistically independent, confidence limits are calculated using the student's t distribution. However, with time series, the observations are generally not independent, so the equivalent number of independent observations or degrees-of-freedom (DOFs) is often far less than the total number of observations in the original time series. For DOFs greater than 120, the t values do not change and no adjustment for serial correlation is required.

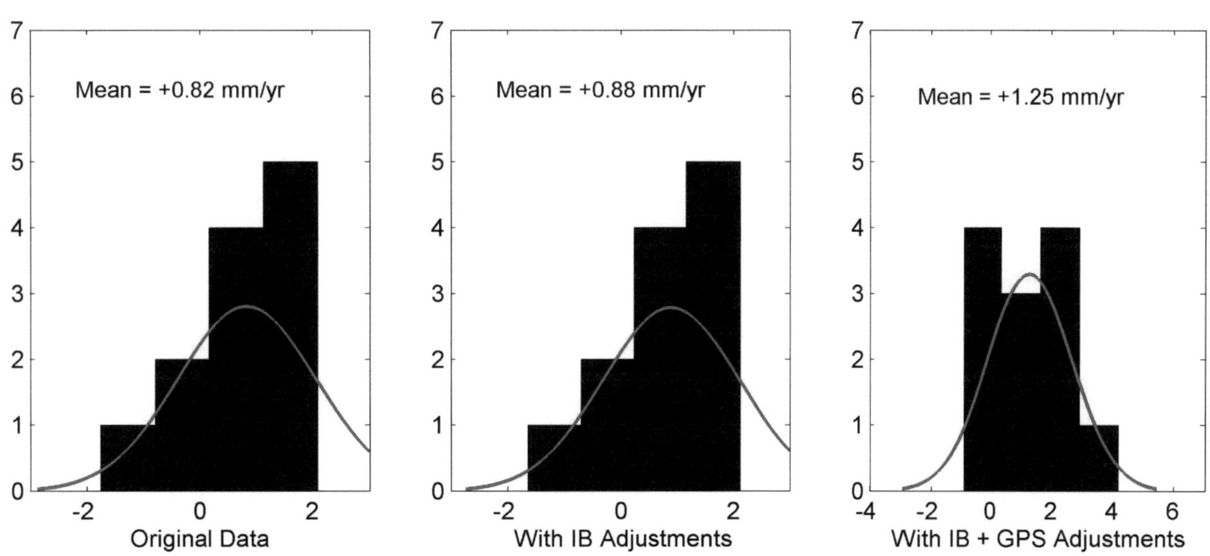

FIGURE A.2 Distributions of slopes for the original relative sea-level data (*left*), with inverse barometer (IB) corrections (*center*), and with IB plus GPS corrections (*right*).

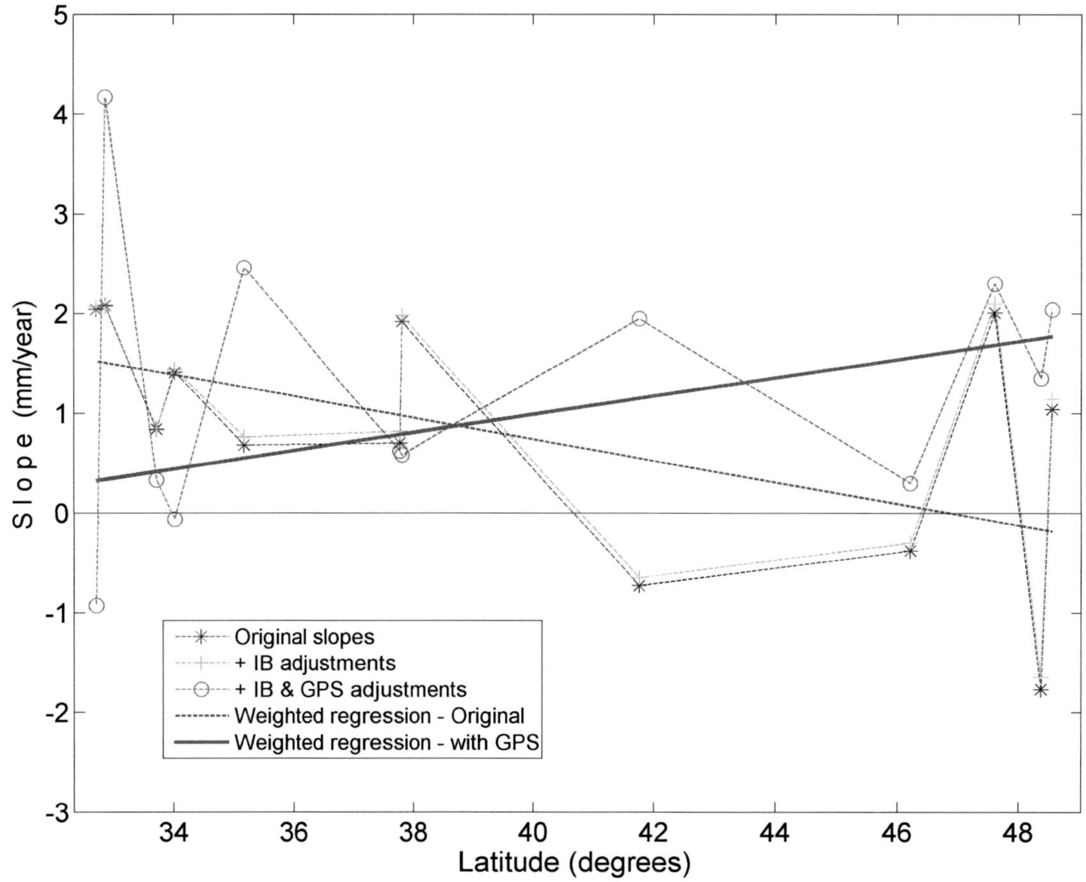

FIGURE A.3 Slopes of linear trends for tide gage records and weighted lines of regression. Slopes for original tide gage records (stars), for IB adjusted records (+), and for IB and GPS adjusted records (o) are shown as a function of latitude from north to south, together with weighted linear trends for the entire coast for the original data (dotted red line) and for the GPS-adjusted data (solid red line).

Several approaches are used to estimate the DOFs when the data are serially correlated, as they are in this case. The committee used the Mitchell et al. (1966) approach to calculate the DOFs for each record, and found that the DOFs were greater than 120 in all cases. Calculating the 95 percent confidence limits reduces to the slope ± 1.96 (from the *t* table) times the square root of the estimated variance of the trend. The confidence interval is thus equal to the difference between the upper and lower confidence limits.

Estimating the confidence limits for the GPS-adjusted trends requires not only a measure of uncertainty associated with the tide gage records but also a measure of uncertainty associated with the GPS data, in this case, the standard deviations. Assuming that the uncertainties for the slopes of the tidal records and the GPS data are independent, the combined uncertainty is equal to the square root of the sum of the variances associated with each. Multiplying this value by 1.96 yields the 95 percent confidence limits. The upper and lower confidence limits are included in Table A.2. In some cases, the uncertainties associated with the GPS data far exceed those associated with the tide gage data.

REFERENCES

Altamimi, Z., X. Collilieux, J. Legrand, B. Garayt, and C. Boucher, 2007, ITRF2005: A new release of the International Terrestrial Reference Frame based on time series of station positions and Earth orientation parameters, *Journal of Geophysical Research*, **112**, B09401, doi:10.1029/2007JB004949.

Argus, D.F., and R.G. Gordon, 2001, Present tectonic motion across the Coast Ranges and San Andreas Fault System in central California, *Geological Society of America Bulletin*, **113**, 1580-1592.

Argus, D.F., M.B. Heflin, A. Donnellan, F.H. Webb, D. Dong, K.J. Hurst, D.C. Jefferson, G.A. Lyzenga, M.M. Watkins, and J.F. Zumberge, 1999, Shortening and thickening of metropolitan Los Angeles measured and inferred by using geodesy, *Geology*, **27**, 703-706.

Argus, D.F., M.B. Heflin, G. Peltzer, F. Crampe, and F.H. Webb, 2005, Interseismic strain accumulation and anthropogenic motion in metropolitan Los Angeles, *Journal of Geophysical Research*, **110**, B04401, doi:10.1029/2003JB002934.

Ashkenazi, V., R.M. Bingley, G.M. Whitmore, and T.F. Baker, 1993, Monitoring changes in mean sea-level to millimetres using GPS, *Geophysical Research Letters*, **20**, 1951-1954.

Bennett, R.A., J.L. Davis, and B.P. Wernicke, 1999, Present-day pattern of Cordilleran deformation in the western United States, *Geology*, **27**, 371-374.

Bromirski, P.D., A.J. Miller, R.E. Flick, and G. Auad, 2011, Dynamical suppression of sea level rise along the Pacific Coast of North America: Indications for imminent acceleration, *Journal of Geophysical Research*, **116**, C07005, doi:10.1029/2010JC006759.

Brooks, B.A., M.A. Merrifield, J. Foster, C.L. Werner, F. Gomez, M. Bevis, and S. Gill, 2007, Space geodetic determination of spatial variability in relative sea level change, Los Angeles Basin, *Geophysical Research Letters*, **34**, L01611, doi:10.1029/2006GL028171.

Dong, D., T.A. Herring, and R.W. King, 1998, Estimating regional deformation from a combination of space and terrestrial geodetic data, *Journal of Geodesy*, **72**, 200-214.

Donnellan, A., B.H. Hager, and R.W. King, 1993a, Discrepancy between geological and geodetic deformation rates in the Ventura Basin, *Nature*, **366**, 333-336.

Donnellan, A., B. Hager, R. King, and T. Herring, 1993b, Geodetic measurement of deformation in the Ventura Basin region, southern California, *Journal of Geophysical Research*, **98**, 21,727-21,739.

Holland, P.W., and R.E. Welsch, 1977, Robust regression using iteratively reweighted least-squares, *Communications in Statistics: Theory and Methods*, **A6**, 813-827.

Huber, P.J., 1981, *Robust Statistics*, John Wiley & Sons, New York, 308 pp.

Mazzotti, S., A. Lambert, N. Courtier, L. Nykolaishen, and H. Dragert, 2007, Crustal uplift and sea level rise in northern Cascadia from GPS, absolute gravity, and tide gauge data, *Geophysical Research Letters*, **34**, L15306, doi:10.1029/2007GL030283.

Mazzotti, S., C. Jones, and R.E. Thomson, 2008, Relative and absolute sea level rise in western Canada and northwestern United States from a combined tide gauge-GPS analysis, *Journal of Geophysical Research*, **113**, C11019, doi:10.1029/2008JC004835.

Mitchell, Jr., J.M., B. Dzerdzeevskii, H. Flohn, W.L. Hofmeyr, H.H. Lamb, K.N. Rao, and C.C. Walléen, 1966, *Climate Change*, Report of a working group of the Commission for Climatology, Technical Note 79, World Meteorological Organization, WMO No. 195, Geneva, Switzerland, 79 pp.

Nerem, R.S., T.M. van Dam, and M.S. Schenewerk, 1997, A GPS network for monitoring absolute sea-level in the Chesapeake Bay: BAYONET, in *Proceedings of the Workshop on Methods for Monitoring Sea Level: GPS and Tide Gauge Benchmark Monitoring and Altimeter Calibration*, Jet Propulsion Laboratory Publication 97-017, Pasadena, CA, pp. 107-115.

Nikolaidis, R., 2002, *Observation of Geodetic and Seismic Deformation with the Global Positioning System*, Ph.D Thesis, University of California, San Diego, 249 pp.

Ponte, R.M., D.A. Salstein, and R.D. Rosen, 1991, Sea level response to pressure forcing in a barotropic numerical model, *Journal of Physical Oceanography*, **21**, 1043-1957.

Spinler, J.C., R.A. Bennett, M.L. Anderson, S.F. McGill, S. Hreinsdóttir, and A. McCallister, 2010, Present-day strain accumulation and slip rates associated with southern San Andreas and eastern California shear zone faults from GPS, *Journal of Geophysical Research*, **115**, B11407, doi:10.1029/2010JB007424.

Strang, G., and K. Borre, 1997, *Linear Algebra, Geodesy, and GPS*, Wellesley-Cambridge Press, Wellesley, MA, 624 pp.

Williams, S.D.P., 2003, The effect of coloured noise on the uncertainties of rates estimated from geodetic time series, *Journal of Geodesy*, **76**, 483-494.

Williams, S.D.P., Y. Bock, P. Fang, P. Jamason, R.M. Nikolaidis, L. Prawirodirdjo, M. Miller, and D.J. Johnson, 2004, Error analysis of continuous GPS position time series, *Journal of Geophysical Research*, **109**, B03412, doi:10.1029/2003JB002741.

Woodworth, P.L., and R. Player, 2003, The permanent service for mean sea level: An update to the 21st century, *Journal of Coastal Research*, **19**, 287-295.

Wöppelmann, G., B.M. Miguez, M.N. Bouin, and Z. Altamimi, 2007, Geocentric sea-level trend estimates from GPS analyses at relevant tide gauges world-wide, *Global Planetary Change*, **57**, 396-406.

Wunsch, C., and D. Stammer, 1997, Atmospheric loading and the oceanic "inverted barometer" effect, *Reviews of Geophysics*, **35**, 79-107.

Zerbini, S., 1997, The SELF II project, in *Proceedings of the Workshop on Methods for Monitoring Sea Level: GPS and Tide Gauge Benchmark Monitoring and Altimeter Calibration*, Jet Propulsion Laboratory Publication 97-017, Pasadena, CA, pp. 93-95.

Appendix B

Sea-Level Rise in the Northeast Pacific Ocean

Sea-level rise along the west coast of the United States can be estimated using tide gage records from coastal stations or satellite altimetry data from the northeast Pacific Ocean. Appendix A estimated sea-level rise along the coast by correcting tide gage records for total vertical land motions—including tectonics, glacial isostatic adjustment (GIA), and fluid withdrawal and recharge—using Global Positioning System data. This appendix estimates sea-level rise using tide gage records corrected for the GIA component of vertical land motion, then compares the estimates with regional altimetry data, which are also corrected for GIA.

It is well established that sea-level rise varies geographically and that this geographical pattern can be measured using satellite altimetry. However, satellite altimetry records are short (< 20 years), and the estimated sea-level trend is significantly influenced by ocean variability at interannual or longer timescales. On the other hand, satellite altimetry is much less affected by GIA than tide gages, and it is unaffected by vertical land motions caused by tectonics or fluid extraction or recharge. This appendix examines the extent to which tide gages and satellite altimetry measure the same sea-level variations at the seasonal and interannual scales in the northeast Pacific. This analysis differs from previous studies, which compared satellite altimetry with island tide gage data for calibration and validation purposes (e.g., Chambers et al., 2003; Mitchum et al., 2010). Such studies show that sea-level trends determined from tide gages are similar to trends determined from altimetry measurements made near the tide gage station (averaged within 220–330 km), as long as the vertical motion of the tide gage benchmark is known. The analysis below compares records from 21 tide gages and altimetry data from the northeast Pacific Ocean over the same timespan, 1992–2008. This exercise is intended (1) to assess confidence in both types of sea-level measurements, (2) to provide an independent check on the GPS-corrected tide gage sea-level estimates presented in Appendix A, and (3) to possibly discern the source of vertical land motion at the tide gage benchmarks along the coasts of California, Oregon, and Washington.

CORRECTIONS

Sea-level data from the TOPEX/Poseidon, Jason-1, and Jason-2 altimeters were obtained for 1992–2010 from JPL PODAAC,[1] and tide gage data were obtained from the Permanent Service for Mean Sea Level (PSMSL). To compare the tide gage and regional altimetry sea-level trends, the altimetry data were first corrected for instrument, media, and geophysical effects, and the gage data were corrected for atmosphere barotropic effects using the inverted barometer assumption. Both types of data were then corrected for the effect of GIA, assuming that GIA is the only geophysical process affecting motion of either the land or the seafloor, for tide gage or altimetry records, respectively. The corrections to the data and the conversion to geocentric sea-level trends are described below.

[1] See <http://podaac.jpl.nasa.gov>.

The sea-level data from TOPEX/Poseidon (ground tracks from the original mission and the tandem mission phase), Jason-1, and Jason-2 were processed by subtracting a mean sea surface from the corrected sea surface topography measurements, and applying cross-track and along-track gradient corrections. Relative altimeter biases between the three different altimeters were estimated globally with respect to TOPEX then applied to the regional study. The altimeter sea surface topography measurements were corrected for media (ionosphere, dry, and wet troposphere), instrument, and geophysical (solid Earth, pole and ocean tides, sea state bias, and atmosphere barotropic response) conditions.

The altimeter sea-level time series were corrected for dynamic atmosphere response using the AVISO dynamic atmosphere response model (Carrère and Lyard, 2003) and for atmospheric pressure (inverse barometer [IB]) using the European Centre for Medium-Range Weather Forecasts surface pressure model. These corrections were applied to the along-track sea-level data from each altimeter, and the data were averaged around each tide gage site into monthly measurements. The monthly tide gage sea-level data were corrected for IB effects using the National Oceanic and Atmospheric Administration Earth System Research Laboratory (NOAA ESRL) 20th Century Reanalysis V2 (Table B.1).[2]

Semiannual and annual signals for both the altimetry and tide gage sea-level time series were removed. Removal of these signals improved the correlation between the altimetry and tide gage sea-level data and reduced their root mean square (RMS) differences. Application of the atmosphere correction further reduced the sea-level variability of both the altimetry and tide gage data. The IB corrections had little impact on the linear sea-level trend estimated from the tide gage records, however they slightly reduced the correlation between the tide gage and altimetry sea-level data from 0.79 to 0.64 (Table B.1).

Both the tide gage and the altimetry sea-level data were corrected for GIA, and thus their estimated trends are directly comparable. For the tide gages, the sea-level trend was calculated by subtracting the GIA predicted relative sea-level trend from the tide gage estimated linear trend. The GIA predicted relative sea-level trend (absolute sea-level change minus the height change of the solid earth surface) at the gage benchmark is effectively a predicted vertical land motion estimate, with an opposite sign, if GIA is the only geophysical process. For altimetry, the geocentric sea-level trend was calculated by subtracting the GIA-model predicted absolute sea-level change from the altimeter sea-level trend (Peltier, 2001, Tamisiea, 2011). Table B.2 shows long-term tide gage estimated sea-level trends, corrected using the van der Wal GIA model, and the GIA model corrections from various GIA models.

RESULTS

Figure B.1 shows the GIA-model predicted relative sea-level change (computed by subtracting the crustal uplift from absolute sea-level change) from an ensemble of eight GIA models in western North America. Predicted values differ significantly from one another in Washington and Canada, but are similar along the coast. In the study region where the tide gages are located, the spread of GIA predicted values is between 1 mm yr^{-1} and 2 mm yr^{-1}, mostly dominated by the difference in the predicted uplift rate of the solid earth.

Figure B.2 shows the tide gage estimated long-term sea-level trends, ~1900–2009, corrected for GIA using an ensemble of 17 models, as a function of latitude. The discrepancy in sea-level trend due to the choice of different GIA models is approximately 1 mm yr^{-1} for the southern tide gages and up to 2 mm yr^{-1} for the Washington gages (Figure B.2).

Figure B.3 compares the estimated sea-level trends for both tide gages and satellite altimetry for 1992–2008. Both records were corrected for IB and for GIA using the van der Wal GIA model (van der Wal et al., 2009). The sea-level trend determined from satellite altimetry (background color) and from the tide gages (color-coded circles) is in reasonable agreement along the coast. Figure B.3 shows that the sea-level trend is nonuniform geographically, and that satellite altimetry can measure the sea-level trends much further away from the coastal regions where tide gages reside.

---

[2] The National Centers for Environmental Prediction (NCEP) Reanalysis Derived Data (Kalnay et al., 1996), which is valid for 1948–2011 and has a spatial resolution of 2.5° × 2.5°, can also be used to correct for inverse barometer. Tests showed that the performance of both the NOAA ESRL and NCEP models was almost identical.

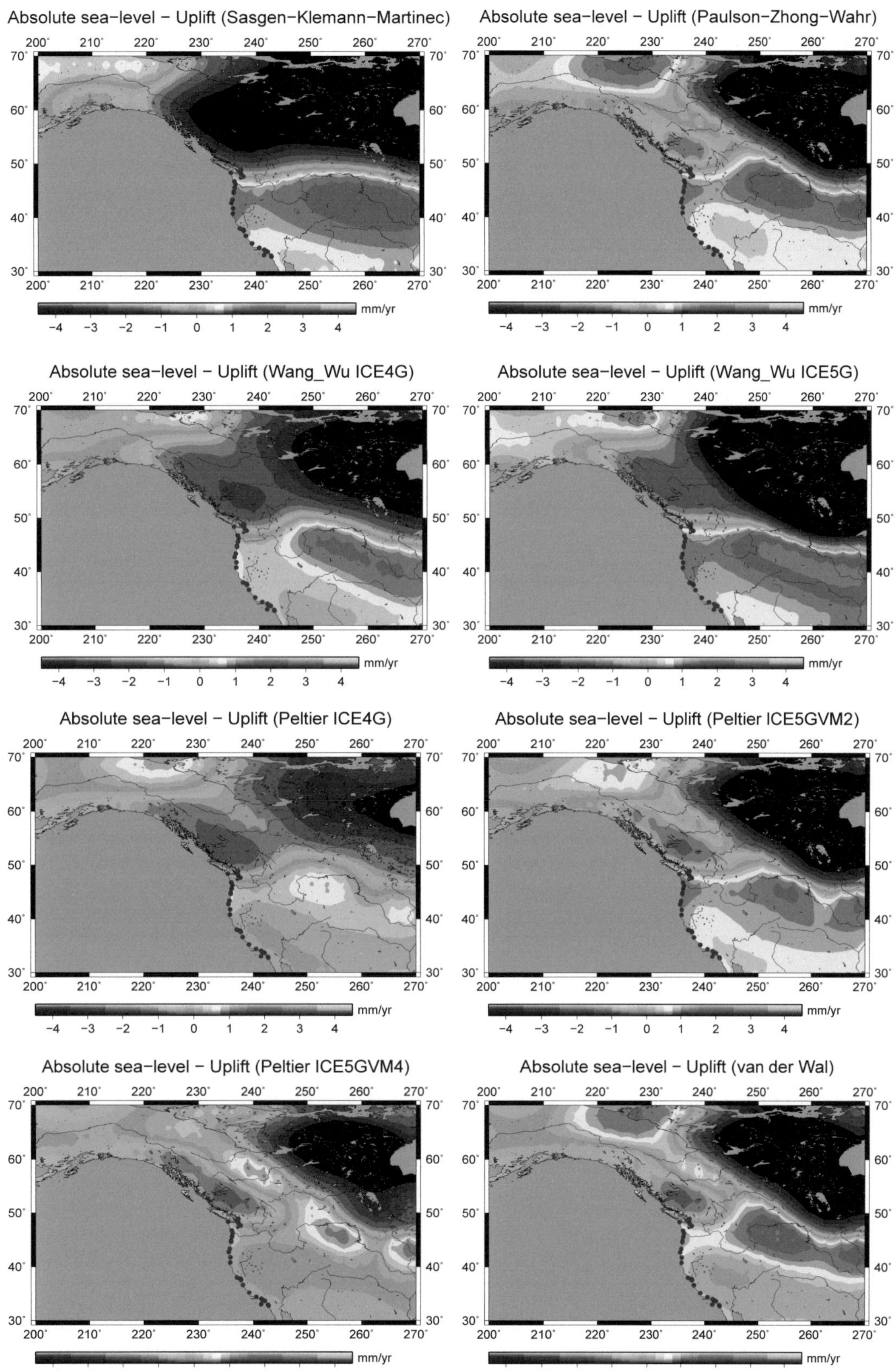

FIGURE B.1 Selected GIA model predicted relative sea-level rise (absolute sea-level change minus height change) in western North America. Tide gage locations are shown as blue dots. For the Paulson-Zhang-Wahr model, only the absolute sea level is available, so a theoretical predicted relationship between absolute sea-level change and uplift (Wahr et al., 1995) was used to compute the associated relative sea-level change.

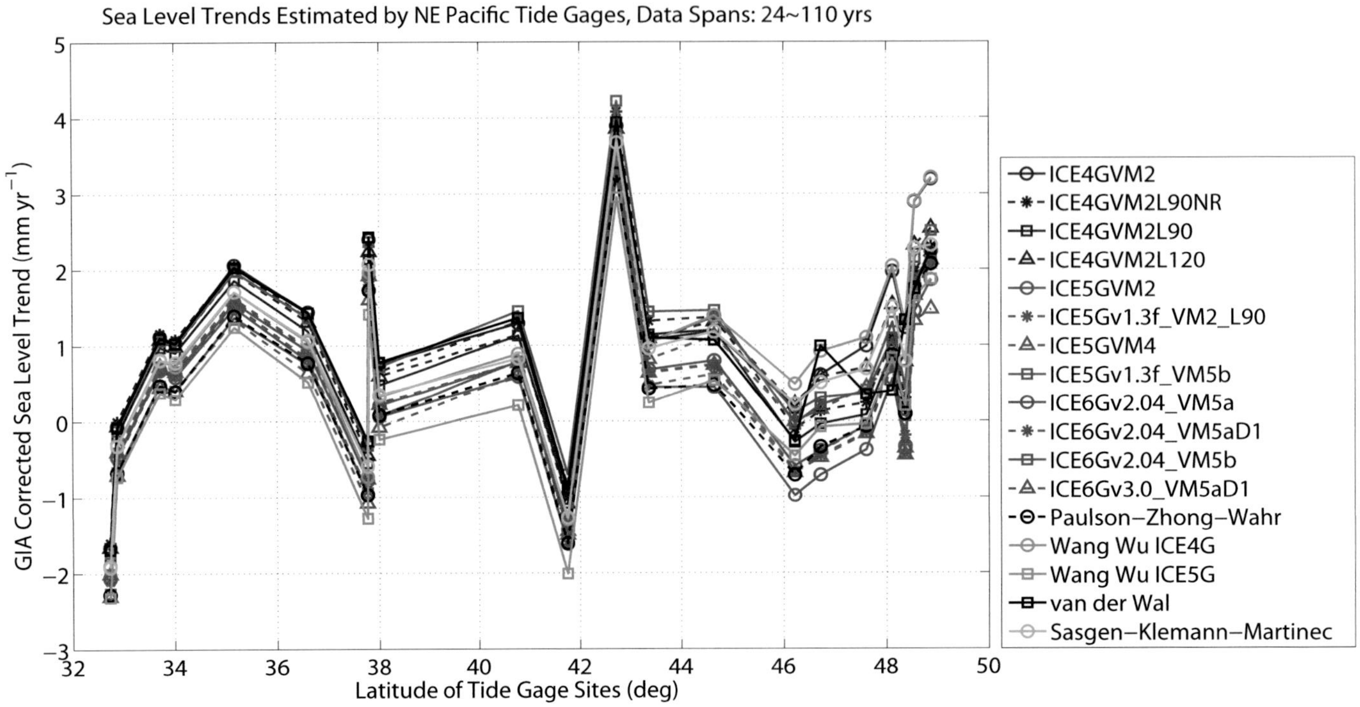

FIGURE B.2 Tide gage estimated long-term sea-level trends, corrected for GIA using an ensemble of 17 models at 21 locations from south (San Diego, 32.72°N) to north (Friday Harbor, 48 55°N).

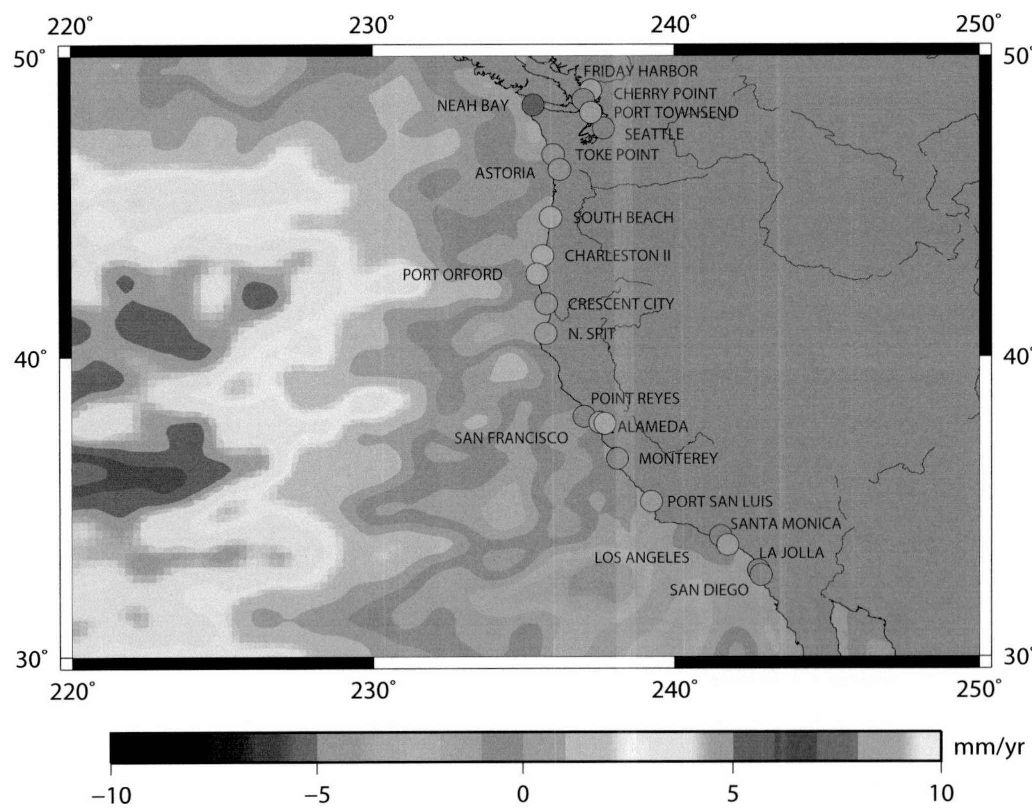

FIGURE B.3 Comparison of geocentric sea-level trends, in mm yr^{-1}, for 1992–2008 from 21 tide gages along the coast of California, Oregon, and Washington (color-coded circles) and from TOPEX/Poseidon, Jason-1, and Jason-2 satellite altimetry missions (background colors).

TABLE B.1 Effect of Inverted Barometer Corrections on Estimated Sea-Level Trends in the Northeast Pacific Ocean

	Without IB Correction to Tide Gage or Altimetry Data	IB Correction Using NOAA ESRL 20th Century Reanalysis V2 Model
Linear sea-level trend averaged from 21 tide gages, 1906–2008	0.71 ± 0.04 mm yr^{-1}	0.76 ± 0.04 mm yr^{-1}
RMS difference of average sea level, altimetry minus tide gage, 1992–2008	3.3 cm	3.3 cm
Correlation, 1992–2008	79.2%	63.7%

Figure B.4 compares the tide gage and altimetry sea-level time series for 1992–2008 at each tide gage site in the northeast Pacific study region. Note than the trend estimates for both data types are probably in error, primarily because of the short data spans and the large sea-level variability at interannual or longer timescales. RMS differences and correlations are given at the top of each figure, and the sea-level trends are indicated by the straight lines. The figures show that seasonal variations between the altimetry and tide gage sea-level time series are largely coherent (as indicated by RMS differences and correlations), indicating that the two independent sea-level measurement systems observed the same sea-level variations at seasonal and interannual timescales during 1992–2008. However, some estimated trends of individual tide gages are substantially different from the altimetry trend, likely because of the poor trend determination due to short data spans and/or because vertical land motion at some of these tide gage sites is dominated by non-GIA processes, such as tectonics.

Next, the estimated gage and altimetry trends were averaged to further examine whether these two independent types of data measure the same sea level at the same location. The top panel of Figure B.5 shows the averaged tide gage and altimetry sea-level time series for 1992–2008. The estimated trends are -2.0 ± 0.7 mm yr^{-1} for the tide gages and 0.0 ± 0.4 mm yr^{-1} for altimetry. Although these trend estimates are significantly different (because of short data spans and possibly non-GIA related uplift or subsidence at the tide gage benchmarks), both time series show good coherence at seasonal and interannual timescales. They have an averaged RMS difference of 3.2 cm and an averaged correlation of 65.9 percent, indicating that they measure essentially the same sea level. The bottom panel of Figure B.5 compares the short-term (last few decades) variation in sea level determined from tide gages and altimetry with the long-term (extending back to 1900) variation in sea level determined from tide gages. The stated reasons for the trend difference between altimetry and gage is further confirmed by the differences in the short-term (1992–2008) and long-term (1900–2009) gage trend estimates, at -2.0 ± 0.7 mm yr^{-1} and 0.76 ± 0.04 mm yr^{-1}, respectively. The latter represents a more robust sea-level trend estimate over the study region.

The above analysis indicates that both altimetry and tide gage observed sea-level variations are coherent at seasonal and interannual timescales. The difference in trend estimates of ~2 mm yr^{-1} over the timespan (1992–2008), correcting for vertical motion assuming an ongoing GIA process, could be caused by the short data span or by non-GIA related vertical motion at some of tide gages in the study region.

REFERENCES

Carrère, L., and F. Lyard, 2003, Modelling the barotropic response of the global ocean to atmospheric wind and pressure forcing – Comparisons with observations, *Geophysical Research Letters*, **30**, 1275.

Chambers, D., S.A. Hayes, J.C. Ries, and T. Urban, 2003, New TOPEX sea state bias models and their effect on global mean sea level, *Journal of Geophysical Research*, **108**, 3305, doi:10.1029/2003JC001839.

Kalnay, E., M. Kanamitsu, R. Kistler, W. Collins, D. Deaven, L. Gandin, M. Iredell, S. Saha, G. White, J. Woollen, Y. Zhu, M. Chelliah, W. Ebisuzaki, W. Higgins, J. Janowiak, K.C. Mo, C. Ropelewski, J. Wang, A. Leetmaa, R. Reynolds, R. Jenne, and D. Joseph, 1996, The NCEP/NCAR 40-year reanalysis project, *Bulletin of the American Meteorological Society*, **77**, 437-470.

Mitchum, G., R. Nerem, M. Merrifield, and W. Gehrels, 2010, Modern sea level estimates, in *Understanding Sea-Level Rise and Variability*, J. Church, P. Woodworth, T. Aarup, and W. Wilson, eds., Wiley-Blackwell, UK, pp. 122-428.

Paulson, A., S.J. Zhong, and J. Wahr, 2007, Inference of mantle viscosity from GRACE and relative sea level data, *Geophysical Journal International*, **171**, 497-508.

TABLE B.2 Sea-Level Trends and Different GIA Corrections for 21 Tide Gages

Tide Gage					van der Wal[a] (mm yr^{-1})		Peltier ICE4G VM2 (mm yr^{-1})	
Name	Latitude	Longitude	Period	Geocentric Sea-Level Trend[b] (mm yr^{-1})	Absolute Sea-Level Trend	Relative Sea-Level Trend[c]	Absolute Sea-Level Trend	Relative Sea-Level Trend
Cherry Point	48.867	-122.750	1985–2009	-0.03	-0.23	-0.50	0.05	-1.10
Friday Harbor	48.550	-123.000	1934–2009	1.13	-0.25	-0.17	0.03	-0.83
Neah Bay	48.367	-124.617	1934–2009	-1.63	-0.28	0.32	0.00	-0.40
Port Townsend	48.117	-122.750	1972–2009	1.48	-0.26	0.12	0.01	-0.53
Seattle	47.760	-122.333	1899–2009	2.11	-0.28	0.43	-0.01	-0.21
Toke Point	46.717	-123.967	1973–2009	1.04	-0.33	1.04	-0.06	0.50
Astoria	46.217	-123.767	1925–2009	-0.29	-0.35	1.07	-0.07	0.64
South Beach	44.633	-124.050	1967–2009	2.42	-0.38	0.91	-0.11	0.80
Charleston II	43.350	-124.317	1970–2009	0.82	-0.40	0.58	-0.13	0.61
Port Orford	42.733	-124.500	1985–2009	1.49	-0.40	0.45	-0.14	0.50
Crescent City	41.750	-124.200	1933–2009	-0.67	-0.41	0.22	-0.15	0.29
North Spit	40.767	-124.217	1985–2009	4.39	-0.41	0.13	-0.16	0.20
Point Reyes	38.000	-122.983	1975–2009	1.68	-0.43	0.04	-0.18	0.06
San Francisco	37.800	-122.467	1854–2009	1.98	-0.43	-0.01	-0.17	0.02
Alameda	37.767	-122.300	1939–2009	0.81	-0.43	-0.03	-0.17	0.01
Monterey	36.600	-121.883	1973–2009	1.11	-0.44	0.05	-0.18	0.04
Port San Luis	35.167	-120.750	1945–2009	0.78	-0.45	0.08	-0.19	0.05
Santa Monica	34.017	-118.500	1933–2009	1.44	-0.45	0.01	-0.19	0.00
Los Angeles	33.717	-118.267	1923–2009	0.83	-0.45	0.03	-0.19	0.01
La Jolla	32.867	-117.250	1924–2009	2.06	-0.45	0.05	-0.19	0.03
San Diego	32.717	-117.167	1906–2009	2.09	-0.45	0.06	-0.19	0.03
Average				1.19	-0.38	0.23	-0.11	0.03

SOURCE: ICE4G VM2 (Peltier, 2002), ICE5G VM2, and ICE5G VM4 (Peltier, 2004) models are from <http://www.sbl.statkart.no/projects/pgs/authors>. ICE5G predictions were computed by R. Peltier (personal communication, 2011). GIA model results (Wang and Wu, 2006; Paulson et al., 2007; van der Wal et al., 2009; Sasgen et al., 2012; H. Wang, personal communication, 2011) were provided by the respective authors.

[a] Model includes rotational feedback.

[b] Tide gage geocentric sea-level trend = tide gage relative sea-level trend minus GIA predicted relative sea-level trend. The van der Wal model was used for the GIA correction. Tide gage relative sea-level trend was estimated by fitting a trend simultaneously with terms modeling semiannual and annual variations.

[c] GIA predicted relative sea-level trend = absolute sea-level change minus height change of the solid earth surface.

Peltier, W., 2001, Global glacial isostatic adjustment and modern instrumental records of relative sea level history, in *Sea Level Rise: History and Consequences*, B. Douglas, M. Kearney, and S. Leatherman, eds., International Geophysics Series, **75**, Academic Press, pp. 65-95.

Peltier, W.R., 2002, On eustatic sea level history: Last Glacial Maximum to Holocene, *Quaternary Science Reviews*, **21**, 377-396.

Peltier, W., 2004, Global glacial isostasy and the surface of the ice-age Earth: The ICE-5G (VM2) model and GRACE, *Annual Review of Earth and Planetary Sciences*, **32**, 111-149.

Sasgen, I., V. Klemann, and Z. Martinec, 2012, Towards the joint inversion of GRACE gravity fields for present-day ice-mass changes and glacial-isostatic adjustment in North America and Greenland, *Journal of Geodynamics*, in press.

Tamisiea, M., 2011, Ongoing glacial isostatic contributions to observations of sea level change, *Geophysical Journal International*, **186**, 1036-1044.

van der Wal, W., A. Braun, P. Wu, and M. Sideris, 2009, Prediction of decadal slope changes in Canada by glacial isostatic adjustment modeling, *Canadian Journal of Earth Sciences*, **46**, 587-595.

Wahr, J.M., D. Han, and A. Trupin, 1995, Predictions of vertical uplift caused by changing polar ice volumes on a viscoelastic Earth, *Geophysical Research Lett*ers, **22**, 977-980.

Wang, H., and P. Wu, 2006, Effects of lateral variations in lithospheric thickness and mantle viscosity on glacially induced surface motion on a spherical, self-gravitating Maxwell Earth, *Earth and Planetary Science Letters*, **244**, 576-589.

Peltier ICE5G VM2[a] (mm yr^{-1})		Peltier ICE5G VM4[a] (mm yr^{-1})		Wang-Wu ICE4G (mm yr^{-1})		Wang-Wu ICE5G (mm yr^{-1})		Sasgen-Klemann-Martinec (mm yr^{-1})	
Absolute Sea-Level Trend	**Relative Sea-Level Trend**	**Absolute Sea-Level Trend**	**Relative Sea-Level Trend**	**Absolute Sea-Level Trend**	**Relative Sea-Level Trend**	**Absolute Sea-Level Trend**	**Relative Sea-Level Trend**	**Absolute Sea-Level Trend**	**Relative Sea-Level Trend**
-0.91	0.23	-0.81	-0.42	0.25	-1.12	0.37	0.22	0.08	-0.23
-0.93	0.61	-0.83	-0.09	0.22	-0.84	0.33	0.42	0.03	-0.22
-0.96	1.18	-0.85	0.40	0.18	-0.45	0.26	0.63	-0.04	0.04
-0.96	0.88	-0.85	0.14	0.20	-0.60	0.31	0.61	-0.01	-0.06
-0.99	1.16	-0.87	0.37	0.18	-0.32	0.28	0.82	-0.06	0.09
-1.04	1.82	-0.91	0.90	0.11	0.19	0.17	1.19	-0.16	0.60
-1.06	1.79	-0.92	0.85	0.09	0.32	0.15	1.27	-0.17	0.59
-1.09	1.53	-0.95	0.52	0.03	0.59	0.06	1.44	-0.19	0.80
-1.10	1.23	-0.96	0.24	-0.01	0.65	-0.01	1.44	-0.18	0.72
-1.11	1.15	-0.96	0.17	-0.03	0.67	-0.03	1.44	-0.18	0.72
-1.11	0.95	-0.96	0.03	-0.05	0.62	-0.06	1.34	-0.16	0.60
-1.12	0.90	-0.97	0.04	-0.07	0.60	-0.09	1.28	-0.15	0.66
-1.13	0.76	-1.00	0.10	-0.10	0.49	-0.14	1.06	-0.12	0.47
-1.13	0.69	-1.00	0.05	-0.10	0.45	-0.14	1.01	-0.12	0.35
-1.13	0.67	-1.00	0.03	-0.10	0.44	-0.14	0.99	-0.12	0.30
-1.14	0.68	-1.01	0.12	-0.11	0.43	-0.15	0.96	-0.11	0.41
-1.14	0.61	-1.02	0.13	-0.12	0.40	-0.17	0.87	-0.10	0.40
-1.14	0.45	-1.03	0.04	-0.11	0.31	-0.17	0.75	-0.09	0.26
-1.14	0.45	-1.03	0.06	-0.12	0.31	-0.17	0.75	-0.08	0.32
-1.14	0.44	-1.03	0.09	-0.12	0.27	-0.17	0.70	-0.08	0.29
-1.14	0.44	-1.03	0.10	-0.12	0.27	-0.17	0.69	-0.08	0.29
-1.08	0.89	-0.95	0.19	0.01	0.18	0.02	0.95	-0.10	0.35

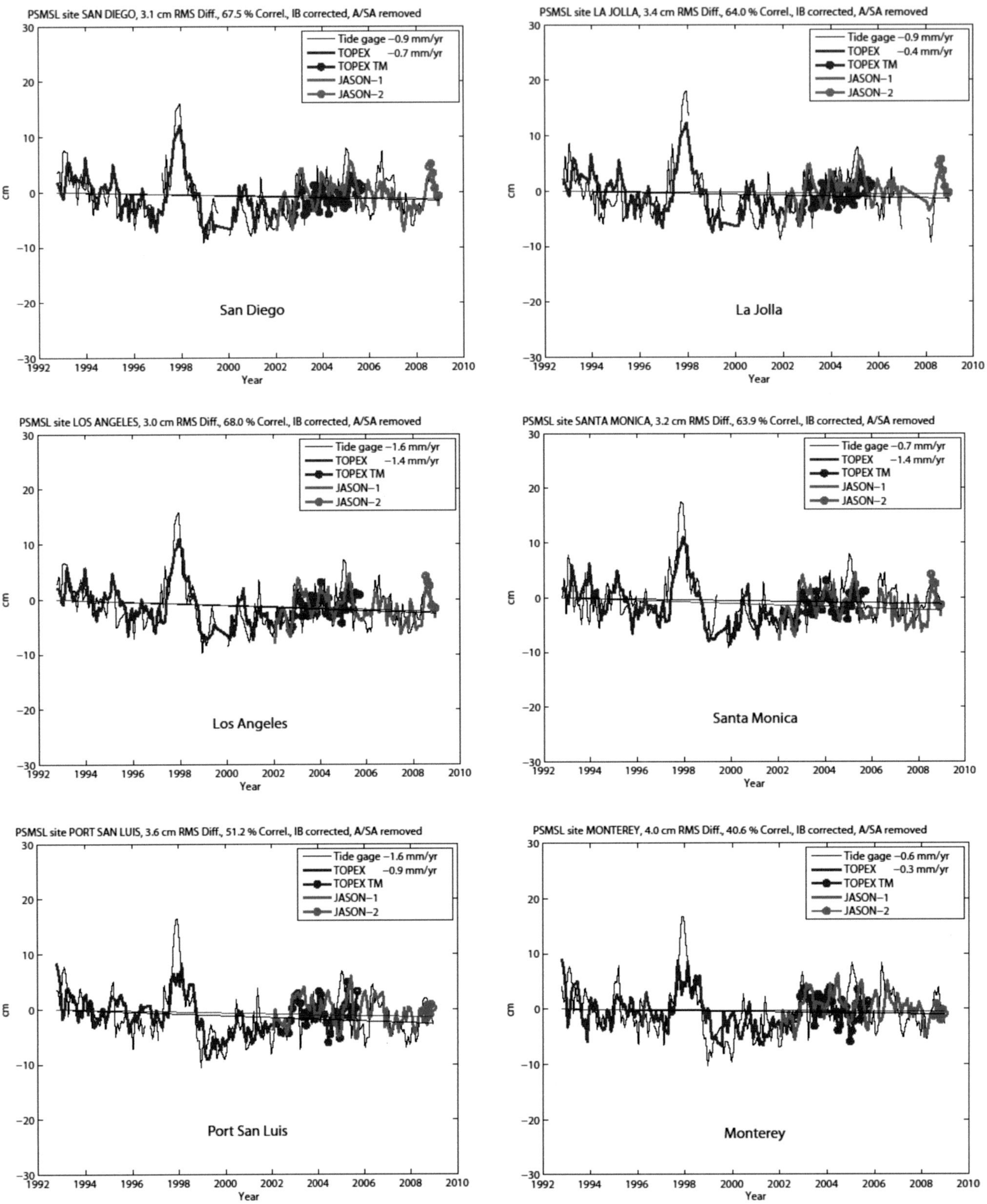

Figure continues

FIGURE B.4 Comparisons for 1992–2008 between the sea-level time series for 21 tide gages and averaged altimetry (TOPEX, Jason-1, and Jason-2) in the northeast Pacific study region.

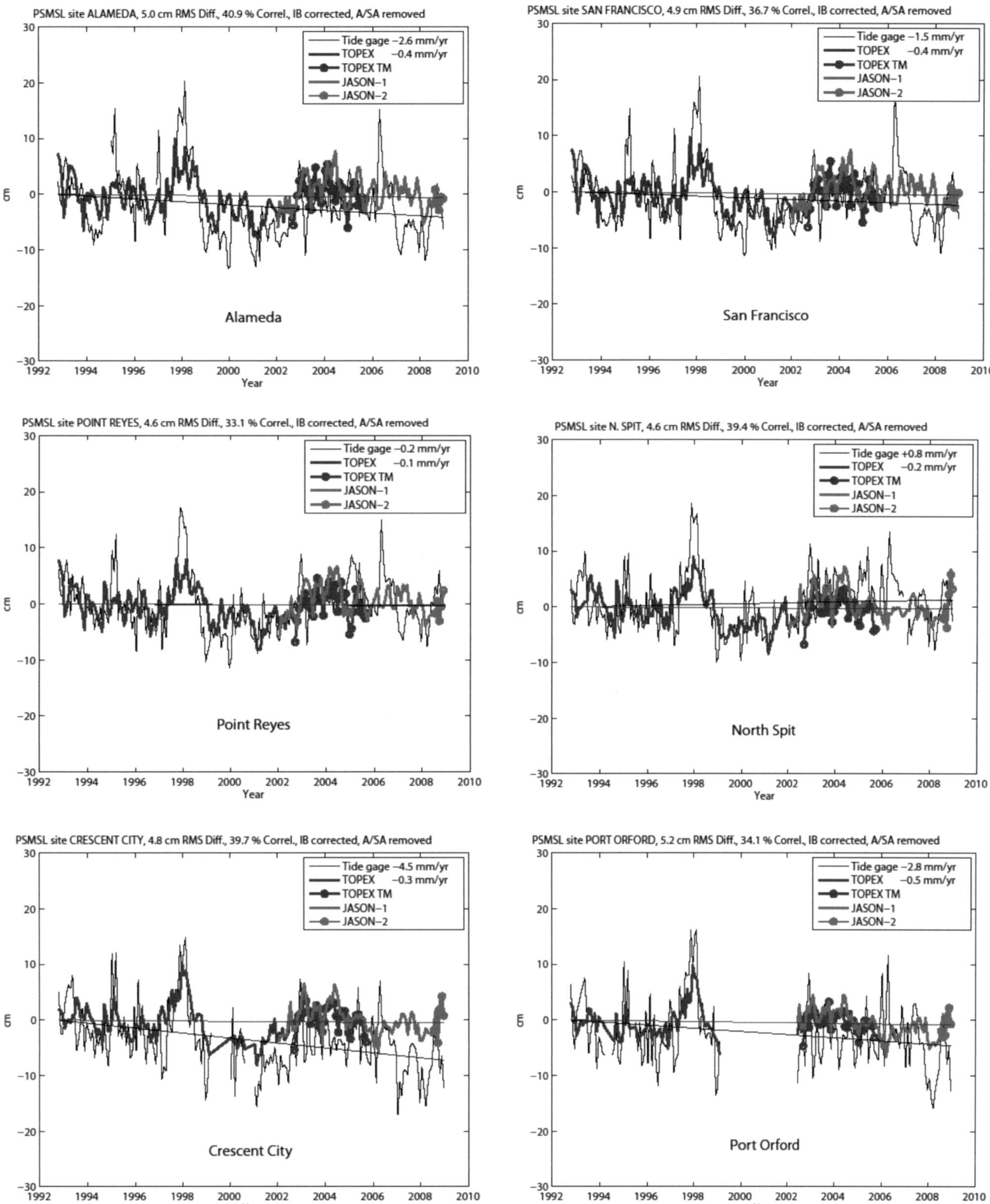

Figure continues

FIGURE B.4 Continued

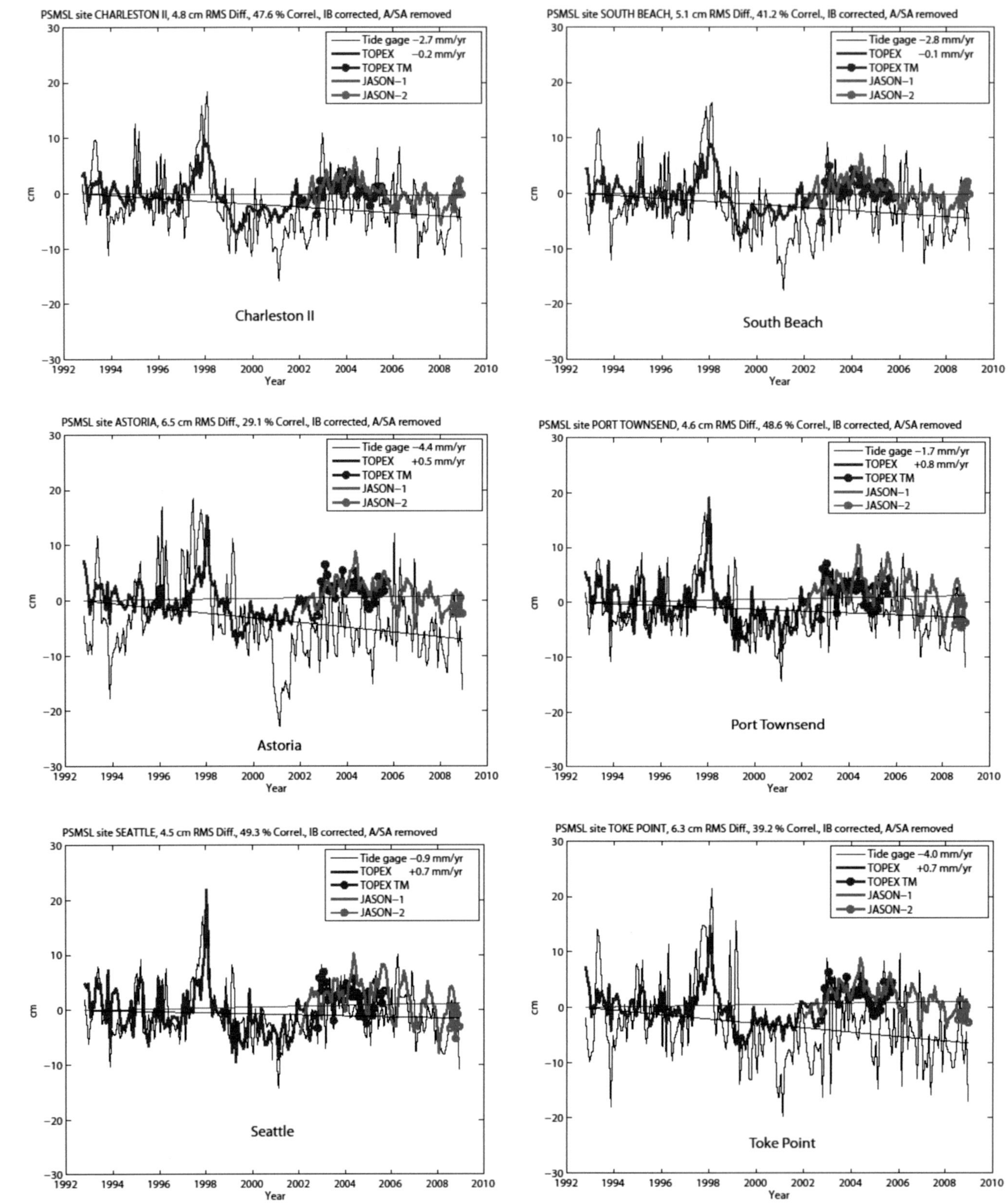

Figure continues

FIGURE B.4 Continued

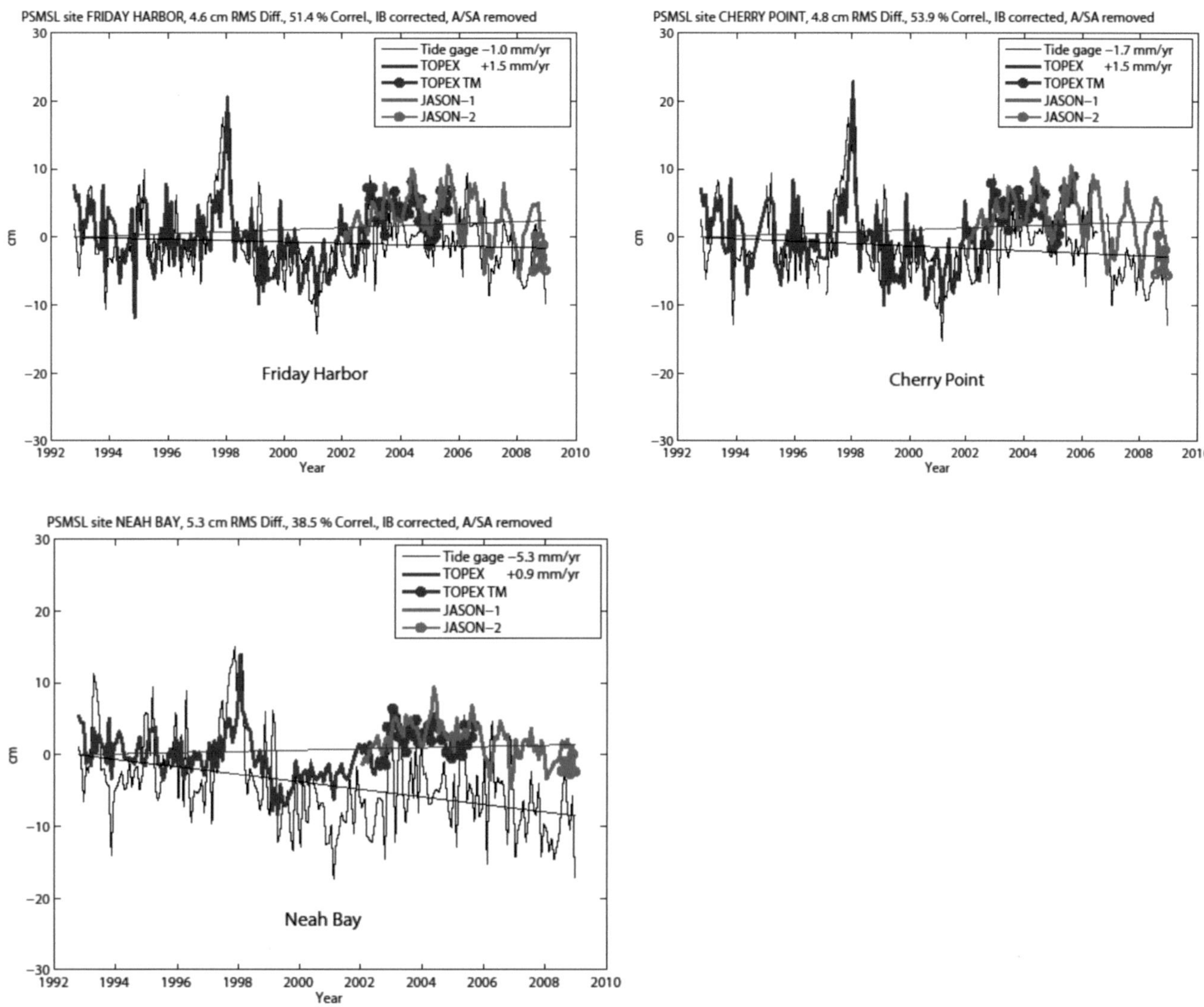

FIGURE B.4 Continued

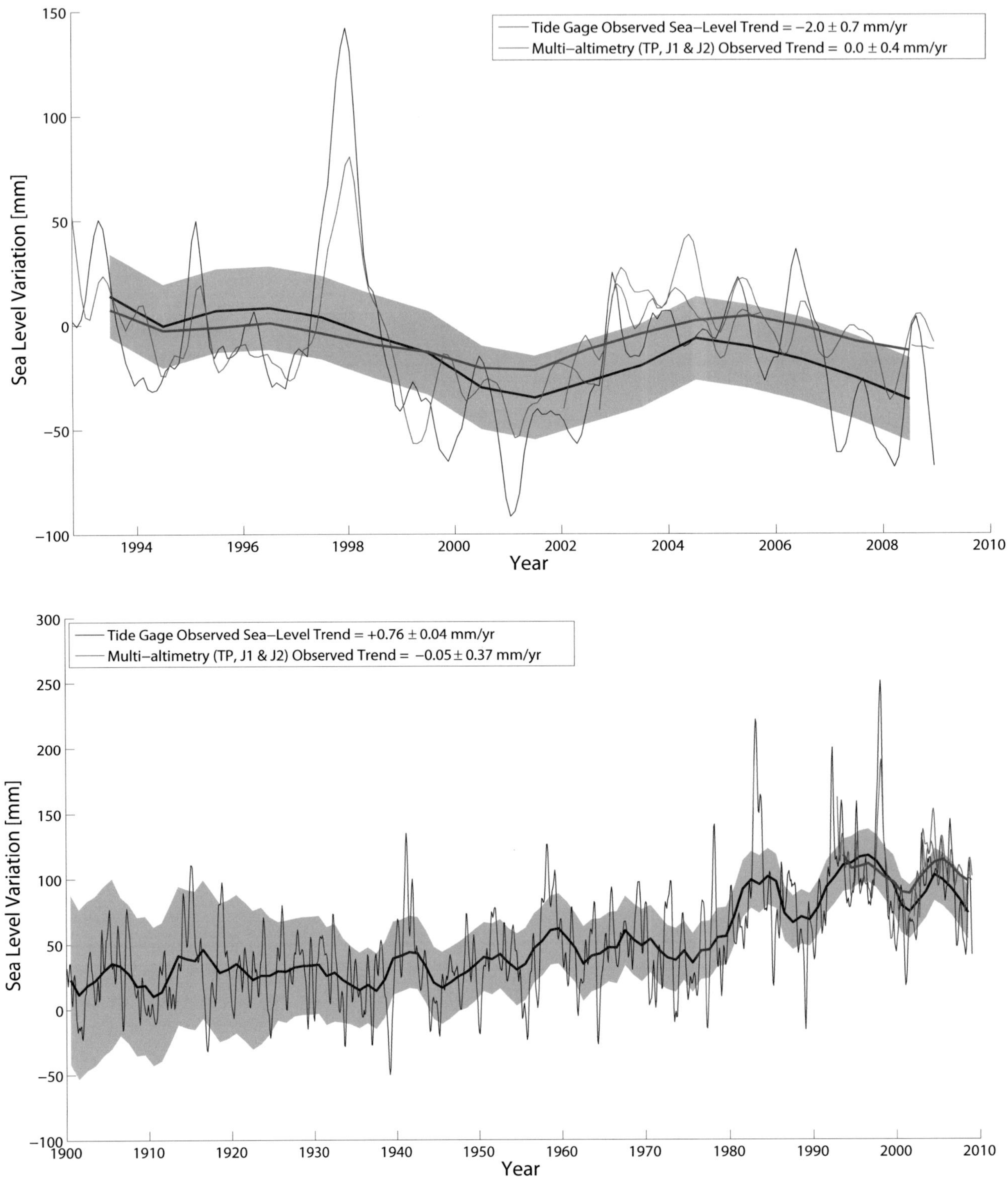

FIGURE B.5 Averaged tide gage and altimetry sea-level time series for 1992–2008 (*top*), and the averaged tide gage time series for 1900–2008 shown also with altimetry sea-level time series (*bottom*). Inverted barometer and GIA corrections have been applied to both tide gage and altimetry sea-level records. The bottom panel also shows yearly averages, solid blue and red lines for tide gage and altimetry sea level, respectively, and estimated tide gage sea-level uncertainty in gray shades.

Appendix C

Analysis of Sea-Level Fingerprint Effects

The effect of the Alaska, Greenland, and Antarctic sea-level fingerprints on relative sea level off the coasts of California, Oregon, and Washington can be calculated by scaling the rate of rise from each source by the appropriate factor (colored contours) indicated in Figure 4.9 and adding the contributions:

$$R_{\Sigma,p}(t) = \sum_{k=1}^{3} s_{k,p} R_k(t),$$

where R is the ice loss rate in mm yr^{-1} or GT yr^{-1}, k indicates the source of new water entering the ocean (Alaska, Greenland, and Antarctica), p indicates the destination of the water (north coast, central coast, or south coast), and $s_{k,p}$ is the fingerprint scale factor (derived from Figure 4.9) for source k delivering water to destination p. Loss rates R for Alaska, Greenland, and Antarctica, as reported in the literature, are given in Table C.1. The 1992–2009 period (1992–2008 for Alaska) was chosen because it was the longest and most nearly common period of availability of the largest number of records for all three regions. Averages were weighted according to the assessed reliability of the individual estimates.

The adjusted rate of sea-level rise is determined by multiplying the ice loss rate R for each of the three sources by the fingerprint scale factor s for each of the three regions along the coast, then summing (see equation). The result is given in Table C.2.

The effect of uncertainties in the ice loss rates on the adjusted rate of relative sea-level rise is shown in Table C.3. The mid-range estimate is the mean estimate, also given in the right column of Table C.2, and the low- and high-range estimates are plus and minus the uncertainties.

REFERENCES

Arendt, A.A., K.A. Echelmeyer, W.D. Harrison, C.S. Lingle, and V.B. Valentine, 2002, Rapid wastage of Alaska glaciers and their contribution to rising sea level, *Science*, **297**, 382-386.

Baur, O., M. Kuhn, and W.E. Featherstone, 2009, GRACE-derived ice-mass variations over Greenland by accounting for leakage effects, *Journal of Geophysical Research*, **114**, B06407, doi:10.1029/2008JB006239.

Berthier, E., E. Schiefer, G.K.C. Clarke, B. Menounos, and F. Remy, 2010, Contribution of Alaskan glaciers to sea level rise derived from satellite imagery, *Nature Geoscience*, **3**, 92-95.

Cazenave, A., K. Dominh, S. Guinehut, E. Berthier, W. Llovel, G. Ramillien, M. Ablain, and G. Larnicol, 2009, Sea level budget over 2003-2008: A reevaluation from GRACE space gravimetry, satellite altimetry and Argo, *Global and Planetary Change*, **65**, 83-88.

Chen, J.L., C.R. Wilson, D. Blankenship, and B.D. Tapley, 2009, Accelerated Antarctic ice loss from satellite gravity measurements, *Nature Geoscience*, **2**, 859-862.

Chen, J.L., C.R. Wilson, and B.D. Tapley, 2011, Interannual variability of Greenland ice losses from satellite gravimetry, *Journal of Geophysical Research*, **116**, B07406, doi:10.1029/2010JB007789.

Cogley, J.G., 2012, The future of the world's glaciers, in *Future Climates of the World*, 2nd edition, A. Henderson-Sellers and K. McGuffie, eds., Elsevier, Waltham, MA, pp. 197-222.

Dong-Chen, E., Y.-D. Yang, and D.-B. Chao, 2009, The sea level change from the Antarctic Ice Sheet based on GRACE, *Chinese Journal of Geophysics*, **52**, 936-942.

Dyurgerov, M.B., 2010, *Reanalysis of Glacier Changes: From the IGY to the IPY, 1960–2008*, Data of Glaciological Studies, Publication 108, Moscow, 116 pp.

Horwath, M., and R. Dietrich, 2009, Signal and error in mass change inferences from GRACE: The case of Antarctica, *Geophysical Journal International*, **177**, 849-864.

TABLE C.1 Ice Mass Loss Rates, in Terms of Sea-Level Equivalent, Measured or Inferred for Alaska, Greenland, and Antarctica

Source	Period	Ice Loss Rates (mm yr^{-1} SLE)
Alaska		
Berthier et al. (2010)	1962–2006	-0.12 ± 0.02
Cogley (2012)	1990–2005	-0.09 ± 0.00
	1995–2005	-0.08 ± 0.00
	2000–2005	-0.15 ± 0.02
Dyurgerov (2010)	1992–2006	-0.20 ± 0.02
Arendt et al. (2002)	1992–2002	-0.27 ± 0.10
Tamisiea et al. (2005)	2002–2003	-0.31 ± 0.09
Luthcke et al. (2008)	2003–2007	-0.23 ± 0.01
Pritchard et al. (2010)	2003–2008	-0.18 ± 0.14
Greenland Ice Sheet		
Wu et al. (2010)	2002–2009	-0.29 ± 0.06
Sørensen et al. (2011)	2004–2008	-0.58 ± 0.06
Schrama and Wouters (2011)	2003–2010	-0.56 ± 0.05
Cazenave et al. (2009)	2003–2008	-0.38 ± 0.05
Zwally et al. (2011)	1992–2002	-0.02 ± 0.01
	2003–2007	-0.47 ± 0.01
Velicogna (2009)	2002–2009	-0.62 ± 0.09
Pritchard et al. (2010)	2004–2010	-0.54 ± 0.06
Baur et al. (2009)	2003–2009	-0.49 ± 0.03
Slobbe et al. (2009)	2003–2008	-0.59 ± 0.22
	2003–2007	-0.38 ± 0.19
Rignot et al. (2011)	1992–2010	-0.43 ± 0.14
Chen et al. (2011)	2002–2005	-0.43 ± 0.10
	2005–2010	-0.68 ± 0.10
Antarctic Ice Sheet		
Wu et al. (2010)	2002–2009	-0.24 ± 0.12
Wingham et al. (2006)	1993–2003	0.07 ± 0.19
Velicogna (2009)	2002–2009	-0.40 ± 0.20
Chen et al. (2009)	2002–2006	-0.40 ± 0.16
	2006–2009	-0.61 ± 0.25
Rignot et al. (2011)	1992–2010	-0.23 ± 0.25
Horwath and Dietrich (2009)	2002–2008	-0.30 ± 0.13
Moore and King (2008)	2002–2006	-0.45 ± 0.22
Cazenave et al. (2009)	2003–2008	-0.55 ± 0.06
Dong-Chen et al. (2009)	2003–2008	-0.22 ± 0.10
Shi et al. (2011)	2003–2008	-0.21 ± 0.01
Zwally et al. (2005)	1992–2001	-0.08 ± 0.14
Ivins et al. (2011)	2003–2009	-0.11 ± 0.02

TABLE C.2 Ice Loss Rates, Sea-Level Fingerprint Scale Factors, and Adjusted Rates of Sea-Level Rise for three U.S. West Coast Locations

Ice Loss Rate	Alaska 0.16 mm yr^{-1} SLE		Greenland 0.35 mm yr^{-1} SLE		Antarctica 0.28 mm yr^{-1} SLE		Sum of Sources 0.79 mm yr^{-1} SLE
Area	Scale Factor	Adjusted Sea-Level Rise (mm yr^{-1})	Scale Factor	Adjusted Sea-Level Rise (mm yr^{-1})	Scale Factor	Adjusted Sea-Level Rise (mm yr^{-1})	Total Adjusted Sea-Level Rise (mm yr^{-1})
North coast	-0.80	-0.13	0.75	0.26	1.17	0.33	0.46
Central coast	-0.20	-0.03	0.87	0.30	1.17	0.33	0.60
South coast	0.20	0.03	0.92	0.32	1.17	0.33	0.68

TABLE C.3 Adjusted Rates of Relative Sea-Level Rise for High, Medium, and Low Ice Loss Rates

Region	Low-Range Estimate (mm yr^{-1})	Mid-Range Estimate (mm yr^{-1})	High-Range Estimate (mm yr^{-1})
North coast	0.07	0.46	0.86
Central coast	0.07	0.60	1.14
South coast	0.06	0.68	1.30

Ivins, E.R., M.M. Watkins, D.N. Yuan, R. Dietrich, G. Casassa, and A. Rulke, 2011, On-land ice loss and glacial isostatic adjustment at the Drake Passage: 2003-2009, *Journal of Geophysical Research*, **116**, B02403, doi:10.1029/2010JB007607.

Luthcke, S.B., A.A. Arendt, D.D. Rowlands, J.J. McCarthy, and C.F. Larsen, 2008, Recent glacier mass changes in the Gulf of Alaska region from GRACE mascon solutions, *Journal of Glaciology*, **54**, 767-777.

Moore, P., and M.A. King, 2008, Antarctic ice mass balance estimates from GRACE: Tidal aliasing effects, *Journal of Geophysical Research*, **113**, F02005, doi:10.1029/2007JF000871.

Pritchard, H.D., S.B. Luthcke, and A.H. Fleming, 2010, Understanding ice sheet mass balance: Progress in satellite altimetry and gravimetry, *Journal of Glaciology*, **56**, 1151-1161.

Rignot, E., I. Velicogna, M.R. van den Broeke, A. Monaghan, and J. Lenaerts, 2011, Acceleration of the contribution of the Greenland and Antarctic ice sheets to sea level rise, *Geophysical Research Letters*, **38**, L05503, doi:10.1029/2011GL046583.

Schrama, E.J.O., and B. Wouters, 2011, Revisiting Greenland Ice Sheet mass loss observed by GRACE, *Journal of Geophysical Research*, **116**, B02407, doi:10.1029/2009JB006847.

Shi, H.L., Y. Lu, Z.L. Du, L.L. Jia, Z.Z. Zhang, and C.X. Zhou, 2011, Mass change detection in Antarctic Ice Sheet using ICESat block analysis techniques from 2003 similar to 2008, *Chinese Journal of Geophysics*, **54**, 958-965.

Slobbe, D.C., P. Ditmar, and R.C. Lindenbergh, 2009, Estimating the rates of mass change, ice volume change and snow volume change in Greenland from ICESat and GRACE data, *Geophysical Journal International*, **176**, 95-106.

Sørensen, L.S., S.B. Simonsen, K. Nielsen, P. Lucas-Picher, G. Spada, G. Adalgeirsdottir, R. Forsberg, and C.S. Hvidberg, 2011, Mass balance of the Greenland Ice Sheet (2003-2008) from ICESat data – the impact of interpolation, sampling and firn density, *The Cryosphere*, **5**, 173-186.

Tamisiea, M.E., E.W. Leuliette, J.L. Davis, and J.X. Mitrovica, 2005, Constraining hydrological and cryospheric mass flux in southeastern Alaska using space-based gravity measurements, *Geophysical Research Letters*, **32**, L20501, doi:10.1029:2005GL023961.

Velicogna, I., 2009, Increasing rates of ice mass loss from the Greenland and Antarctic ice sheets revealed by GRACE, *Geophysical Research Letters*, **36**, L19503, doi:10.1029/2009GL040222.

Wingham, D.J., A. Shepherd, A. Muir, and G.J. Marshall, 2006, Mass balance of the Antarctic Ice Sheet, *Philosophical Transactions of the Royal Society A*, **364**, 1627-1635.

Wu, X., M.B. Heflin, H. Schotman, B.L.A. Vermeersen, D. Dong, R.S. Gross, E.R. Ivins, A.W. Moore, and S.E. Owen, 2010, Simultaneous estimation of global present-day water transport and glacial isostatic adjustment, *Nature Geoscience*, **3**, 642-646.

Zwally, H.J., M.B. Giovinetto, J. Li, H.G. Cornejo, M.A. Beckley, A.C. Brenner, J.L. Saba, and D. Yi, 2005, Mass changes of the Greenland and Antarctic ice sheets and shelves and contributions to sea level rise: 1992-2002, *Journal of Glaciology*, **51**, 509-527.

Zwally, H.J., L.I. Jun, A.C. Brenner, M. Beckley, H.G. Cornejo, J. Dimarzio, M.B. Giovinetto, T.A. Neumann, J. Robbins, J.L. Saba, Y.I. Donghui, and W. Wang, 2011, Greenland Ice Sheet mass balance: Distribution of increased mass loss with climate warming; 2003-07 versus 1992-2002, *Journal of Glaciology*, **57**, 88-102.

Appendix D

Long-Term Tide Gage Stability From Leveling Data

James Foster, University of Hawai'i at Manoa

Note: The committee commissioned the following discussion paper. Dr. Foster's views, as expressed below, may not always reflect the views of the Committee on Sea Level Rise in California, Oregon, and Washington, or vice versa.

Abstract

Leveling observations from tide gage benchmark networks can be used to assess the long-term stability of the tide gage with respect to the benchmarks, to provide estimates on the degree to which short-term (e.g., decadal timescale) estimates of vertical motion can be inferred to represent longer term rates, and to provide a means for detecting and estimating possible steps in the sea-level record due to changes in tide gage instrumentation, monument instabilities or local ground motion. Leveling data from six California tide gages confirm that long-term vertical motions of the tide gages are small, typically less than 0.25 mm yr^{-1} with respect to the stable local coastline, and represent only a very minor portion of the relative sea-level change budget. Decadal timescale estimates of benchmark vertical motions mostly range between ± 0.5 mm yr^{-1} from their long-term values, indicating that decadal scale estimates of geodetic motion are generally a reasonable approximation to longer-term rates. The formal errors on the rate estimates however, based on an assumption of white noise, are over optimistically tight and should be scaled up by a factor of 2.33 to provide more realistic bounds. This analysis also suggests that the tide gage at Point Reyes may be experiencing recent subsidence, despite the long-term estimates indicating relative stability. Jumps were evident in the leveling data from all tide gages except Los Angeles: some are clearly due to changes in equipment, others are less clearly attributable to mechanical issues at the sites and may indicate corresponding steps remain uncorrected in the sea-level time series data.

INTRODUCTION

The observations recorded by tide gages are relative sea-level values, and are a linear combination of real changes in local sea level as well as any local vertical motions, due to either regional tectonics, local subsidence/uplift of the ground, and possibly offsets due to motions of the tide gage itself with respect to its mount. Routine leveling at tide gages has been performed for many decades, and provides a resource with which to explore the issues of local vertical land motions (Shinkle and Dokka, 2004) and tide gage stability over longer timescales than is possible with more modern geodetic techniques such as Global Navigation Satellite Systems (GNSS, such as GPS) measurements or satellite-based Interferometric Synthetic Aperture Radar (InSAR). With this data set it is possible to assess two key issues of importance to understanding long-term relative sea-level rise: (1) What proportion of the observed sea-level change is due to local vertical motions of the tide gage itself, or its immediate vicinity? (2) How representative are the vertical land motions estimated from GNSS/InSAR data over the last 5–15 years of multidecadal to century scale rates? Leveling data from 6 California tide gage networks (Figure D.1) were used to address these questions.

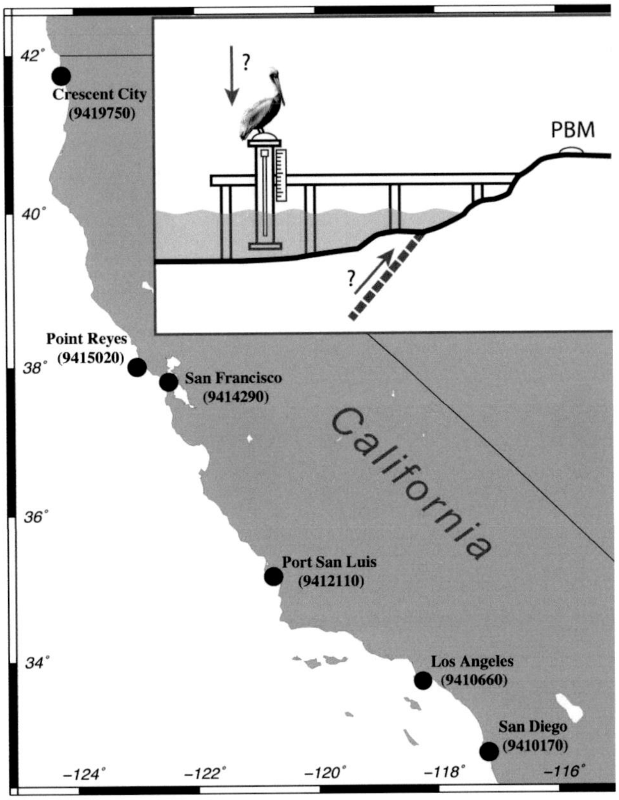

FIGURE D.1 Location of tide gages examined with inset showing schematic of a tide gage and benchmark leveling network illustrating fundamental ambiguity that may exist between motion of the tide gage instrument and local land motion.

Tide gages measure the height of the sea surface with respect to their own internal reference level. As the gages require relatively deep water to operate in, away from shore wave breaks and currents, they are often installed on manmade structures such as piers, built out from the coast. Although tide gage locations are chosen carefully in an effort to ensure long-term stability and a minimum of vertical motion, it is impossible to guarantee these qualities in a site. In order to maintain an accessible external reference mark from which the local sea-level datums can be assigned, to allow for tide gage equipment changes, and to confirm the stability of the gage, each tide gage is supported by a network of leveling benchmarks, with one of these designated as the primary benchmark. This is normally chosen to be both as close to the tide gage as possible, and installed in as stable a location as can be found (Hicks et al., 1987). As the local sea-level datums are defined with respect to this primary benchmark, each tide gage has a correction term that is applied to the raw ranges it observes to the sea-surface in order to map them to the vertical datum defined by the primary benchmark. Every time the gage is either changed, upgraded, or experiences some possible vertical offset—due, for example, to being hit by a harbor vehicle—a new correction term is calculated by performing leveling between the primary benchmark and the reference point on the new (or newly offset) tide gage. This term is applied to the data stream so that is it transparent to the end-user of the data, maintaining what is, in theory, a continuous record of sea level with respect to the primary benchmark, rather than the gage itself. Although these constants are recorded as they are recalculated and programmed into the tide gage data logger, they are not all archived digitally, and were not available for this study.

Current recommended operating procedures require a network of at least 10 benchmarks be established and monitored in support of tide gage sea-level observations (Woodworth, 2002). Historically, however, significantly fewer marks have been regularly observed—some of which may no longer exist due to construction (or other reasons) around the typically extremely changeable industrial environment of the harbors, where most tide gages are installed. Leveling of the west coast network is currently performed using 2nd Order Levels standards approximately annually, though historically it was often done much less frequently and to a lower standard. The leveling data are used to determine the general stability of each benchmark as well as the tide gage, and identify any tide gage offsets (any jump greater than 6 mm) necessitating a site correction factor adjustment. The long-term relative vertical velocity of the tide gage with respect to the primary benchmark is not applied as a correction to the sea-level time series.

DATA AND METHODS

The leveling data, corrected for atmospheric refraction, from all occupations of the tide gage benchmark networks for San Diego, Los Angeles, Port San Luis, San Francisco, Point Reyes, and Crescent City, were provided by the National Oceanographic and Atmospheric Administration's (NOAA's) Center for Operational Oceanographic Products and Services (CO-OPS). Each benchmark network data set consisted of a spreadsheet listing the date of each leveling project and the height determined for each benchmark occupied during that project. The recorded heights themselves have little value for the purposes of this study, as they are determined with respect to some arbitrary starting height (typically a predefined value for the height of the primary benchmark) for each leveling run. The observations required for this study are changes in the height differences between benchmarks, so the first step was to determine the differential heights measured during each project with respect to the primary benchmark. A network solution was then performed to estimate and remove the mean differential height for each benchmark from the observations, generating a time series of observed changes in relative height for each benchmark in the network. Bad data points, due to incorrect readings or transcriptions of values, contaminate this initial analysis, and were identified and removed from the data set. These points are typically obvious as single observation outliers, displaced by several centimeters from the trend defined by the remainder of that benchmark's data.

To interpret observed vertical motions and assess whether they are likely due to benchmark instability (Karcz et al., 1976) or rather to local land motions, the approximate spatial locations of the benchmarks are required. These locations were determined either by digitizing the survey benchmark sketches or, if available, extracted from the National Geodetic Survey (NGS) benchmark data sheet database. Some of the older, discontinued benchmarks have no recorded location and could not contribute to this aspect of the analysis. In most cases, the digitized locations of the benchmarks are estimated to be accurate to 40 m or better, while those determined by the NGS using GPS or other methods are considerably more accurate. This accuracy is sufficient to allow us to examine and interpret spatial patterns of vertical motions. Benchmarks interpreted as being unstable are identified from their time series, either objectively by virtue of very high variance about their mean trend (typically greater than 10 mm^2, in contrast to most benchmarks which have variances typically less than 2 mm^2), or more subjectively due to having high apparent relative vertical velocities, uncorrelated with their neighboring marks, by comparison with the rest of the network. In most cases, those benchmarks identified as unstable have been given a stability rating of *C* or *D* by the NGS, indicating unknown or doubtful long-term stability. However, in some cases even benchmarks with the highest *A* rating were found to have velocities and/or variances at odds with the neighboring marks.

Both the formal NGS/CO-OPS datum definition and, at this stage in the processing, this analysis, define the primary benchmark as a fixed reference. There is clearly a danger, however, that the primary benchmark itself might experience vertical motions, due either to local benchmark instability or real land motions. In order to assess and account for this, a subset of the benchmarks in the network that show no sign of either monument instability or anomalous rates of motion was used to define a combined vertical reference datum for the network, which should provide a more robust datum. The choice of benchmarks is somewhat subjective, and is unable to correct for vertical deformation occurring on spatial scales greater than the width of the network. It turns out, however, that the specific choice of reference benchmarks has only a small impact on the final estimates of rates of vertical motion for the tide gage and primary benchmark.

The rates of vertical motion were estimated using a robust linear fit that uses an iterative reweighting scheme to reduce the impact of outlier observations on the final estimation of slope. To assess the degree to which estimates of vertical motions based on decadal scale windows of data can be trusted as reasonable approximations to the longer-term trends, a moving window was applied to the data set, and vertical rates for each window were estimated. A minimum of 5 observations and at least a 10-year time span were required for each window location in order to generate an estimate for any given period in order to prevent the results being merely a reflection of the measurement errors and/or sparse data.

RESULTS

San Diego

Three benchmarks—9, N 57, and RIVET—showed anomalous vertical motions and were excluded from the spatial analysis. Benchmarks RIVET and N 75 were given a *D* stability code by the National Geodetic Survey, indicating they might be expected to be unstable. Benchmark 9, on the other hand, has an *A* rating, indicating highest level of confidence in the benchmark stability. It is, however, located in the middle of downtown San Diego, and it is possible that this location has over the decades experienced significant vertical motion due to construction and settling, unrelated to broader underlying tectonic trends. For the remaining network (Figure D.2), vertical motions suggest a gentle northwest to southeast trend, with the majority of benchmarks nearest to the tide gage location in close agreement. Defining a network vertical reference using these sites implies a vertical motion for the primary benchmark of -0.03 mm yr^{-1}, and for the tide gage reference marks of +0.07 mm yr^{-1}. Estimating the apparent decadal scale vertical velocities by calculating rates for each 10 consecutive measurements shows that these rates range from -0.4 mm yr^{-1} to nearly 1 mm yr^{-1}, though mostly clustered between ± 0.2 mm yr^{-1}. The exceptional rates in the 1980s are possibly due to a small, unmodeled residual step in the tide gage leveling time series, either at the tide gage itself, or possibly at the primary benchmark, as the other benchmark time series also show excursions at this time. With less consistent and rigorous field operational procedures, data prior to 1970 are noisier, and this is reflected in the increased range and errors for the velocity estimates for this period.

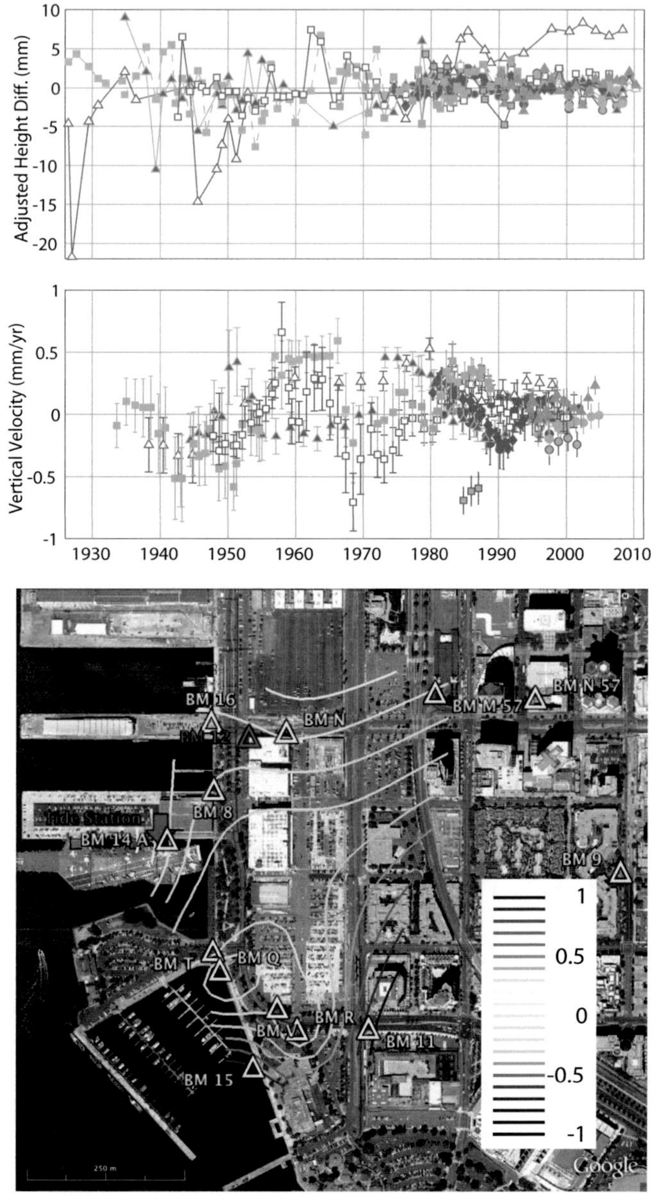

FIGURE D.2 (*Top*) Time series of vertical displacements for San Diego (9410170) tide gage leveling network. All benchmarks are shown in gray except tide gage marks RM OF ETG (yellow) and AQUATRAK (cyan). (*Middle*) Time series of apparent vertical velocities based on at least 5 consecutive observations covering at least 10 years (approximately decadal). All benchmarks are shown in gray except tide gage marks RM OF ETG (yellow) and AQUATRAK (cyan). (*Bottom*) Contours of best-fit linear vertical deformation rates for San Diego benchmarks with respect to network vertical reference frame. Contours every 0.1 mm yr^{-1}.

Los Angeles

Los Angeles is recognized as having complex temporal and spatial variations in vertical deformation (Brooks et al., 2007), and this complexity is evident in the range of vertical motions calculated for the tide gage leveling network and their spatial pattern (Figure D.3; for which the limited and inconsistent 4 measurements for a STAFF STOP tide gage mark were ignored). Although it appears that a large portion of the pier on which the tide gage is installed is experiencing significant subsidence relative to the coast, the end nearest the tide gage and primary benchmark sees less of this. The choice of benchmarks for the network vertical reference is particularly difficult for this location, as there is no obvious subset of benchmarks demonstrating broadly similar motion that might be interpreted as "stable." However, excluding benchmarks with the most extreme motions results in a reference

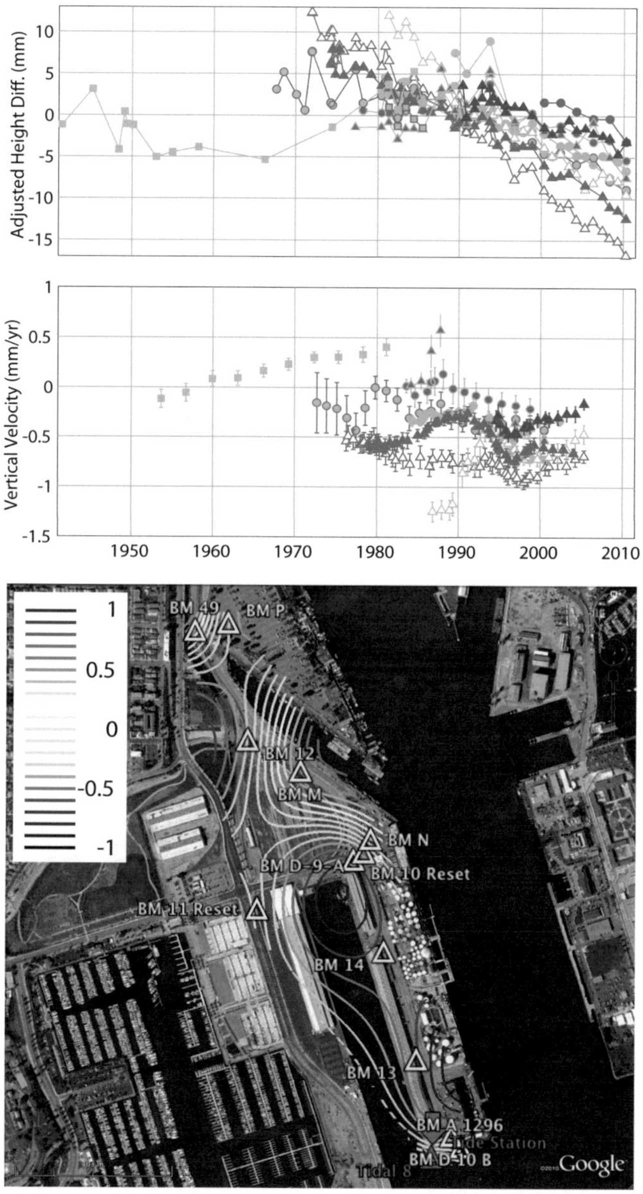

FIGURE D.3 (*Top*) Time series of vertical displacements for Los Angeles (9410660) tide gage leveling network. All benchmarks shown in gray except tide gage mark AQUATRAK (green). (*Middle*) Time series of apparent decadal-scale vertical velocities. All benchmarks shown in gray except tide gage mark AQUATRAK (green). (*Bottom*) Contours of best-fit linear vertical deformation rates for Los Angeles benchmarks with respect to network vertical reference frame. Contours every 0.1 mm yr^{-1}.

frame in which the primary benchmark is experiencing 0.14 ± 0.14 mm yr^{-1} uplift, and the tide gage subsiding at 0.17 ± 0.12 mm yr^{-1}. The large errors probably reflect the strong quasi-seasonal signals recognized in Los Angeles (Bawden et al., 2001) that will be strongly aliased into the roughly annual leveling campaigns. The vertical velocity time series suggests relatively large variations in apparent velocities on the decadal scale, with a strong perturbation in the 1990s due largely to a leveling run that recorded unusual, but not obviously incorrect, values.

Port San Luis

The benchmark network at Port San Luis appears particularly stable, with the exception of two benchmarks (B and H 828), which were removed from the analysis. The network vertical reference implies a small uplift of 0.08 ± 0.05 mm yr^{-1} for the primary benchmark, and slight subsidence of 0.05 ± 0.05 mm yr^{-1} for the tide gage, which for this location is located at the end of a long pier (Figure D.4). Notably, there is a distinct downward step in the leveling time series for both tide gage marks in 1996 with a mean value of -5.25 ± 0.59 mm. It is not clear whether this step is a result of equipment change or some other source; however, it should be noted that when corrected, the sign of the tide gage motion changes, with slight uplift indicated (0.29 ± 0.05 mm yr^{-1}). All plots in Figure D.4 show the data after these steps have been estimated and removed from the time series. If left in, this step dominates the analysis of the decadal-scale velocity estimate stability. If removed, however, the short-term velocities indicate that all (stable) benchmarks are well described over their lifetime by a consistent vertical velocity.

San Francisco

San Francisco has a very long record of benchmark leveling, with the earliest measurements dating from the mid 1920s. Several of the benchmarks show significant excursions over that time period (Figure D.5), yet the long-term pattern is of relative consistency. Steps were estimates and removed from benchmarks 173 (in 1943) and 175 (in 1980). Benchmarks M and 175 were considered outlier time series and not included in the spatial analysis. The network vertical reference

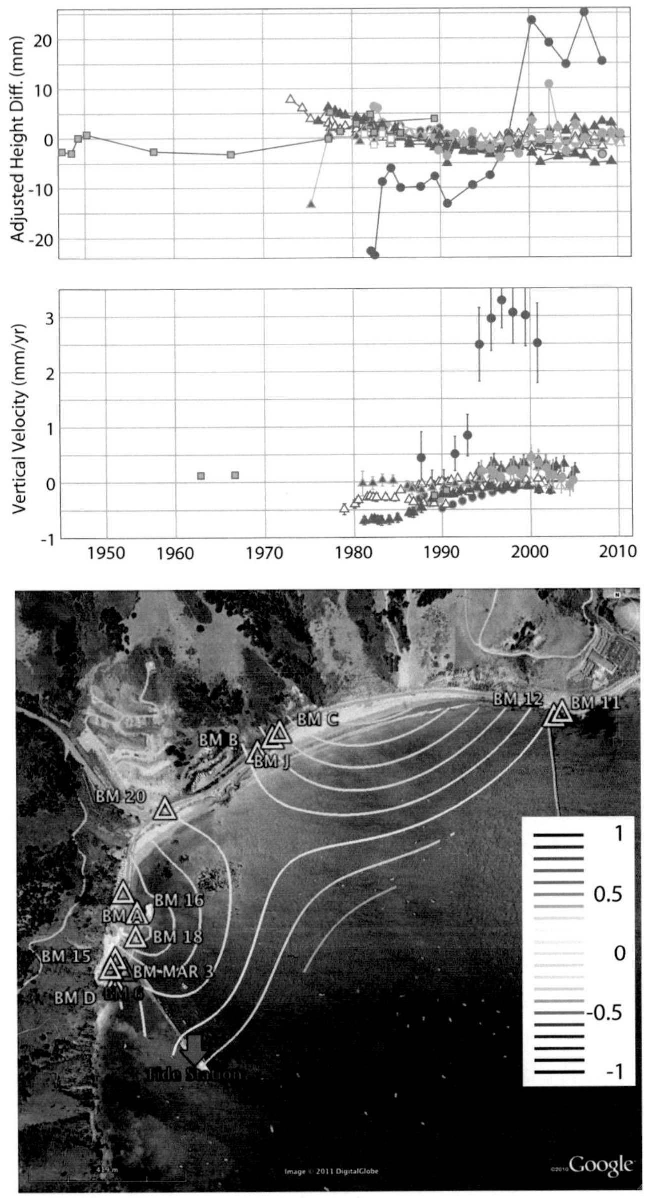

FIGURE D.4 (*Top*) Time series of vertical displacements for Port San Luis (9412110) tide gage leveling network. All benchmarks shown in gray except tide gage marks AQUAREF (yellow) and AQUATRAK (red). Steps in the tide gage time series have been removed. Unstable benchmark BM B clearly visible as outlier time series. (*Middle*) Time series of apparent decadal-scale vertical velocities. All benchmarks shown in gray except tide gage marks AQUAREF (yellow) and AQUATRAK (red). (*Bottom*) Contours of best-fit linear vertical deformation rates for Port San Luis benchmarks with respect to network vertical reference frame. Contours every 0.1 mm yr^{-1}.

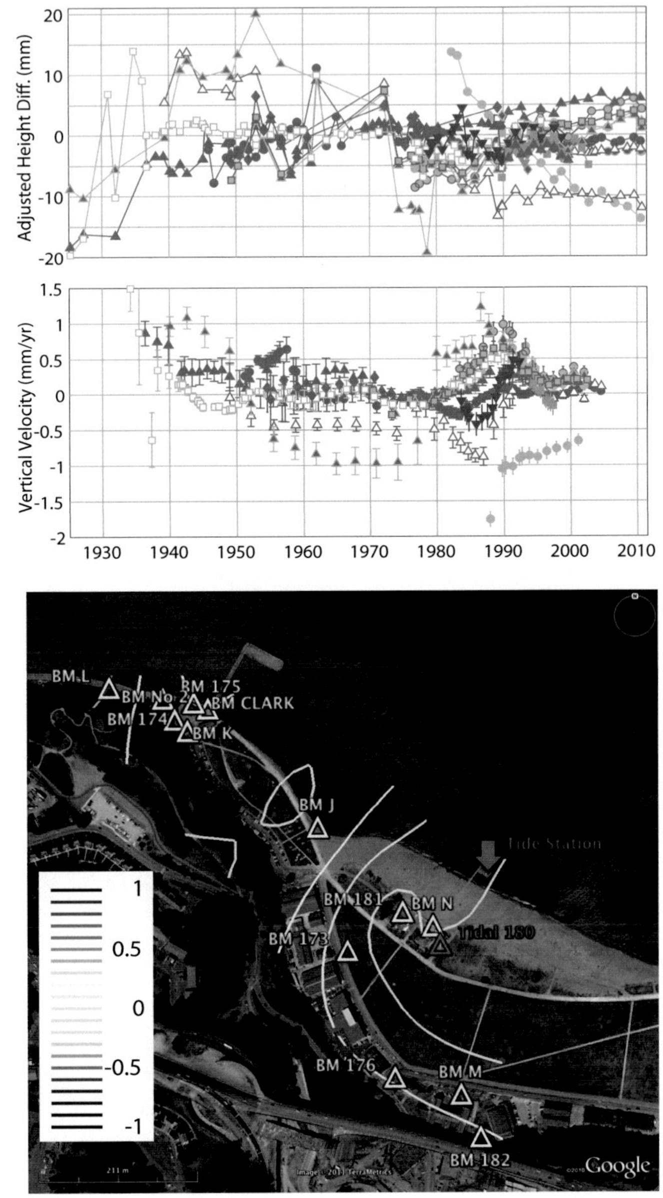

FIGURE D.5 (*Top*) Time series of vertical displacements for San Francisco (9414290) tide gage leveling network. All benchmarks shown in gray except tide gage marks AQUATRAK REF (red), AQUATRAK (cyan), RM OF ETG (green) and STAFF STOP (orange). (*Middle*) Time series of apparent decadal-scale vertical velocities. All benchmarks shown in gray except tide gage marks AQUATRAK (cyan) and RM OF ETG (green) (AQUATRAK REF and STAFF STOP did not have long enough time series to generate decadal velocity estimates). (*Bottom*) Contours of best-fit linear vertical deformation rates for San Francisco benchmarks with respect to network vertical reference frame. Contours every 0.1 mm yr^{-1}.

analysis suggests that in late 1989—around the time of the Loma Prieta earthquake—the primary benchmark moved with respect to the rest of the network. Correcting the data for this motion does not entirely remove the apparent upward motions of several benchmarks at this time, or shortly afterward, suggesting that the monuments or the local ground around them moved either during the shaking of the earthquake itself or during a period of post-seismic adjustment. Despite this, the rate of motion calculated for the primary benchmark is small (0.05 ± 0.07 mm yr^{-1}) and the mean tide gage motion is negligible (0.01 ± 0.07 mm yr^{-1}). As there is little spatial coherence in the vertical velocities, with adjacent benchmarks often exhibiting opposite senses of motion, the contour map results in only small amplitude variability in vertical motion rates. The decadal-scale vertical velocities are strongly affected by the 1989 events, but nonetheless generally indicate that estimates of long-term vertical motions based on any 10-year period would be accurate to better than 0.5 mm yr^{-1}.

Point Reyes

Although the time series from Point Reyes is relatively short, with the first regular observations starting in 1973 (an initial leveling run in 1930 was too removed from the rest of the data set to contribute usefully to the analysis), it shows relatively complex behavior (Figure D.6). The AQUAREF mark, observed between 1990 and 1999, shows strong linear downward motion, at odds with the other tide gage mark, AQUATRAK, and were ignored during the spatial analysis, as were outlier records from benchmarks 1 FMK and 11. The benchmark leveling observations, which were otherwise largely stable prior to ~2001, show strong heterogeneous motions after this point, with many sites uplifted for several years before subsiding back to their long-term baselines. The recent AQUATRAK record, however, shows continued subsidence. The contour plot highlights an area experiencing significant uplift to the southeast of the tide gage, with all other benchmarks appearing relatively stable over the full time window of the observations. This uplift may be related to motion on a mapped fault segment (Clark and Brabb, 1997) near the eastern edge of the network. Although there is no evidence to suggest this fault is affecting the

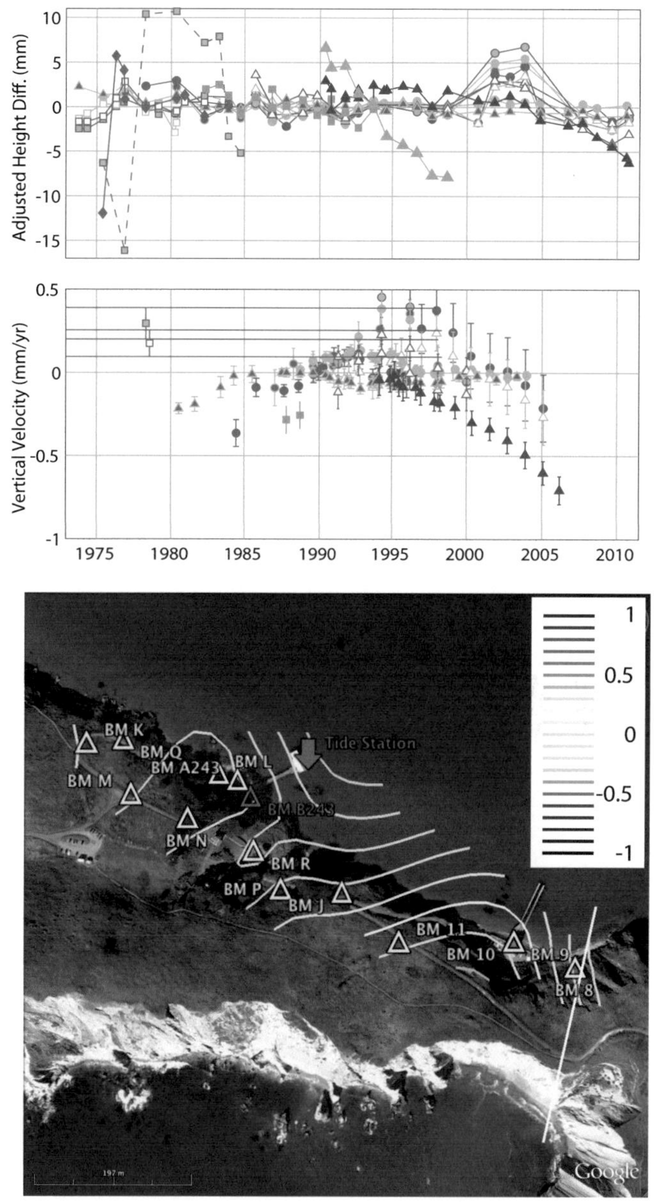

FIGURE D.6 (*Top*) Time series of vertical displacements for Point Reyes (9415020) tide gage leveling network. All benchmarks shown in gray except tide gage marks AQUAREF (orange), AQUATRAK (magenta) and RM OF ETG (blue). (*Middle*) Time series of apparent decadal-scale vertical velocities. All benchmarks shown in gray except tide gage marks AQUAREF (orange), AQUATRAK (magenta) and RM OF ETG (blue). (*Bottom*) Contours of best-fit linear vertical deformation rates for Point Reyes benchmarks with respect to network vertical reference frame. Contours every 0.1 mm yr^{-1}. White line indicates mapped fault trace.

tide gage itself, it emphasizes the tectonic complexity of many locations along the California coast. The velocity stability analysis highlights the AQUAREF outlier velocities, suggesting that these observations are most likely indicating monument instability. The other benchmarks have decadal velocity estimates consistent to ± 0.5 mm yr^{-1}, until the impact of the events of 2001–2011, which indicate some possibly transient deformation event affecting a subset of the benchmark network. Interestingly, the vertical velocity rate estimates for the AQUATRAK suggest continued, or even accelerating subsidence, with the most recent decadal rate approaching -1 mm yr^{-1}.

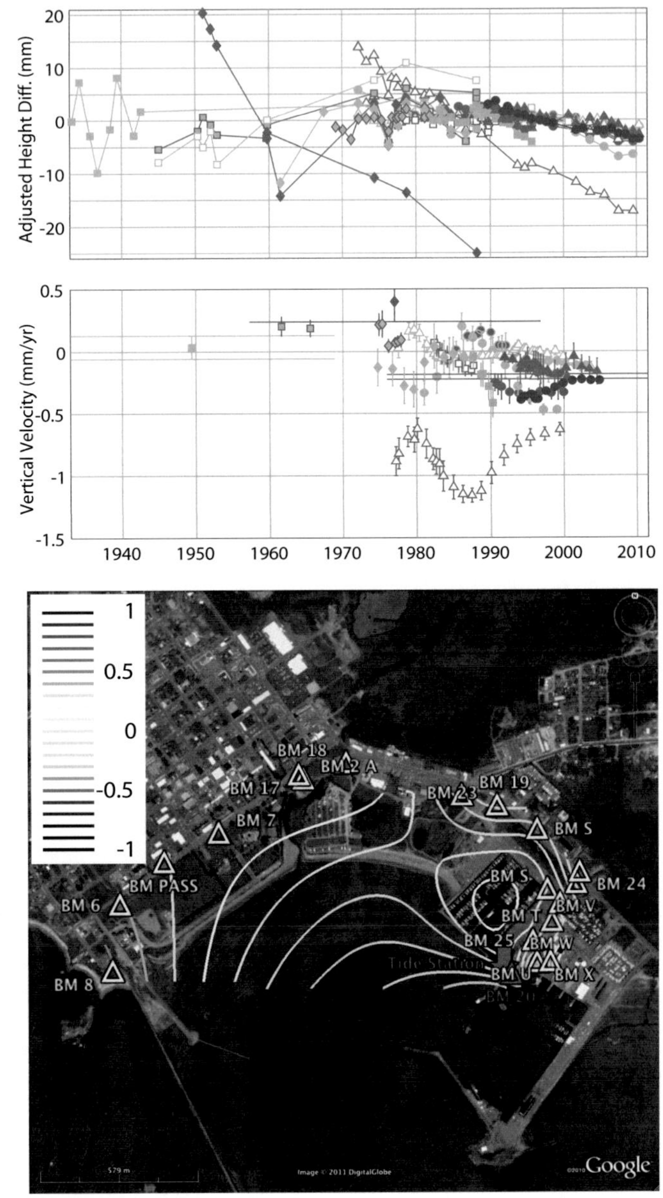

FIGURE D.7 (*Top*) Time series of vertical displacements for Crescent City (9419750) tide gage leveling network. All benchmarks shown in gray except tide gage marks AQUAREF (magenta), AQUATRAK (red) and RM OF ETG (yellow). (*Middle*) Time series of apparent decadal-scale vertical velocities. All benchmarks shown in gray except tide gage marks AQUAREF (magenta), AQUATRAK (red) and RM OF ETG (yellow). (*Bottom*) Contours of best-fit linear vertical deformation rates for Crescent City benchmarks with respect to network vertical reference frame. Contours every 0.1 mm yr^{-1}.

Crescent City

Long-term relative height residuals are generally small, though several benchmarks show unusually large motions compared to the rest of the network, presumably due to benchmark instability (Figure D.7). The pattern of the contours suggest that the benchmarks nearest the main downtown area are being uplifted slightly relative to the benchmarks around the port. It is perhaps more likely that the port area is subsiding; however, the relative rates are small. The velocity analysis highlights one of the more unstable benchmarks (NO 24), which was excluded from the contouring solution, along with outliers PASS and 18. It also, however, highlights a change in behavior of the RM OF ETG tide gage mark in the late 1980s. Following a relatively small possible step in the time series sometime between September 1986 and September 1987, there is a strong change in decadal vertical velocity. A large step in November 1988 indicates a change in equipment. However, the apparent motion continues past this date, perhaps indicating a problem with the tide gage monument, as the AQUATRAK mark, observed from 1989, does not show the same motion. For the other marks, decadal velocity estimates are generally within ± 0.4 mm yr^{-1}, indicating long-term rates can be reasonably reliably inferred from relatively short-term observation records.

DISCUSSION

The locations of tide gages are chosen to be stable, and so it is unsurprising—though reassuring—that in general they show only small vertical motions with respect to their benchmark leveling networks (Table D.1). Some of the motions approach 0.25 mm yr^{-1}, however, which is large enough that they should be accounted for when assessing the long-term rates of relative sea-level rise. The data indicate that, although the long-term vertical motion rates are small, there are significant spatial and temporal variations in these rates within several of the networks. Although several instances of apparent monument instability are evident, in most cases these motions appear to reflect real local ground motions. Some of these spatial patterns appear to be consistent over the time window of leveling observations, but others show changes in behavior over time. In particular, the San Francisco tide gage network appears to have been significantly affected by the Loma Prieta earthquake, while Point Reyes appears to have experienced a longer period event, whose impact might still be influencing the tide gage.

Another issue of concern is the impact of tide gage monument instability on estimates of relative sea-level change. Standard procedures for tide gage operations and maintenance of the local vertical reference datum include programming the tide gage data logger with a correction constant that takes into account changes in the instrument height during equipment changes or other events. In particular, any change of more than 6 mm is assumed to reflect instrument monument instability and should be accounted for by a change in the correction constant. Unfortunately, although the values of these constants and when they were changed have been recorded and archived, they are mostly in

TABLE D.1 Long-Term Vertical Motion Estimates for Primary Benchmarks and Tide Gage Reference Marks, with Respect to a Network Defined Vertical Reference

Tide Gage Location	Benchmark	Vertical Velocity (mm yr^{-1})	Velocity Error (mm yr^{-1})[a]	Data Range
San Diego (9410170)	12 (primary benchmark)	-0.03	0.02/0.05	1917–2011
	AQUATRAK	+0.06	0.05/0.12	1990–2011
	RM of ETG	+0.08	0.05/0.12	1966–1994
	Mean	+0.07	0.02/0.05	
Los Angeles (9410660)	8–14 FT ABOVE MLLW (primary benchmark)	+0.14	0.14/0.33	1920–2011
	AQUATRAK	-0.17	0.14/0.33	1990–2011
	STAFF STOP	*+126*	*144/335*	*1980–1984*
	Mean	-0.17	0.14/0.33	
Port San Luis[b] (9412110)	6 (primary benchmark)	+0.08	0.05/0.12	1933–2011
	AQUAREF	(-0.10) +0.24	(0.11) 0.09/0.21	1989–2011
	AQUATRAK	(-0.03) +0.32	(0.08) 0.07/0.16	1989–2011
	Mean	(-0.05) +0.29	(0.05) 0.05/0.12	
San Francisco (9414290)	180 (primary benchmark)	-0.05	0.07/0.16	1925–2011
	AQUATRAK REF	-0.06	0.12/0.28	1990–2000
	AQUATRAK	-0.06	0.12/0.28	1989–2003
	RM of ETG	-0.01	0.10/0.23	1978–1999
	STAFF STOP	+0.16	0.17/0.40	1991–1999
	Mean	-0.01	0.07/0.16	
Point Reyes (9415020)	B243 (primary benchmark)	+0.03	0.02/0.05	1930–2011
	AQUAREF	-1.78	0.11/0.26	1990–1999
	AQUATRAK	-0.28	0.05/0.12	1990–2011
	RM of ETG	-0.22	0.07/0.16	1982–1994
	Mean	-0.22	0.07/0.16	
Crescent City (9419750)	TIDAL 20 1959 RESET (primary benchmark)	+0.05	0.03/0.07	1933–2011
	AQUAREF	-0.24	0.04/0.09	1989–2011
	AQUATRAK	-0.11	0.05/0.12	1989–2011
	RM of ETG	-0.23	0.09/0.21	1982–1996
	Mean	-0.20	0.06/0.14	

[a] Two error estimates are listed: the formal error and the formal error multiplied by the empirically determined 2.33 scale factor.
[b] Values for Port San Luis in parentheses are calculated before a ~6.5 mm step in 1996 is removed from the time series.

analog form and were not available to this study. Without them it becomes impossible to say with complete confidence whether any particular step detected in the leveling data for a tide gage has been accurately removed a priori from the sea-level time series by adjustment of the on-site correction value. Each of San Diego, Point Reyes and Crescent City tide gages have significant steps in the leveling (Table D.2) that appear to match detectable steps in the sea-level record (Larry Breaker, personal communication). The smaller of these (e.g., 11.2 mm at San Diego) would have only a minor impact on long-term sea-level change estimates; however the larger ones, if entirely uncorrected, could contribute significantly to apparent rates of change. It is unclear whether it is possible to retroactively determine whether these steps have been entirely, partly, or not corrected. This problem increases the ambiguity of whether a leveling step indicates a recognized change or step in the tide gage instrumentation, a problem with the mounting of the tide gage instrument, or motion of the entire pier, due to either settling or, possibly, local ground motion.

The analysis of the stability of decadal-scale velocity estimates suggests that, for most tide gages and benchmark networks, although there is significant variation over the full time window, the limited range suggests it is largely reasonable to extrapolate vertical rates based on a limited time window of observations. Figure D.8 shows the rate differences between the decadal-scale rate estimates and the final "correct" long-term estimate. These results confirm that the decadal-scale rate estimates are tightly clustered around their long-term values with $2\sigma = 0.48$ mm yr^{-1}. The same data can be mapped into differences expressed as multiples of the formal errors determined for the decadal-scale rate estimates (Figure D.8 right). Presented this way, the 2σ limits indicate that the formal errors for the decadal rate estimates are significantly too optimistic, with the 2σ value being 4.66. Interestingly this 2.33 multiple of the formal error is close to the 2.5 empirical scale factor often adopted in GPS geodesy to take account of red noise in the time series produced by quasi-seasonal and other long-term signals. Detailed analyses of the noise character of GPS coordinate time series (Mao et al., 1999; Williams et al., 2004) find that the errors are typically well described by a power-law function. The results presented here suggest that the same power-law functions describing noise in decadal-scale GPS vertical time series may also be applicable to multi-decadal leveling measurements.

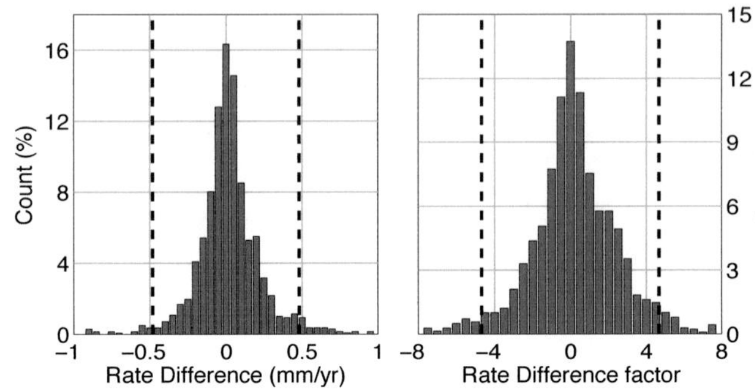

FIGURE D.8 Histograms of differences between decadal-scale estimates of vertical motion rates and the long-term rate determined from the full time series. (*Left*) Rate differences observed in mm yr^{-1}. (*Right*) Rate differences expressed as multiples of the formal error determined for the decadal velocity fit.

REFERENCES

Bawden, G.W., W.R. Thatcher, R.S. Stein, K.W. Hudnut, and G. Peltzer, 2001, Tectonic contraction across Los Angeles after removal of groundwater pumping effects, *Nature*, **412**, 812-815.

TABLE D.2 Steps Detected in Tide Gage Leveling Marks

Location	Step Size (mm)	Step Date
San Diego	+11.2	Early 1974
Port San Luis	-6.5	Between late 1995 and late 1996
San Francisco	+15.0	May 9, 1979
San Francisco	-5.0	Between November 17, 1983, and May 31, 1984
Point Reyes	+42.3	1994
Crescent City	+84.5	Between November 2 and 9, 1988
Crescent City	+3.0	Between September 1986 and September 1987

Brooks, B.A., M.A. Merrifield, J. Foster, C.L. Werner, F. Gomez, M. Bevis, and S. Gill, 2007, Space geodetic determination of spatial variability in relative sea level change, Los Angeles Basin, *Geophysical Research Letters*, **34**, L01611, doi:10.1029/2006GL028171.

Clark, J.C., and E.E. Brabb, 1997, *Geology of Point Reyes National Seashore and Vicinity, California: A Digital Database*, U.S. Geological Survey Open File Report 97-456, available at <http://pubs.usgs.gov/of/1997/of97-456/>.

Hicks, S.D., P.C. Morris, H.A. Lippincott, and M.C. O'Hargan, 1987, *User's Guide for the Installation of Bench Marks and Leveling Requirements for Water Level Stations*, NOAA National Ocean Service, Rockville, MD, 73 pp.

Karcz, I., J. Morreale, and F. Porebski, 1976, Assessment of benchmark credibility in the study of recent vertical crustal movements, *Tectonophysics*, **33**, T1-T6.

Mao, A., C.G.A. Harrison, and T.H. Dixon, 1999, Noise in GPS coordinate time series, *Journal of Geophysical Research*, **104**, 2797-2816.

Shinkle, K.D., and R.K. Dokka, 2004, *Rates of Vertical Displacement at Benchmarks in the Lower Mississippi Valley and the Northern Gulf Coast*, NOAA Technical Report NOS/NGS 50, 135 pp.

Williams, S.D.P., Y. Bock, P. Fang, P. Jamason, R.M. Nikolaidis, L. Prawirodirdjo, M. Miller, and D.J. Johnson, 2004, Error analysis of continuous GPS position time series, *Journal of Geophysical Research*, **109**, B03412, doi:10.1029/2003JB002741.

Woodworth, P.L., 2002, *Manual on Sea Level Measurement and Interpretation, Volume III - Reappraisals and Recommendations as of the Year 2000*, Intergovernmental Oceanographic Commission, Manuals and Guides, **14**, 49 pp.

ACKNOWLEDGMENTS

Thanks to Todd Ehret of NOAA CO-OPS for providing the leveling data and Clyde Kakazu, Marti Ikehara, and Stephen Gill for valuable discussion about the NGS/CO-OPS leveling procedures and data handling and archiving.

GLOSSARY

AQUAREF/ AQUATRAK REF	Physical point on tide gage used for leveling
AQUATRAK	Physical point on tide gage used for leveling
Primary Bench Mark (PBM)	Benchmark which defines the local vertical datum
RM OF ETG	Physical point on tide gage used for leveling (reference mark of electric tape gage)
STAFF STOP	Physical point on tide gage used for leveling

Appendix E

Cryosphere Extrapolations

EXTRAPOLATION BY GENERALIZED LINEAR MODEL

Extrapolations into the future, based on observational records, were done separately for the three categories of ice sources (glaciers and ice caps, Greenland Ice Sheet, and Antarctic Ice Sheet) using a generalized linear model (GLM) approach. The data were assumed to have a normal distribution, and the parameters of the model were estimated using a weighted least squares approach, which allows data uncertainty to be incorporated (McCullagh and Nelder, 1989). This is especially important because there are multiple data sets for each category with different estimates of data error. Traditionally, the data uncertainty is discarded in fitting the models, thus exaggerating the trends in the face of uncertain information.

For each data set in a category, the weighted least squares method was applied to obtain a robust linear model. The linear model is

$$Y = X\beta + \varepsilon,$$

where Y is the vector of response variable (ice mass loss rate per year), X is the matrix of the dependent variables (in this case, the intercept represented as a column of 1 and year), β is the vector of model coefficients (in this case β_0 and β_1), $X\beta$ is the model estimate ($\hat{Y}$), and ε is the error assumed to be normally distributed with mean zero and variance σ^2. Using weighted least squares, β is estimated as

$$\beta = (X^T W X)^{-1}(X^T W Y).$$

Here W is a diagonal matrix containing the uncertainty of each observation. It is populated as $1/d_i$, where d_i is the error associated with each observation point i.

The variance of the error is obtained from standard linear model theory (McCullagh and Nelder, 1989):

$$\sigma^2 = \frac{\sum_{i=1}^{N}\left(Y - \hat{Y}\right)^2}{(N - p)}$$

where N is the number of observations used in the model fitting and p is the number of model coefficients.

With the fitted model, estimate of the response variable for any time i is obtained as $\hat{Y}_i = X_i\beta$, and the 95 percent prediction interval (or prediction uncertainty) is obtained as

$$\hat{Y}_i \pm 1.96\left\{\sigma\sqrt{1 + X_i^T\left(X^T X\right)^{-1} X_i}\right\}.$$

It is apparent from the above equation that the intervals tend to widen as the extrapolation domain extends further from the observations. The term within the curly bracket is the standard error of the estimate $\hat{Y}_i$.

Multi-Data Averaging

Several independent data sets are available for loss rates from glaciers and ice caps as well as for the Greenland and Antarctic ice sheets (data sources are described below). The following steps were performed to obtain a multi-data averaged estimate and standard error for each category of ice source. First, a weighted least squares based linear model was fit for each data set for a given category of ice source c following the method described above. Second, the multiple estimates were combined using a weighted averaging approach in which each estimate was weighted according to its standard error. In the following discussion, the subscript *ic* refers to a variable within category *c* at time *i*. Suppose that, for a category, *K* data sets are available; thus, for any year *i* in the future, *K* estimates $\left(\hat{Y}_{ij}, j=1,2,\cdots K\right)$ can be obtained along with their standard errors $\left(e_{ij}, j=1,2,\cdots K\right)$. A normalized set of weights is computed as

$$m=\left(\frac{1}{e_{i1}}+\frac{1}{e_{i2}}+\cdots+\frac{1}{e_{iK}}\right)$$

and

$$\hat{w}_j=\left(\frac{1}{e_{ij}}\right)m^{-1}$$

for j = 1,2···K. The weighted estimate is obtained as

$$\overline{Y}_{ic}=\sum_{j=1}^{K}\hat{Y}_{ij}\hat{w}_j$$

and its standard error as

$$\overline{e}_{ic}=\sqrt{\sum_{j=1}^{KN}e_{ij}^2\hat{w}_j^2}\,.$$

The 95 percent confidence interval of this weighted estimate is provided as

$$\overline{Y}_{ic}=\pm 1.96\overline{e}_{ic}.$$

These calculations were applied for each of the three ice categories, yielding a multi-data averaged estimate and standard error for each.

Third, the global ice mass loss rate for any year *i* and error were calculated. The global ice mass loss rate was estimated as

$$\overline{Y}_i=\sum_{c=1}^{C}\overline{Y}_{ic}$$

and the standard error as

$$\overline{e}_c=\sqrt{\sum_{c=1}^{C}\overline{e}_{ic}^2}\,.$$

The 95 percent confidence interval of this multi-data averaged global mass loss rate is

$$\overline{Y}_{ic}=\pm 1.96\overline{e}_{ic}.$$

The second and third steps were repeated for all of the projection years. The mass loss rates and the interval were subsequently converted to sea-level rates and then cumulatively summed.

RAPID DYNAMIC RESPONSE

Simple extrapolation of existing trends will not capture the effect of rapid dynamic response that begins after the period of observation. The committee calculated the effects of both acceleration and deceleration in ice discharge relative to observed present-day rates, as described below. The term "rapid dynamic response" is defined here as mass changes in a glacier or ice sheet that occur at rates faster than accompanying climatic mass balance and which force glacier or ice sheet conditions further away from equilibrium with climate.

Increases in Dynamic Discharge

To simulate the effect of rapid dynamics, supplementary ice fluxes were added to the loss rates determined by extrapolation. The parameters for the added rapid dynamic response are summarized in Box E.1. The choice of dynamic variations was intended to capture the general magnitude of plausible changes. Although these particular events may not occur, the calculations provide a means to quantitatively estimate the influence of rapid dynamic response on sea-level rise and to translate ranges of plausible future glaciological changes into equivalent sea-level changes.

The range of added rapid dynamic response for each ice source for each projection period is given in Table E.1, and the effect of the simulated rapid dynamic response on the projections, summed for all three sources, is shown in Table E.2. The top rows ("base values") of Table E.2 show the integrated cumulative sea-level rise from the extrapolation and the low and high values based on uncertainties in the extrapolation. The middle rows of Table E.2 show the effect of additional rapid dynamic response on the projections of sea-level rise. The bottom rows ("percentage effect") in Table E.2 show the effect of added dynamics expressed as a percentage of total sea-level rise. Rapid dynamic response is not an insignificant factor in future sea-level rise, but according to this simple analysis, it is also not a "wild card" variable that will swamp all other sources if it comes into play.

BOX E.1
Parameters for Added Rapid Dynamic Response

Added dynamics (additional discharge assigned to each land ice source for simulation of increased rapid dynamic contribution)

- Glaciers and ice caps: 50 percent of 324.8 GT yr^{-1} = 162.4 GT yr^{-1}
- Greenland Ice Sheet: Increase outlet glacier speed by 2 km yr^{-1} = 375.1 GT yr^{-1}
- Antarctic Ice Sheet: Double outlet glacier discharge to 264 GT yr^{-1}

Variations (perturbations to base values selected for discharge used in the sensitivity calculation)

- Glaciers and ice caps: Use 30 percent and 70 percent of 324.8 GT yr^{-1}
- Greenland Ice Sheet: Use 80 percent and 120 percent of 375.1 GT yr^{-1}
- Antarctic Ice Sheet: Use 80 percent and 120 percent of 264 GT yr^{-1}

TABLE E.1 Range of Added Rapid Dynamic Response (cm) for the Cryosphere Components of Sea-Level Rise

Term	2030	2050	2100
Glaciers and ice caps	0–0.5	0–1.4	0–3.7
Greenland	0–1.2	0–3.3	0–8.4
Antarctica	0–0.8	0–2.3	0–5.9
Total cryosphere	0–2.5	0–7.0	0–18.0

TABLE E.2 Effect of Rapid Dynamic Response and Uncertainty on Future Cumulative Sea-Level Rise

	2030	2050	2100
Base values: Projected sea-level rise Z with uncertainty δZ (cm)			
Z	6.6	17.7	57.0
$Z - \delta Z$	5.9	14.1	44.2
$Z + \delta Z$	7.3	19.0	69.6
Projected sea-level rise with added dynamics Z_d (cm)			
Z_d	2.5	7.0	18.0
$Z + Z_d$	9.1	24.7	75.0
$Z - \delta Z + Z_d$	8.4	21.1	62.2
$Z + \delta Z + Z_d$	9.8	26.0	87.6
Percentage effect of added dynamics			
$Z + Z_d$	38%	40%	32%
$Z - \delta Z + Z_d$	42%	50%	41%
$Z + \delta Z + Z_d$	34%	37%	26%

Sensitivity of the Added Dynamics Analysis

The choice of fluxes for added rapid dynamic response (Box E.1) was guided by the analogy to the doubling of the Greenland mass balance deficit in 2000–2006 but was otherwise rather arbitrary. To investigate how sensitive this calculation is to the choice of added discharge flux, the input variables (discharge from glaciers and ice caps, Greenland, and Antarctica) were varied by ± 20 percent individually (no two inputs were varied at the same time), and the calculation was repeated to determine the response in the output (additional sea-level rise). The result of this sensitivity test is summarized in Table E.3. For variations of 20 percent magnitude in the inputs, output magnitudes varied by no more than 7 percent.

TABLE E.3 Sensitivity of Rapid Dynamic Response Estimate to the Choice of Parameters

	Percentage Change	Δ mm Sea-Level Rise
Glaciers and ice caps		
30 percent of 324.8 GT yr^{-1}	0.99	179
70 percent of 324.8 GT yr^{-1}	1.01	182
Greenland Ice Sheet		
80 percent of 375.1 GT yr^{-1}	0.9	165
120 percent of 375.1 GT yr^{-1}	1.09	197
Antarctic Ice Sheet		
80 percent of 264 GT yr^{-1}	0.93	169
120 percent of 264 GT yr^{-1}	1.07	192

NOTE: Input parameters were varied by ± 20 percent individually (only one parameter was varied at a time).

TABLE E.4 Effect of Reduced Greenland Dynamic Discharge on Sea-Level Rise Projections

Year	Total Sea-Level Rise (mm)	Cryosphere Component (mm)
Base-Rate Projection (From Table 5.2)		
2030	135	81
2050	280	180
2100	827	584
50 Percent Slowdown in Greenland Dynamic Discharge		
2030	128	76
2050	273	168
2100	774	535
Percent Change		
2030	-5%	-6%
2050	-3%	-7%
2100	-6%	-8%

Decreases in Dynamic Discharge

To test the effect of decreased dynamic discharge, the projected output of the Greenland Ice Sheet was reduced by 25 percent from its projected base value and all other cryosphere terms were left unchanged. The results are summarized in Table E.4. The table shows the cumulative sea-level rise (central value only) for 2030, 2050, and 2100 for both the base rate projection (Table 5.2) and for a 50 percent reduction in Greenland calving discharge (equivalent to a 25 percent reduction in overall Greenland discharge). The cryosphere component totals are for the Greenland and Antarctica ice sheets, glaciers, and ice caps, and the total sea-level rise includes the steric component. The percent change rows show that reducing the Greenland discharge affects sea-level projections by a maximum of 8 percent.

EFFECT OF SEA-LEVEL FINGERPRINT

The influence of melting from Alaska, Greenland, and Antarctica on regional sea level was described in "Sea-Level Fingerprints of Modern Land Ice Change" in Chapter 4. To estimate this effect on projected future sea-level rise, land ice loss rates were subdivided into Alaska, Greenland, Antarctic, and all other glacier and ice cap losses other than Alaska. The fingerprint scale factors for Alaska, Greenland, and Antarctica (specified in Table 4.1) were then applied to losses from those regions, on a year-by-year basis, and the losses from other glacier and ice cap regions were carried forward without adjustment. The projected sea level ΔZ at destination region p and time t is then, using the notation defined in Appendix C:

$$\Delta Z_p(t)=\sum_{k=1}^{3} s_{k,p}\int_0^t R_k(\tau)d\tau+\int_0^t R_{GIC-AK}(t),$$

where R is rate of ice loss (e.g., in GT yr^{-1}), s is the fingerprint scale factor, k =1,2,3 is the set of source locations (Alaska, Greenland, and Antarctica), p = 1,2,3 is the set of destination locations (north coast, central coast, south coast), and t is time. The term R_{GIC-AK} designates the loss rate from all glacier and ice cap regions with the exception of Alaska.

REFERENCE

McCullagh, P., and J.A. Nelder, 1989, *Generalized Linear Models*, 2nd edition, Chapman and Hall, London, 532 pp.

Appendix F

Biographical Sketches of Committee Members

Robert A. Dalrymple, *chair*, is the Willard and Lillian Hackerman Professor of Civil Engineering at Johns Hopkins University. His research interests are in coastal engineering, water wave mechanics, high-performance computing, fluid mechanics, littoral processes, and tidal inlets. Dr. Dalrymple has chaired several National Research Council (NRC) committees, including the Committee on Review of the Louisiana Coastal Protection and Restoration Program, and served on others including the Committee on Responding to Sea Level: Engineering Implications. He also has held leadership positions in professional societies including president of the Association of Coastal Engineers, president of the American Society of Civil Engineer's (ASCE's) Coasts, Oceans, Ports, and Rivers Institute, and chair of the Coastal Engineering Research Council. Dr. Dalrymple is a member of the National Academy of Engineering and a recipient of ASCE's International Coastal Engineering Award for his achievements and contributions to the advancement of coastal engineering through research, teaching, and professional leadership. He received a B.A. in engineering sciences from Dartmouth College, an M.S. in ocean engineering from the University of Hawai'i, and a Ph.D. in civil and coastal engineering from the University of Florida.

Laurence C. Breaker is an adjunct professor at the Moss Landing Marine Laboratories at San Jose State University. Prior to joining the laboratory in 2001, he spent 13 years as a senior research physical scientist at the National Oceanic and Atmospheric Administration's (NOAA's) National Centers for Environmental Prediction. Dr. Breaker's research focuses on the analysis of long-term observations of sea-level rise along the California coast, modeling of both global and local sea-level rise, physical oceanography, and satellite remote sensing. His recent papers have examined the 154-year record of monthly sea level at San Francisco and sea-level responses to large earthquakes in California and Alaska. Dr. Breaker was awarded NOAA's Bronze Medal for major contributions to the Coastal Marine Demonstration Project, which tested the state of the art in marine forecasting and evaluated the potential benefits of experimental higher resolution predictions. He received a B.S. in mechanical engineering from Bucknell University, an M.S. in applied marine physics from the University of Miami, and a Ph.D. in oceanography (minor in meteorology) from the Naval Postgraduate School.

Benjamin A. Brooks is an associate researcher (tenured) and director of the Pacific GPS Facility in the Hawai'i Institute of Geophysics and Planetology at the University of Hawai'i. His research interests are in tectonic geodesy and active tectonics with a recent focus on relative sea-level change as a result of subsidence in the Los Angeles Basin and the San Francisco Bay Delta. Dr. Brooks is a member of the advisory committee for the Plate Boundary Observatory, which collects geodetic data on active deformation across the western United States. He is a Fulbright Fellow. He received a B.S. in earth science from the University of California, Santa Cruz, and a Ph.D. in geological sciences from Cornell University.

Daniel R. Cayan is a research meteorologist at the Scripps Institution of Oceanography and is also a researcher in the U.S. Geological Survey. His work is directed at understanding climate variability and changes over the Pacific Ocean and North America and climate impacts on water, wildfire, health, and agriculture in California and western North America. Among his recent publications are projections of sea-level extremes along the California coast. Dr. Cayan heads two climate research programs aimed at improving climate information and forecasts for decision makers in the California region: the California Nevada Applications Program and the California Climate Change Center. He is a fellow of the American Geophysical Union. He received a B.S. in meteorology and oceanography from the University of Michigan and a Ph.D. in oceanography from the University of California, San Diego.

Gary B. Griggs is a distinguished professor of earth and planetary sciences and the director of the Institute of Marine Sciences at the University of California, Santa Cruz. His research is focused on the coastal zone and ranges from coastal evolution and development to shoreline processes—including the evaluation of long-term shoreline changes and geomorphic evolution of coastlines—to coastal hazards and coastal engineering. Dr. Griggs is the author or coauthor of several books, including *Living with the Changing California Coast* and *Introduction to California's Beaches and Coast*. He served as chair of the University of California Marine Council from 1999 to 2009 and is a current member and past chair of the science advisory team to the Governor's Ocean Protection Council. He is a fellow of the California Academy of Sciences. He received a B.A. in geological sciences from the University of California, Santa Barbara, and a Ph.D. in oceanography from Oregon State University.

Weiqing Han is an associate professor in the Department of Atmospheric and Oceanic Sciences at the University of Colorado. Her research interests are in sea level, ocean circulation and dynamics, air-sea interaction, and climate variability and change. Among her recent work is an analysis of patterns of sea-level change in the Indian Ocean and the influence of North Atlantic circulation on glacial sea-level changes. Dr. Han serves on a panel of the World Climate Research Programme Climate Variability and Predictability project, and she is a recipient of a National Science Foundation Faculty Early CAREER Award. She received a B.S. in meteorology from the Nanjing Institute of Meteorology, an M.S. in meteorology from the Chinese Academy of Meteorological Sciences, and a Ph.D. in physical oceanography from the Nova Southeastern University Oceanographic Center.

Benjamin P. Horton is an associate professor in the Department of Earth and Environmental Science at the University of Pennsylvania. His research focuses on mechanisms of sea-level changes, including climate change, earthquakes, tsunamis, and the coastal sedimentary budget. He also has examined the response of estuaries to sea-level rise. He aims to bridge the gap between instrumental and geological observations of sea-level change. Dr. Horton has worked on sea-level rise in several countries; his U.S. work has focused on the contributions of eustacy and isostacy along the Atlantic coast and earthquakes and ground deformation along the west coast. He is a contributing author to the Intergovernmental Panel on Climate Change (IPCC) Fifth Assessment Report, a member of the steering committee of PALSEA (PALeo-constraints on SEA-level rise), and the project leader of the International Geoscience Programme's Preparing for Coastal Change. He received a B.A. in geography from the University of Liverpool and a Ph.D. in geography from the University of Durham.

Christina L. Hulbe is a professor and chair of the Department of Geology at Portland State University. Her research focuses on understanding and modeling the dynamics of ice sheets, the interactions between ice shelves and ice sheets, and on the role of ice sheets in climate change. New work now under way involves the uncertainty associated with poorly known boundary conditions in mathematical models of ice sheets. Dr. Hulbe is a representative of the NRC's Scientific Committee on Antarctic Research (SCAR) and co-chairs the local organizing committee for the 2012 SCAR Open Science Conference. She received a B.S. in geological engineering from Montana College of Mineral Science and Technology, an M.S. in geology from Ohio State University, and a Ph.D. in geophysics from the University of Chicago.

James C. McWilliams is Louis B. Slichter Professor of Earth Sciences in the Institute of Geophysics and Planetary Physics and the Department of Atmospheric and Oceanic Sciences at the University of California, Los Angeles. He also is a senior research scientist at the National Center for Atmospheric Research. His research interests are in theory and computational modeling of Earth's ocean and atmosphere. In addition to his work in fluid dynamics, he developed a three-dimensional simulation model of the U.S. west coast that incorporates physical oceanographic, biogeochemical, and sediment transport aspects of the coastal circulation and is being used to interpret coastal phenomena, diagnose historical variability in relation to observational data, and assess future possibilities. Dr. McWilliams has served on many NRC climate committees, including the Committee on Science of Climate Change. He is a fellow of the American Geophysical Union and a member of the National Academy of Sciences. He received a B.S. in applied mathematics from Caltech and an M.S. and Ph.D. from Harvard.

Philip W. Mote is a professor in the College of Earth, Ocean, and Atmospheric Sciences at Oregon State University. He also is the director of the Oregon Climate Change Research Institute for the Oregon University System. Before joining Oregon State University, he was a research scientist at the University of Washington and the state climatologist for Washington. Dr. Mote's research interests include climate variability and change in the Pacific Northwest; regional climate modeling; mountain snowpack and its response to climate variability and change; sea-level rise; impacts of climate change on water resources, forests, and shorelands; and adaptation to climate change. Among his publications in these areas is an analysis of sea-level rise in the coastal waters of Washington state. Dr. Mote has served on several committees associated with climate change and sea-level rise, including the NRC Panel on Adapting to the Impacts of Climate Change and the IPCC. He received a B.A. in physics from Harvard University and a Ph.D. in atmospheric sciences from the University of Washington.

William Tad Pfeffer is a professor of civil, environmental, and architectural engineering at the University of Colorado. He also is a fellow of the university's Institute of Arctic and Alpine Research. Dr. Pfeffer's research interests are in modern glacier physics, including ice mechanics and glacier dynamics, heat and mass transfer in snow and ice, atmosphere/glacier and ocean/glacier interactions, and the application of the results to estimates of future sea-level change. He is a member of the executive committee of the American Geophysical Union's Cryospheric Sciences Focus Group. He received a B.A. in geology from the University of Vermont, an M.A. in geology from the University of Maine, and a Ph.D. in geophysics from the University of Washington.

Denise Reed is a professor in the Department of Earth and Environmental Sciences at the University of New Orleans. Her research interests include coastal marsh response to sea-level rise and how this is affected by human activities. She has worked on coastal issues on the Atlantic, Pacific, and Gulf coasts of the United States, as well as other parts of the world, and also is involved in ecosystem restoration planning both in Louisiana and in California. Dr. Reed has served on numerous boards and panels concerning coastal environments and ecosystem restoration, including NRC committees on water and environmental management in the California Bay Delta and on mitigating shore erosion, the Corps of Engineers Environmental Advisory Board, the NOAA Science Advisory Board's Ecosystems Sciences and Management Working Group, the National Science Panel for South Bay Salt Ponds Restoration, and the Strategic Science Review Panel for the Puget Sound Nearshore Ecosystem Restoration Program. She received her B.A. and Ph.D. in geography from the University of Cambridge, United Kingdom.

C.K. Shum is a professor and Distinguished University Scholar in the Division of Geodetic Science, School of Earth Sciences, at the Ohio State University. His research focuses on the accurate measurement of present-day sea-level rise and the improved understanding of the geophysical causes of this rise. He also works on satellite geodesy, temporal gravity field and tide modeling, satellite oceanography, hydrology and geodynamics, ice mass balance, precision satellite orbit determination, GPS meteorology, and space physics. Dr. Shum was a lead author of the chapter on observations of oceanic climate change and sea level in the IPCC Fourth Assessment Report. He is a fellow of the International Association of Geodesy and the American Association for the Advancement of Science. He is a recipient of the European Geosciences Union's Vening Meinesz Medal for distinguished research in geodesy, and of several NASA awards for his work on the TOPEX/POSEIDON and GRACE missions. He received his B.S. and Ph.D. in aerospace engineering from the University of Texas, Austin.

Appendix G

Acronyms and Abbreviations

CGPS	continuous Global Positioning System
CMIP 3	Coupled Model Intercomparison Project Phase 3
CO_2	carbon dioxide
CO-OPS	Center for Operational Oceanographic Products and Services
DCW	Digital Chart of the World
DOFs	degrees-of-freedom
DORIS	Doppler Orbitography and Radio-positioning Integrated by Satellite
ENSO	El Niño-Southern Oscillation
FEMA	Federal Emergency Management Administration
GCM	general circulation model
GIA	glacial isostatic adjustment
GLIMS	Global Land Ice Measurements from Space
GLM	generalized linear model
GNSS	Global Navigation Satellite System
GPS	Global Positioning System
GRACE	Gravity Recovery and Climate Experiment
IB	inverse barometer
InSAR	interferometric synthetic aperture radar
IPCC	Intergovernmental Panel on Climate Change
lidar	light detection and ranging
MBT	mechanical bathythermograph
MTJ	Mendocino Triple Junction
NCEP	National Centers for Environmental Prediction
NGS	National Geodetic Survey
NOAA	National Oceanic and Atmospheric Administration
NRC	National Research Council
PDO	Pacific Decadal Oscillation
PSMSL	Permanent Service for Mean Sea Level
RMS	root mean square
SLA	sea-level anomaly
SLE	sea-level equivalent
SLP	sea-level pressure
SOPAC	Scripps Orbit and Permanent Array Center
SWH	significant wave height
USACE	U.S. Army Corps of Engineers
USGS	U.S. Geological Survey
VLM	vertical land motion
WGI	World Glacier Inventory
XBT	expendable bathythermograph